STUDENT SOLUTIONS MANUAL

Neil Wigley
University of Windsor

to accompany

CALCULUS
Late Transcendentals

Eighth Edition

Howard Anton
Drexel University

Irl C. Bivens
Davidson College

Stephen L. Davis
Davidson College

WILEY

JOHN WILEY & SONS, INC.

Cover Photo: ©Arthur Tilley/Taxi/Getty Images

To order books or for customer service, please call 1-800-CALL-WILEY (225-5945).

ISBN 0-471-67210-6

Printed in the United States of America

10 9 8 7 6 5 4 3 2 1

Printed and bound by Malloy Lithographing, Inc.

CONTENTS

CHAPTER 1
Functions

EXERCISE SET 1.1

1. **(a)** $-2.9, -2.0, 2.35, 2.9$ **(b)** none   **(c)** $y = 0$
     (d) $-1.75 \le x \le 2.15$ **(e)** $y_{\max} = 2.8$ at $x = -2.6$; $y_{\min} = -2.2$ at $x = 1.2$

3. **(a)** yes   **(b)** yes
     (c) no (vertical line test fails)   **(d)** no (vertical line test fails)

5. **(a)** around 1943   **(b)** 1960; 4200
     (c) no; you need the year's population   **(d)** war; marketing techniques
     (e) news of health risk; social pressure, antismoking campaigns, increased taxation

7. **(a)** 1999, \$43,200   **(b)** 1985, \$37,000
     (c) second year; graph has a larger (negative) slope

9. **(a)** $f(0) = 3(0)^2 - 2 = -2; f(2) = 3(2)^2 - 2 = 10; f(-2) = 3(-2)^2 - 2 = 10; f(3) = 3(3)^2 - 2 = 25;$
     $f(\sqrt{2}) = 3(\sqrt{2})^2 - 2 = 4; f(3t) = 3(3t)^2 - 2 = 27t^2 - 2$
     (b) $f(0) = 2(0) = 0; f(2) = 2(2) = 4; f(-2) = 2(-2) = -4; f(3) = 2(3) = 6; f(\sqrt{2}) = 2\sqrt{2};$
     $f(3t) = 1/3t$ for $t > 1$ and $f(3t) = 6t$ for $t \le 1$.

11. **(a)** $x \ne 3$ **(b)** $x \le -\sqrt{3}$ or $x \ge \sqrt{3}$
     (c) $x^2 - 2x + 5 = 0$ has no real solutions so $x^2 - 2x + 5$ is always positive or always negative. If
     $x = 0$, then $x^2 - 2x + 5 = 5 > 0$; domain: $(-\infty, +\infty)$.
     (d) $x \ne 0$ **(e)** $\sin x \ne 1$, so $x \ne (2n + \frac{1}{2})\pi$, $n = 0, \pm 1, \pm 2, \ldots$

13. **(a)** $x \le 3$ **(b)** $-2 \le x \le 2$ **(c)** $x \ge 0$ **(d)** all x **(e)** all x

15. **(a)** Breaks could be caused by war, pestilence, flood, earthquakes, for example.
     (b) C decreases for eight hours, takes a jump upwards, and then repeats.

17.

19. **(a)** $x = 2, 4$ **(b)** none   **(c)** $x \le 2; 4 \le x$ **(d)** $y_{\min} = -1$; no maximum value

21. The cosine of θ is $(L - h)/L$ (side adjacent over hypotenuse), so $h = L(1 - \cos\theta)$.

23. **(a)** If $x < 0$, then $|x| = -x$ so $f(x) = -x + 3x + 1 = 2x + 1$. If $x \ge 0$, then $|x| = x$ so
     $f(x) = x + 3x + 1 = 4x + 1$;
$$f(x) = \begin{cases} 2x + 1, & x < 0 \\ 4x + 1, & x \ge 0 \end{cases}$$

(b) If $x < 0$, then $|x| = -x$ and $|x - 1| = 1 - x$ so $g(x) = -x + 1 - x = 1 - 2x$. If $0 \le x < 1$, then $|x| = x$ and $|x - 1| = 1 - x$ so $g(x) = x + 1 - x = 1$. If $x \ge 1$, then $|x| = x$ and $|x - 1| = x - 1$ so $g(x) = x + x - 1 = 2x - 1$;

$$g(x) = \begin{cases} 1 - 2x, & x < 0 \\ 1, & 0 \le x < 1 \\ 2x - 1, & x \ge 1 \end{cases}$$

25. (a) $V = (8 - 2x)(15 - 2x)x$

(b) $0 \le x \le 4$

(c) $0 \le V \le 91$

(d) As x increases, V increases and then decreases; the maximum value could be approximated by zooming in on the graph.

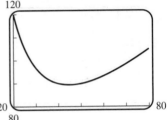

27. (a) The side adjacent to the building has length x, so $L = x + 2y$.

(b) $A = xy = 1000$, so $L = x + 2000/x$.

(c) all $x \ne 0$

(d) $L \approx 89.44$ ft

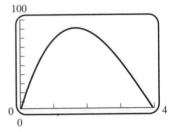

29. (a) $V = 500 = \pi r^2 h$ so $h = \dfrac{500}{\pi r^2}$. Then

$$C = (0.02)(2)\pi r^2 + (0.01)2\pi r h = 0.04\pi r^2 + 0.02\pi r \frac{500}{\pi r^2}$$

$$= 0.04\pi r^2 + \frac{10}{r}; \quad C_{\min} \approx 4.39 \text{ cents at } r \approx 3.4 \text{ cm},$$
$$h \approx 13.8 \text{ cm}.$$

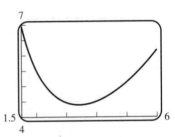

(b) $C = (0.02)(2)(2r)^2 + (0.01)2\pi r h = 0.16r^2 + \dfrac{10}{r}$. Since $0.04\pi < 0.16$, the top and bottom now get more weight. Since they cost more, we diminish their sizes in the solution, and the cans become taller.

(c) $r \approx 3.1$ cm, $h \approx 16.0$ cm, $C \approx 4.76$ cents

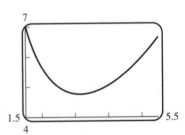

31. (i) $x = 1, -2$ causes division by zero   (ii) $g(x) = x + 1$, all x

33. (a) $25°F$ (b) $13°F$ (c) $5°F$

35. As in the previous exercise, WCT $\approx 1.4157T - 30.6763$; thus $T \approx 15°F$ when WCT $= -10$.

EXERCISE SET 1.2

1. (e) seems best, though only (a) is bad.

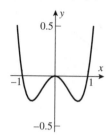

3. (b) and (c) are good; (a) is very bad.

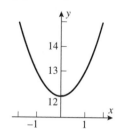

5. $[-3, 3] \times [0, 5]$

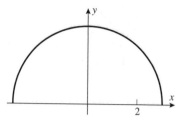

7. **(a)** window too narrow, too short
(b) window wide enough, but too short
(c) good window, good spacing
(d) window too narrow, too short
(e) window too narrow, too short

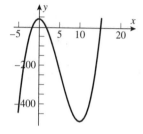

9. $[-5, 14] \times [-60, 40]$

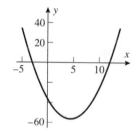

11. $[-0.1, 0.1] \times [-3, 3]$

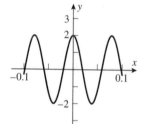

13. $[-250, 1050] \times [-1500000, 600000]$

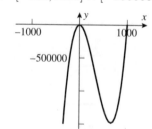

15. $[-2, 2] \times [-20, 20]$

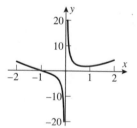

17. depends on graphing utility

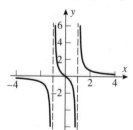

19. (a) $f(x) = \sqrt{16 - x^2}$

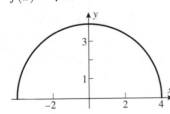

(b) $f(x) = -\sqrt{16 - x^2}$

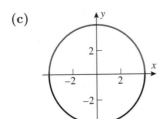

(c)

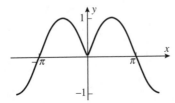

(d)

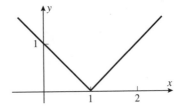

(e) No; the vertical line test fails.

21. (a)

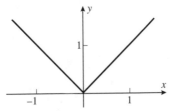

(b)

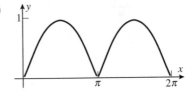

(c)

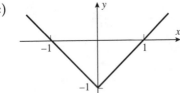

(d)

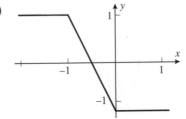

(e)

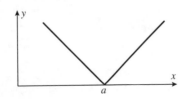

(f)

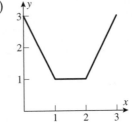

23. The portions of the graph of $y = f(x)$ which lie below the x-axis are reflected over the x-axis to give the graph of $y = |f(x)|$.

25. (a) for example, let $a = 1.1$

(b)

27.

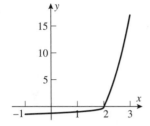

29. (a)

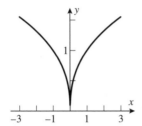

(b)

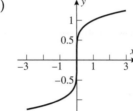

(c)

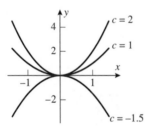

(d)

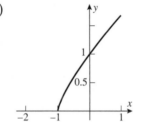

31. (a) stretches or shrinks the graph in the y-direction; reflects it over the x-axis if c changes sign

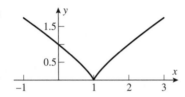

(b) As c increases, the lower part of the parabola moves down and to the left; as c decreases, the motion is down and to the right.

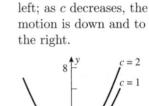

(c) The graph rises or falls in the y-direction with changes in c.

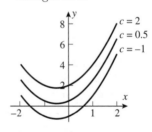

As c increases, the lower part of the parabola moves down and to the left; as c decreases, the motion is down and to the right

33. The curve oscillates between the lines $y = x$ and $y = -x$ with increasing rapidity as $|x|$ increases.

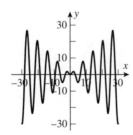

EXERCISE SET 1.3

1. (a)

    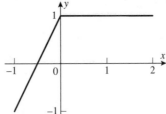

    (b)

    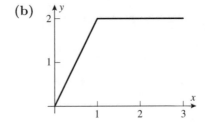

    (c)

    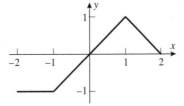

    (d)

    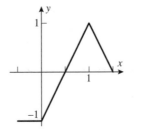

3. (a)

    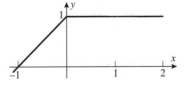

    (b)

    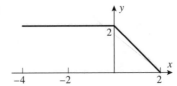

    (c)

    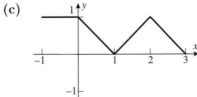

    (d)

    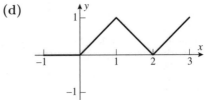

5. Translate left 1 unit, stretch vertically by a factor of 2, reflect over x-axis, translate down 3 units.

    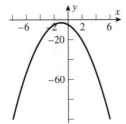

7. $y = (x + 3)^2 - 9$; translate left 3 units and down 9 units.

    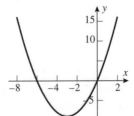

9. $y = -(x-1)^2 + 2$;
translate right 1 unit,
reflect over x-axis,
translate up 2 units.

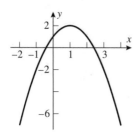

11. Translate left 1 unit,
reflect over x-axis,
translate up 3 units.

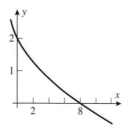

13. Compress vertically
by a factor of $\frac{1}{2}$,
translate up 1 unit.

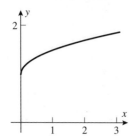

15. Translate right 3 units.

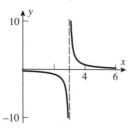

17. Translate left 1 unit,
reflect over x-axis,
translate up 2 units.

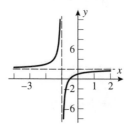

19. Translate left 2 units
and down 2 units.

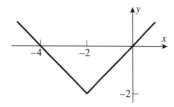

21. Stretch vertically by a factor of 2,
translate right 1/2 unit and up 1 unit.

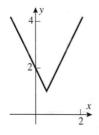

23. Stretch vertically by a factor of 2,
reflect over x-axis, translate up 1 unit.

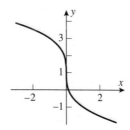

25. Translate left 1 unit and up 2 units.

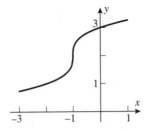

27. **(a)**

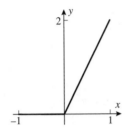

(b) $y = \begin{cases} 0 \text{ if } x \le 0 \\ 2x \text{ if } 0 < x \end{cases}$

29. $(f+g)(x) = 3\sqrt{x-1}, \, x \ge 1; \, (f-g)(x) = \sqrt{x-1}, \, x \ge 1; \, (fg)(x) = 2x - 2, \, x \ge 1;$
$(f/g)(x) = 2, \, x > 1$

31. **(a)** 3   **(b)** 9   **(c)** 2   **(d)** 2

33. **(a)** $t^4 + 1$ **(b)** $t^2 + 4t + 5$ **(c)** $x^2 + 4x + 5$ **(d)** $\dfrac{1}{x^2} + 1$
     (e) $x^2 + 2xh + h^2 + 1$ **(f)** $x^2 + 1$ **(g)** $x + 1$ **(h)** $9x^2 + 1$

35. $(f \circ g)(x) = 1 - x, \, x \le 1; \, (g \circ f)(x) = \sqrt{1 - x^2}, \, |x| \le 1$

37. $(f \circ g)(x) = \dfrac{1}{1 - 2x}, \, x \ne \dfrac{1}{2}, 1; \, (g \circ f)(x) = -\dfrac{1}{2x} - \dfrac{1}{2}, \, x \ne 0, 1$

39. $x^{-6} + 1$

41. **(a)** $g(x) = \sqrt{x}, \, h(x) = x + 2$ **(b)** $g(x) = |x|, \, h(x) = x^2 - 3x + 5$

43. **(a)** $g(x) = x^2, \, h(x) = \sin x$ **(b)** $g(x) = 3/x, \, h(x) = 5 + \cos x$

45. **(a)** $f(x) = x^3, \, g(x) = 1 + \sin x, \, h(x) = x^2$ **(b)** $f(x) = \sqrt{x}, \, g(x) = 1 - x, \, h(x) = \sqrt[3]{x}$

47.

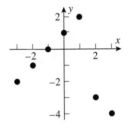

49. Note that
$f(g(-x)) = f(-g(x)) = f(g(x)),$
so $f(g(x))$ is even.

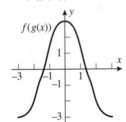

51. $f(g(x)) = 0$ when $g(x) = \pm 2$, so $x = \pm 1.4$; $g(f(x)) = 0$ when $f(x) = 0$, so $x = \pm 2$.

53. $\dfrac{3w^2 - 5 - (3x^2 - 5)}{w - x} = \dfrac{3(w - x)(w + x)}{w - x} = 3w + 3x$

$\dfrac{3(x + h)^2 - 5 - (3x^2 - 5)}{h} = \dfrac{6xh + 3h^2}{h} = 6x + 3h;$

55. $\dfrac{1/w - 1/x}{w - x} = \dfrac{x - w}{wx(w - x)} = -\dfrac{1}{xw}; \quad \dfrac{1/(x + h) - 1/x}{h} = \dfrac{x - (x + h)}{xh(x + h)} = \dfrac{-1}{x(x + h)}$

57. neither; odd; even

59. **(a)** 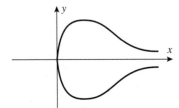   **(b)**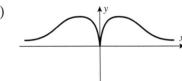

(c)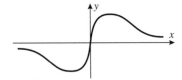

61. **(a)** even   **(b)** odd   **(c)** odd   **(d)** neither

63. **(a)** $f(-x) = (-x)^2 = x^2 = f(x)$, even   **(b)** $f(-x) = (-x)^3 = -x^3 = -f(x)$, odd

(c) $f(-x) = |-x| = |x| = f(x)$, even   **(d)** $f(-x) = -x + 1$, neither

(e) $f(-x) = \dfrac{(-x)^5 - (-x)}{1 + (-x)^2} = -\dfrac{x^5 - x}{1 + x^2} = -f(x)$, odd

(f) $f(-x) = 2 = f(x)$, even

65. In Exercise 64 it was shown that g is an even funcction, and h is odd. Moreover by inspection $f(x) = g(x) + h(x)$ for all x, so f is the sum of an even function and an odd function.

67. **(a)** y-axis, because $(-x)^4 = 2y^3 + y$ gives $x^4 = 2y^3 + y$

(b) origin, because $(-y) = \dfrac{(-x)}{3 + (-x)^2}$ gives $y = \dfrac{x}{3 + x^2}$

(c) x-axis, y-axis, and origin because $(-y)^2 = |x| - 5$, $y^2 = |-x| - 5$, and $(-y)^2 = |-x| - 5$ all give $y^2 = |x| - 5$

69. 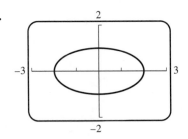   **71.**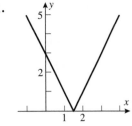

73. (a)

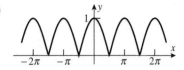

(b)

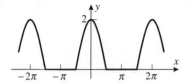

75. Yes, e.g. $f(x) = x^k$ and $g(x) = x^n$ where k and n are integers.

EXERCISE SET 1.4

1. (a) $y = 3x + b$ **(b)** $y = 3x + 6$ **(c)**

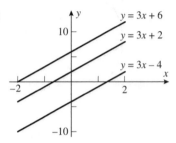

3. (a) $y = mx + 2$

   (b) $m = \tan\phi = \tan 135° = -1$, so $y = -x + 2$ **(c)**

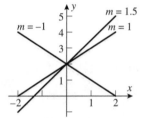

5. Let the line be tangent to the circle at the point (x_0, y_0) where $x_0^2 + y_0^2 = 9$. The slope of the tangent line is the negative reciprocal of y_0/x_0 (why?), so $m = -x_0/y_0$ and $y = -(x_0/y_0)x + b$. Substituting the point (x_0, y_0) as well as $y_0 = \pm\sqrt{9 - x_0^2}$ we get $y = \pm\dfrac{9 - x_0 x}{\sqrt{9 - x_0^2}}$.

7. The x-intercept is $x = 10$ so that with depreciation at 10% per year the final value is always zero, and hence $y = m(x - 10)$. The y-intercept is the original value.

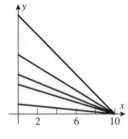

9. **(a)** The slope is -1.   **(b)** The y-intercept is $y = -1$.

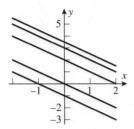   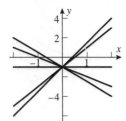

(c) They pass through the point $(-4, 2)$.   **(d)** The x-intercept is $x = 1$.

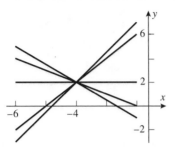   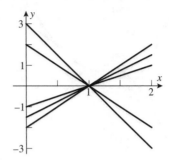

11. **(a)** VI   **(b)** IV   **(c)** III   **(d)** V   **(e)** I   **(f)** II

13. **(a)**

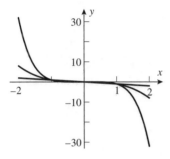   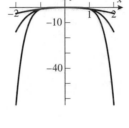

(b)

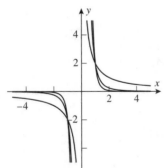

(c)

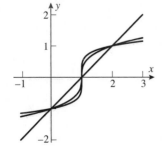   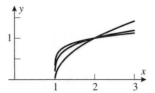

15. (a)

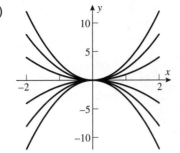

(b)

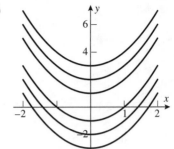

(c)

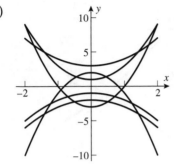

17. (a)

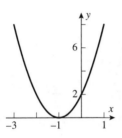

(b)

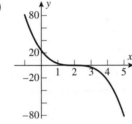

(c)

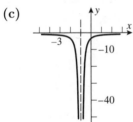

(d)

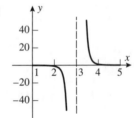

19. (a)

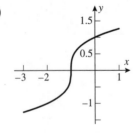

(b)

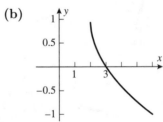

(c)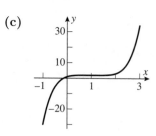

(d)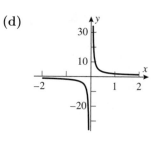

21. $y = x^2 + 2x = (x+1)^2 - 1$

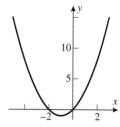

23. (a) N·m

(b) k = 20 N·m

(c)

$V\,(\text{L})$	0.25	0.5	1.0	1.5	2.0
$P\,(\text{N/m}^2)$	80×10^3	40×10^3	20×10^3	13.3×10^3	10×10^3

(d)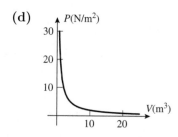

25. (a) $F = k/x^2$ so $0.0005 = k/(0.3)^2$ and $k = 0.000045$ N·m^2.

(b) $F = 0.000005$ N

(c)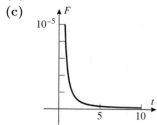

(d) When they approach one another, the force becomes infinite; when they get far apart it tends to zero.

27. (a) II; $y = 1$, $x = -1, 2$

(b) I; $y = 0$, $x = -2, 3$

(c) IV; $y = 2$

(d) III; $y = 0$, $x = -2$

29. (a) $y = 3\sin(x/2)$

(b) $y = 4\cos 2x$

(c) $y = -5\sin 4x$

31. (a) $y = \sin(x + \pi/2)$

(b) $y = 3 + 3\sin(2x/9)$

(c) $y = 1 + 2\sin(2(x - \pi/4))$

33. (a) $3, \pi/2$

(b) $2, 2$

(c) $1, 4\pi$

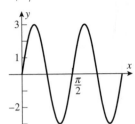

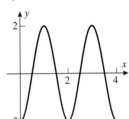

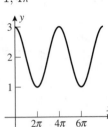

35. (a) $x = A\sin(\omega t + \theta) = A[\sin \omega t \cos \theta + \sin \theta \cos \omega t]$, so let $A\cos \theta = A_1, A\sin \theta = A_2$, then $A_1^2 + A_2^2 = A^2(\cos^2 \theta + \sin^2 \theta) = A^2$, and $A_2/A_1 = A\sin \theta/A\cos \theta = \tan \theta$.

(b) $A^2 = 2 + 6, A = 2\sqrt{2}; \tan \theta = \sqrt{6}/\sqrt{2} = \sqrt{3}, \theta = \pi/3$, and $x = 2\sqrt{2}\sin(2\pi t + \pi/3)$.

EXERCISE SET 1.5

1. (a) $f(g(x)) = 4(x/4) = x$, $g(f(x)) = (4x)/4 = x$, f and g are inverse functions
  (b) $f(g(x)) = 3(3x - 1) + 1 = 9x - 2 \neq x$ so f and g are not inverse functions
  (c) $f(g(x)) = \sqrt[3]{(x^3 + 2) - 2} = x$, $g(f(x)) = (x - 2) + 2 = x$, f and g are inverse functions
  (d) $f(g(x)) = (x^{1/4})^4 = x$, $g(f(x)) = (x^4)^{1/4} = |x| \neq x$, f and g are not inverse functions

3. (a) yes   **(b)** yes   **(c)** no   **(d)** yes   **(e)** no   **(f)** no

5. (a) yes; all outputs (the elements of row two) are distinct
  (b) no; $f(1) = f(6)$

7. (a) f has an inverse because the graph passes the horizontal line test. To compute $f^{-1}(2)$ start at 2 on the y-axis and go to the curve and then down, so $f^{-1}(2) = 8$; similarly, $f^{-1}(-1) = -1$ and $f^{-1}(0) = 0$.
  (b) domain of f^{-1} is $[-2, 2]$, range is $[-8, 8]$ **(c)**

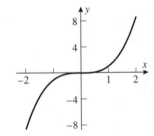

9. $y = f^{-1}(x), x = f(y) = 7y - 6, y = \dfrac{1}{7}(x + 6) = f^{-1}(x)$

11. $y = f^{-1}(x), x = f(y) = 3y^3 - 5, y = \sqrt[3]{(x + 5)/3} = f^{-1}(x)$

13. $y = f^{-1}(x), x = f(y) = \sqrt[3]{2y - 1}, y = (x^3 + 1)/2 = f^{-1}(x)$

15. $y = f^{-1}(x), x = f(y) = 3/y^2, y = -\sqrt{3/x} = f^{-1}(x)$

17. $y = f^{-1}(x), x = f(y) = \begin{cases} 5/2 - y, & y < 2 \\ 1/y, & y \geq 2 \end{cases}, \quad y = f^{-1}(x) = \begin{cases} 5/2 - x, & x > 1/2 \\ 1/x, & 0 < x \leq 1/2 \end{cases}$

19. $y = f^{-1}(x), x = f(y) = (y + 2)^4$ for $y \geq 0$, $y = f^{-1}(x) = x^{1/4} - 2$ for $x \geq 16$

21. $y = f^{-1}(x)$, $x = f(y) = -\sqrt{3 - 2y}$ for $y \le 3/2$, $y = f^{-1}(x) = (3 - x^2)/2$ for $x \le 0$

23. $y = f^{-1}(x)$, $x = f(y) = y - 5y^2$ for $y \ge 1$, $5y^2 - y + x = 0$ for $y \ge 1$,
$y = f^{-1}(x) = (1 + \sqrt{1 - 20x})/10$ for $x \le -4$

25. (a) $y = f(x) = (6.214 \times 10^{-4})x$ (b) $x = f^{-1}(y) = \dfrac{10^4}{6.214}y$
(c) how many meters in y miles

27. (a) $f(f(x)) = \dfrac{3 - \dfrac{3 - x}{1 - x}}{1 - \dfrac{3 - x}{1 - x}} = \dfrac{3 - 3x - 3 + x}{1 - x - 3 + x} = x$ so $f = f^{-1}$

(b) symmetric about the line $y = x$

29. if $f^{-1}(x) = 1$, then $x = f(1) = 2(1)^3 + 5(1) + 3 = 10$

31. $f(f(x)) = x$ thus $f = f^{-1}$ so the graph is symmetric about $y = x$.

33.

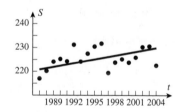

EXERCISE SET 1.6

1. The sum of the squares for the residuals for line I is approximately
$1^2 + 1^2 + 1^2 + 0^2 + 2^2 + 1^2 + 1^2 + 1^2 = 10$, and the same for line II is approximately
$0^2 + (0.4)^2 + (1.2)^2 + 0^2 + (2.2)^2 + (0.6)^2 + (0.2)^2 + 0^2 = 6.84$; line II is the regression line.

3. Least squares line $S = 0.6414t - 1054.517$, correlation coefficient 0.5754

5. (a) Least squares line $p = 0.0146T + 3.98$, correlation coefficient 0.9999
(b) $p = 3.25$ atm   (c) $T = -272°C$

7. (a) $R = 0.00723T + 1.55$ (b) $T = -214°C$

9. (a) $S = 0.50179w - 0.00643$ (b) $S = 8$, $w = 16$ lb

11. (a) Let h denote the height in inches and y the number of rebounds per minute. Then
$y = 0.0174h - 1.0549$, $r = 0.8423$

(b)

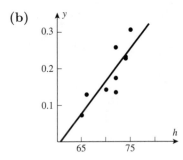

(c) The least squares line is a fair model for these data, since the correlation coefficient is 0.8423.

13. **(a)** $(0 + b - 0)^2 + (1 + b - 0)^2 + (1 + b - 1)^2 = 3b^2 + 2b + 1$

(b) The function of Part (a) is a parabola in the variable b, and has its minimum when $b = -\frac{1}{3}$, hence the desired line is $y = x - \frac{1}{3}$.

15. Let the points be $(A, B), (A, C)$ and (A', B') and let the line of regression have the form $y = m(x - A) + b$. Then the sum of the squares of the residues is $(B - b)^2 + (C - b)^2 + (m(x - A') + b - B')^2$.

Choose m and b so that the line passes through the point (A', B'), which makes the third term zero, and passes through the midpoint between (A, B) and (A, C), which minimizes the sum $(B - b)^2 + (C - b)^2$ (see Exercise 15). These two conditions together minimize the sum of the squares of the residuals, and they determine the slope and one point, and thus they determine the line of regression.

17. **(a)** $H \approx 20000/110 \approx 182$ km/s/Mly

(b) One light year is 9.408×10^{12} km and
$$t = \frac{d}{v} = \frac{1}{H} = \frac{1}{20\text{km/s/Mly}} = \frac{9.408 \times 10^{18}\text{km}}{20\text{km/s}} = 4.704 \times 10^{17} \text{ s} = 1.492 \times 10^{10} \text{ years.}$$

(c) The Universe would be even older.

19. As in Example 4, a possible model is of the form $T = D + A\sin\left[B\left(t - \frac{C}{B}\right)\right]$. Since the longest day has 993 minutes and the shortest has 706, take $2A = 993 - 706 = 287$ or $A = 143.5$. The midpoint between the longest and shortest days is 849.5 minutes, so there is a vertical shift of $D = 849.5$. The period is about 365.25 days, so $2\pi/B = 365.25$ or $B = \pi/183$. Note that the sine function takes the value -1 when $t - \frac{C}{B} = -91.8125$, and T is a minimum at about $t = 0$. Thus the phase shift $\frac{C}{B} \approx 91.5$. Hence $T = 849.5 + 143.5\sin\left[\frac{\pi}{183}t - \frac{\pi}{2}\right]$ is a model for the temperature.

21. $t = 0.445\sqrt{d}$

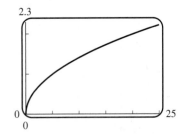

EXERCISE SET 1.7

1. **(a)** $x + 1 = t = y - 1, y = x + 2$

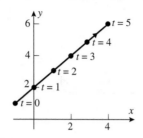

(c)

t	0	1	2	3	4	5
x	-1	0	1	2	3	4
y	1	2	3	4	5	6

3. $t = (x + 4)/3; y = 2x + 10$

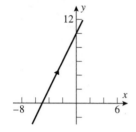

5. $\cos t = x/2, \sin t = y/5;$
$x^2/4 + y^2/25 = 1$

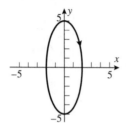

7. $\cos t = (x - 3)/2,$
$\sin t = (y - 2)/4;$
$(x - 3)^2/4 + (y - 2)^2/16 = 1$

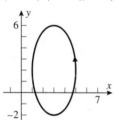

9. $\cos 2t = 1 - 2\sin^2 t;$
$x = 1 - 2y^2,$
$-1 \le y \le 1$

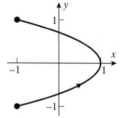

11. $x/2 + y/3 = 1, 0 \le x \le 2, 0 \le y \le 3$

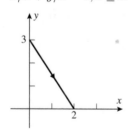

13. $x = 5\cos t, y = -5\sin t, 0 \le t \le 2\pi$

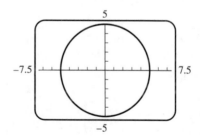

15. $x = 2$, $y = t$

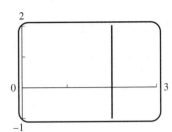

17. $x = t^2$, $y = t$, $-1 \leq t \leq 1$

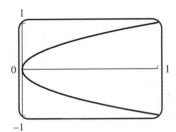

19. (a)

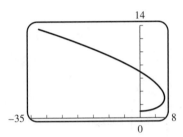

(b)

t	0	1	2	3	4	5
x	0	5.5	8	4.5	-8	-32.5
y	1	1.5	3	5.5	9	13.5

(c) $x = 0$ when $t = 0, 2\sqrt{3}$.

(d) for $0 < t < 2\sqrt{2}$

(e) at $t = 2$

21. (a)

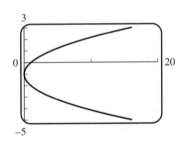

(b)

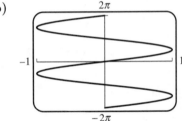

23. (a) Eliminate t to get $\dfrac{x - x_0}{x_1 - x_0} = \dfrac{y - y_0}{y_1 - y_0}$

(b) Set $t = 0$ to get (x_0, y_0); $t = 1$ for (x_1, y_1).

(c) $x = 1 + t$, $y = -2 + 6t$

(d) $x = 2 - t$, $y = 4 - 6t$

25. (a) $|R - P|^2 = (x - x_0)^2 + (y - y_0)^2 = t^2[(x_1 - x_0)^2 + (y_1 - y_0)^2]$ and $|Q - P|^2 = (x_1 - x_0)^2 + (y_1 - y_0)^2$, so $r = |R - P| = |Q - P|t = qt$.

(b) $t = 1/2$

(c) $t = 3/4$

27. (a) IV, because x always increases whereas y oscillates.

(b) II, because $(x/2)^2 + (y/3)^2 = 1$, an ellipse.

(c) V, because $x^2 + y^2 = t^2$ increases in magnitude while x and y keep changing sign.

(d) VI; examine the cases $t < -1$ and $t > -1$ and you see the curve lies in the first, second and fourth quadrants only.

(e) III because $y > 0$.

(f) I; since x and y are bounded, the answer must be I or II; but as t runs, say, from 0 to π, x goes directly from 2 to -2, but y goes from 0 to 1 to 0 to -1 and back to 0, which describes I but not II.

29. The two branches corresponding to $-1 \le t \le 0$ and $0 \le t \le 1$ coincide.

31. (a) $\dfrac{x-b}{a} = \dfrac{y-d}{c}$

(b)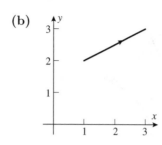

33.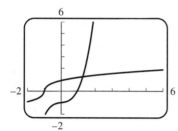

35.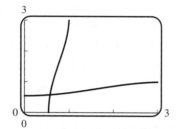

37.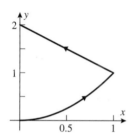

39. (a) $x = 4\cos t,\ y = 3\sin t$

(b) $x = -1 + 4\cos t,\ y = 2 + 3\sin t$

(c)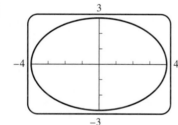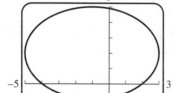

41. (a) From Exercise 40, $x = 400\sqrt{2}\,t,\ y = 400\sqrt{2}\,t - 4.9t^2$.

(b) 16,326.53 m

(c) 65,306.12 m

43. Assume that $a \ne 0$ and $b \ne 0$; eliminate the parameter to get $(x-h)^2/a^2 + (y-k)^2/b^2 = 1$. If $|a| = |b|$ the curve is a circle with center (h, k) and radius $|a|$; if $|a| \ne |b|$ the curve is an ellipse with center (h, k) and major axis parallel to the x-axis when $|a| > |b|$, or major axis parallel to the y-axis when $|a| < |b|$.

(a) ellipses with a fixed center and varying axes of symmetry

(b) (assume $a \ne 0$ and $b \ne 0$) ellipses with varying center and fixed axes of symmetry

(c)   circles of radius 1 with centers on the line $y = x - 1$

45. **(a)**

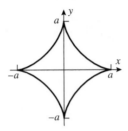

(b) Use $b = a/4$ in the equations of Exercise 44 to get

$x = \frac{3}{4}a \cos \phi + \frac{1}{4}a \cos 3\phi$, $y = \frac{3}{4}a \sin \phi - \frac{1}{4}a \sin 3\phi$;

but trigonometric identities yield $\cos 3\phi = 4 \cos^3 \phi - 3 \cos \phi$, $\sin 3\phi = 3 \sin \phi - 4 \sin^3 \phi$,

so $x = a \cos^3 \phi$, $y = a \sin^3 \phi$.

(c) $x^{2/3} + y^{2/3} = a^{2/3}(\cos^2 \phi + \sin^2 \phi) = a^{2/3}$

REVIEW EXERCISES CHAPTER 1

1.

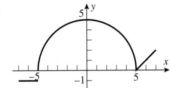

3.

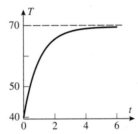

5. **(a)** If the side has length x and height h, then $V = 8 = x^2h$, so $h = 8/x^2$. Then the cost
$C = 5x^2 + 2(4)(xh) = 5x^2 + 64/x$.

(b) The domain of C is $(0, +\infty)$ because x can be very large (just take h very small).

7. **(a)** The base has sides $(10 - 2x)/2$ and $6 - 2x$, and the height is x, so $V = (6 - 2x)(5 - x)x$ ft^3.

(b) From the picture we see that $x < 5$ and $2x < 6$, so $0 < x < 3$.

(c) 3.57 ft $\times$3.79 ft $\times$1.21 ft

9.

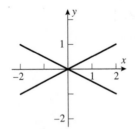

11.

x	-4	-3	-2	-1	0	1	2	3	4
$f(x)$	0	-1	2	1	3	-2	-3	4	-4
$g(x)$	3	2	1	-3	-1	-4	4	-2	0
$(f \circ g)(x)$	4	-3	-2	-1	1	0	-4	2	3
$(g \circ f)(x)$	-1	-3	4	-4	-2	1	2	0	3

13. $f(g(x)) = (3x+2)^2+1, g(f(x)) = 3(x^2+1)+2$, so $9x^2+12x+5 = 3x^2+5, 6x^2+12x = 0, x = 0, -2$

15. When $f(g(h(x)))$ is defined, we require $g(h(x)) \neq 1$ and $h(x) \neq 0$. The first requirement is equivalent to $x \neq \pm 1$, the second is equivalent to $x \neq \pm\sqrt{(2)}$. For all other $x, f \circ g \circ h = 1/(2-x^2)$.

17. (a) even $\times$ odd $=$ odd
     (b) a square is even
     (c) even $+$ odd is neither
     (d) odd $\times$ odd $=$ even

19. (a) The circle of radius 1 centered at (a, a^2); therefore, the family of all circles of radius 1 with centers on the parabola $y = x^2$.

     (b) All parabolas which open up, have latus rectum equal to 1 and vertex on the line $y = x/2$.

21. (a)

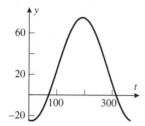

     (b) when $\dfrac{2\pi}{365}(t - 101) = \dfrac{3\pi}{2}$, or $t = 374.75$, which is the same date as $t = 9.75$, so during the night of January 10th-11th

     (c) from $t = 0$ to $t = 70.58$ and from $t = 313.92$ to $t = 365$ (the same date as $t = 0$), for a total of about 122 days

23. When $x = 0$ the value of the green curve is higher than that of the blue curve, therefore the blue curve is given by $y = 1 + 2\sin x$.
The points A, B, C, D are the points of intersection of the two curves, i.e. where
$1+2\sin x = 2\sin(x/2)+2\cos(x/2)$. Let $\sin(x/2) = p, \cos(x/2) = q$. Then $2\sin x = 4\sin(x/2)\cos(x/2)$ (basic trigonometric identity, so the equation which yields the points of intersection becomes $1 + 4pq = 2p + 2q$,
$4pq - 2p - 2q + 1 = 0, (2p-1)(2q-1) = 0$; thus whenever either $\sin(x/2) = 1/2$ or $\cos(x/2) = 1/2$, i.e. when $x/2 = \pi/6, 5\pi/6, \pm\pi/3$. Thus A has coordinates $(-2\pi/3, 1 - \sqrt{3})$, B has coordinates $(\pi/3, 1 + \sqrt{3})$, C has coordinates $(2\pi/3, 1 + \sqrt{3})$, and D has coordinates $(5\pi/3, 1 - \sqrt{3})$.

25. (a) $f(g(x)) = x$ for all x in the domain of g, and $g(f(x)) = x$ for all x in the domain of f.
     (b) They are reflections of each other through the line $y = x$.
     (c) The domain of one is the range of the other and vice versa.
     (d) The equation $y = f(x)$ can always be solved for x as a function of y. Functions with no inverses include $y = x^2, y = \sin x$.

27. (a)

1.90	1.92	1.94	1.96	1.98	2.00	2.02	2.04	2.06	2.08	2.10
3.4161	3.4639	3.5100	3.5543	3.5967	3.6372	3.6756	3.7119	3.7459	3.7775	3.8068

     (b) $y = 1.9589x - 0.2910$

     (c) $y - 3.6372 = 1.9589(x - 2)$, or $y = 1.9589x - 0.2806$

(d) As one zooms in on the point $(2, f(2))$ the two curves seem to converge to one line.

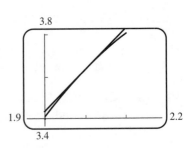

29. The data are periodic, so it is reasonable that a trigonometric function might approximate them. A possible model is of the form $T = D + A \sin \left[B \left(t - \dfrac{C}{B} \right) \right]$. Since the highest level is 1.032 meters and the lowest is 0.045, take $2A = 1.032 - 0.042 = 0.990$ or $A = 0.495$. The midpoint between the lowest and highest levels is 0.537 meters, so there is a vertical shift of $D = 0.537$. The period is about 12 hours, so $2\pi/B = 12$ or $B = \pi/6$. The phase shift $\dfrac{C}{B} \approx 6.5$. Hence $T = 0.537 + 0.495 \sin \left[\dfrac{\pi}{6} (t - 6.5) \right]$ is a model for the temperature.

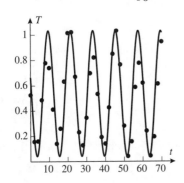

31. $x(t) = \sqrt{2} \cos t, \quad y(t) = -\sqrt{2} \sin t, \quad 0 \le t \le 3\pi/2$

33.

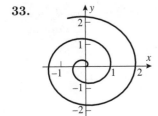

CHAPTER 2
Limits and Continuity

EXERCISE SET 2.1

1. (a) 0   (b) 0   (c) 0   (d) 3

3. (a) $-\infty$ (b) $-\infty$ (c) $-\infty$ (d) 1

5. for all $x_0 \neq -4$

13. (a)

2	1.5	1.1	1.01	1.001	0	0.5	0.9	0.99	0.999
0.1429	0.2105	0.3021	0.3300	0.3330	1.0000	0.5714	0.3690	0.3367	0.3337

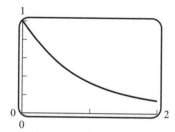

The limit is 1/3.

(b)

2	1.5	1.1	1.01	1.001	1.0001
0.4286	1.0526	6.344	66.33	666.3	6666.3

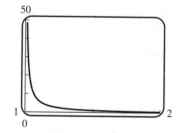

The limit is $+\infty$.

(c)

0	0.5	0.9	0.99	0.999	0.9999
-1	-1.7143	-7.0111	-67.001	-667.0	-6667.0

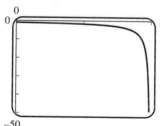

The limit is $-\infty$.

15. (a)

-0.25	-0.1	-0.001	-0.0001	0.0001	0.001	0.1	0.25
2.7266	2.9552	3.0000	3.0000	3.0000	3.0000	2.9552	2.7266

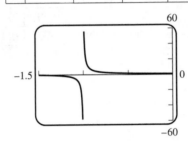

The limit is 3.

(b)

0	−0.5	−0.9	−0.99	−0.999	−1.5	−1.1	−1.01	−1.001
1	1.7552	6.2161	54.87	541.1	−0.1415	−4.536	−53.19	−539.5

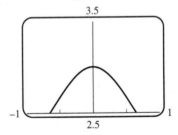

The limit does not exist.

17. $m_{sec} = \dfrac{x^2 - 1}{x + 1} = x - 1$ which gets close to −2 as x gets close to −1, thus $y - 1 = -2(x + 1)$ or $y = -2x - 1$

19. $m_{sec} = \dfrac{x^4 - 1}{x - 1} = x^3 + x^2 + x + 1$ which gets close to 4 as x gets close to 1, thus $y - 1 = 4(x - 1)$ or $y = 4x - 3$

21. **(a)** The limit appears to be 3.   **(b)** The limit appears to be 3.

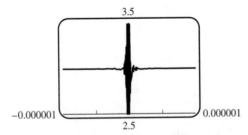

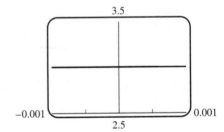

(c) The limit does not exist.

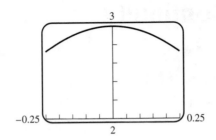

23. **(a)** The plot over the interval $[-a, a]$ becomes subject to catastrophic subtraction if a is small enough (the size depending on the machine).

   (c) It does not.

25. **(a)** The length of the rod while at rest

   (b) The limit is zero. The length of the rod approaches zero.

EXERCISE SET 2.2

1. (a) -6 **(b)** 13 **(c)** -8 **(d)** 16 **(e)** 2 **(f)** $-1/2$
  (g) The limit doesn't exist because the denominator tends to zero but the numerator doesn't.
  (h) The limit doesn't exist because the denominator tends to zero but the numerator doesn't.

3. 6 **5.** $3/4$ **7.** 4 **9.** $-4/5$

11. -3 **13.** $3/2$ **15.** $+\infty$ **17.** does not exist

19. does not exist   **21.** $+\infty$ **23.** does not exist   **25.** $+\infty$

27. $+\infty$ **29.** 6

31. (a) 2 **(b)** 2 **(c)** 2

33. (a) 3 **(b)**

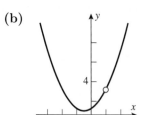

35. (a) Theorem 2.2.2(a) doesn't apply; moreover one cannot add/subtract infinities.
  (b) $\displaystyle \lim_{x\to 0^+}\left(\frac{1}{x}-\frac{1}{x^2}\right)=\lim_{x\to 0^+}\left(\frac{x-1}{x^2}\right)=-\infty$

37. $\displaystyle \lim_{x\to 0}\frac{x}{x\left(\sqrt{x+4}+2\right)}=\frac{1}{4}$

39. The left and/or right limits could be plus or minus infinity; or the limit could exist, or equal any preassigned real number. For example, let $q(x)=x-x_0$ and let $p(x)=a(x-x_0)^n$ where n takes on the values $0,1,2$.

EXERCISE SET 2.3

1. (a) $-\infty$ **(b)** $+\infty$

3. (a) 0 **(b)** -1

5. (a) -12 **(b)** 21 **(c)** -15 **(d)** 25
  (e) 2 **(f)** $-3/5$ **(g)** 0
  (h) The limit doesn't exist because the denominator tends to zero but the numerator doesn't.

7. $-\infty$ **9.** $+\infty$ **11.** $3/2$ **13.** 0

15. 0 **17.** $-5^{1/3}/2$ **19.** $-\sqrt{5}$ **21.** $1/\sqrt{6}$

23. $\sqrt{3}$ **25.** $-\infty$ **27.** $-1/7$

29. It appears that $\lim\limits_{t\to+\infty} n(t) = +\infty$, and $\lim\limits_{t\to+\infty} e(t) = c$.

31. **(a)** $+\infty$ **(b)** -5

33. $\lim\limits_{x\to+\infty} (\sqrt{x^2+3} - x)\dfrac{\sqrt{x^2+3}+x}{\sqrt{x^2+3}+x} = \lim\limits_{x\to+\infty} \dfrac{3}{\sqrt{x^2+3}+x} = 0$

35. $\lim\limits_{x\to+\infty} \left(\sqrt{x^2+ax} - x\right)\dfrac{\sqrt{x^2+ax}+x}{\sqrt{x^2+ax}+x} = \lim\limits_{x\to+\infty} \dfrac{ax}{\sqrt{x^2+ax}+x} = a/2$

37. $\lim\limits_{x\to+\infty} p(x) = (-1)^n\infty$ and $\lim\limits_{x\to-\infty} p(x) = +\infty$

39. If $m > n$ the limits are both zero. If $m = n$ the limits are both equal to a_m, the leading coefficient of p. If $n > m$ the limits are $\pm\infty$ where the sign depends on the sign of a_m and whether n is even or odd.

41. If $m > n$ the limit is 0. If $m = n$ the limit is -3. If $m < n$ and $n - m$ is odd, then the limit is $+\infty$; if $m < n$ and $n - m$ is even, then the limit is $-\infty$.

43. $\lim\limits_{x\to+\infty} f(x) = \lim\limits_{x\to-\infty} f(x) = L$

45. $f(x) = x + 2 + \dfrac{2}{x-2}$,

so $\lim\limits_{x\to\pm\infty} (f(x) - (x+2)) = 0$

and $f(x)$ is asymptotic to $y = x + 2$.

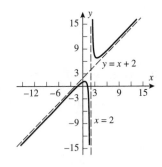

47. $f(x) = -x^2 + 1 + 2/(x-3)$

so $\lim\limits_{x\to\pm\infty} [f(x) - (-x^2+1)] = 0$

and $f(x)$ is asymptotic to $y = -x^2 + 1$.

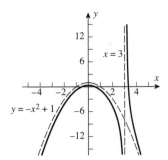

49. $f(x) - \sin x = 0$ and $f(x)$ is asymptotic to $y = \sin x$.

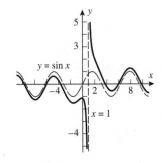

EXERCISE SET 2.4

1. **(a)** $|f(x) - f(0)| = |x + 2 - 2| = |x| < 0.1$ if and only if $|x| < 0.1$

   (b) $|f(x) - f(3)| = |(4x - 5) - 7| = 4|x - 3| < 0.1$ if and only if $|x - 3| < (0.1)/4 = 0.0025$

   (c) $|f(x) - f(4)| = |x^2 - 16| < \epsilon$ if $|x - 4| < \delta$. We get $f(x) = 16 + \epsilon = 16.001$ at $x = 4.000124998$, which corresponds to $\delta = 0.000124998$; and $f(x) = 16 - \epsilon = 15.999$ at $x = 3.999874998$, for which $\delta = 0.000125002$. Use the smaller δ: thus $|f(x) - 16| < \epsilon$ provided $|x - 4| < 0.000125$ (to six decimals).

3. **(a)** $x_1 = (1.95)^2 = 3.8025, x_2 = (2.05)^2 = 4.2025$

   (b) $\delta = \min\left(|4 - 3.8025|, |4 - 4.2025|\right) = 0.1975$

5. $|(x^3 - 4x + 5) - 2| < 0.05,$
   $-0.05 < (x^3 - 4x + 5) - 2 < 0.05,$
   $1.95 < x^3 - 4x + 5 < 2.05;$
   $x^3 - 4x + 5 = 1.95$ at $x = 1.0616,$
   $x^3 - 4x + 5 = 2.05$ at $x = 0.9558;$
   $\delta = \min(1.0616 - 1, 1 - 0.9558) = 0.0442$

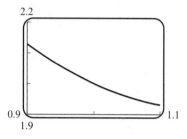

7. With the TRACE feature of a calculator we discover that (to five decimal places) $(0.87000, 1.80274)$ and $(1.13000, 2.19301)$ belong to the graph. Set $x_0 = 0.87$ and $x_1 = 1.13$. Since the graph of $f(x)$ rises from left to right, we see that if $x_0 < x < x_1$ then $1.80274 < f(x) < 2.19301$, and therefore $1.8 < f(x) < 2.2$. So we can take $\delta = 0.13$.

9. $|2x - 8| = 2|x - 4| < 0.1$ if $|x - 4| < 0.05$, $\delta = 0.05$

11. $\left|\dfrac{x^2 - 9}{x - 3} - 6\right| = |x + 3 - 6| = |x - 3| < 0.05$ if $|x - 3| < 0.05, \delta = 0.05$

13. On the interval $[1, 3]$ we have $|x^2 + x + 2| \leq 14$, so $|x^3 - 8| = |x - 2||x^2 + x + 2| \leq 14|x - 2| < 0.001$ provided $|x - 2| < 0.001 \cdot \dfrac{1}{14}$; but $0.00005 < \dfrac{0.001}{14}$, so for convenience we take $\delta = 0.00005$ (there is no need to choose an 'optimal' δ).

15. if $\delta \leq 1$ then $|x| > 3$, so $\left|\dfrac{1}{x} - \dfrac{1}{5}\right| = \dfrac{|x - 5|}{5|x|} \leq \dfrac{|x - 5|}{15} < 0.05$ if $|x - 5| < 0.75, \delta = 3/4$

17. **(a)** $\lim\limits_{x \to 4} f(x) = 3$

   (b) $|10f(x) - 30| = 10|f(x) - 3| < 0.005$ provided $|f(x) - 3| < 0.0005$, which is true for $|x - 3| < 0.0001, \delta = 0.0001$

19. It suffices to have $|10f(x) + 2x - 38| \leq |10f(x) - 30| + 2|x - 4| < 0.01$, by the triangle inequality. To ensure $|10f(x) - 30| < 0.005$ use Exercise 17 (with $\epsilon = 0.0005$) to get $\delta = 0.0001$. Then $|x - 4| < \delta$ yields $|10f(x) + 2x - 38| \leq 10|f(x) - 3| + 2|x - 4| \leq (10)\dot{0}.0005 + (2)\dot{0}.0001 \leq 0.005 + 0.0002 < 0.01$

21. $|3x - 15| = 3|x - 5| < \epsilon$ if $|x - 5| < \frac{1}{3}\epsilon, \delta = \frac{1}{3}\epsilon$

23. $\left|\dfrac{2x^2 + x}{x} - 1\right| = |2x| < \epsilon$ if $|x| < \frac{1}{2}\epsilon, \delta = \frac{1}{2}\epsilon$

25. $|f(x) - 3| = |x + 2 - 3| = |x - 1| < \epsilon$ if $0 < |x - 1| < \epsilon, \delta = \epsilon$

27. **(a)** $|(3x^2 + 2x - 20 - 300| = |3x^2 + 2x - 320| = |(3x + 32)(x - 10)| = |3x + 32| \cdot |x - 10|$

   (b) If $|x - 10| < 1$ then $|3x + 32| < 65$, since clearly $x < 11$

   (c) $\delta = \min(1, \epsilon/65); \quad |3x + 32| \cdot |x - 10| < 65 \cdot |x - 10| < 65 \cdot \epsilon/65 = \epsilon$

29. if $\delta < 1$ then $|2x^2 - 2| = 2|x - 1||x + 1| < 6|x - 1| < \epsilon$ if $|x - 1| < \frac{1}{6}\epsilon$, $\delta = \min(1, \frac{1}{6}\epsilon)$

31. If $\delta < \frac{1}{2}$ and $|x - (-2)| < \delta$ then $-\frac{5}{2} < x < -\frac{3}{2}$, $x + 1 < -\frac{1}{2}$, $|x + 1| > \frac{1}{2}$; then
$$\left|\frac{1}{x + 1} - (-1)\right| = \frac{|x + 2|}{|x + 1|} < 2|x + 2| < \epsilon \text{ if } |x + 2| < \frac{1}{2}\epsilon, \delta = \min\left(\frac{1}{2}, \frac{1}{2}\epsilon\right)$$

33. $|\sqrt{x} - 2| = \left|(\sqrt{x} - 2)\frac{\sqrt{x} + 2}{\sqrt{x} + 2}\right| = \left|\frac{x - 4}{\sqrt{x} + 2}\right| < \frac{1}{2}|x - 4| < \epsilon \text{ if } |x - 4| < 2\epsilon, \delta = 2\epsilon$

35. **(a)** $|f(x) - L| = \frac{1}{x^2} < 0.1 \text{ if } x > \sqrt{10}, N = \sqrt{10}$

   (b) $|f(x) - L| = \left|\frac{x}{x + 1} - 1\right| = \left|\frac{1}{x + 1}\right| < 0.01 \text{ if } x + 1 > 100, N = 99$

   (c) $|f(x) - L| = \left|\frac{1}{x^3}\right| < \frac{1}{1000} \text{ if } |x| > 10, x < -10, N = -10$

   (d) $|f(x) - L| = \left|\frac{x}{x + 1} - 1\right| = \left|\frac{1}{x + 1}\right| < 0.01 \text{ if } |x + 1| > 100, -x - 1 > 100, x < -101,$
   $N = -101$

37. **(a)** $\frac{x_1^2}{1 + x_1^2} = 1 - \epsilon, x_1 = -\sqrt{\frac{1 - \epsilon}{\epsilon}}; \quad \frac{x_2^2}{1 + x_2^2} = 1 - \epsilon, x_2 = \sqrt{\frac{1 - \epsilon}{\epsilon}}$

   (b) $N = \sqrt{\frac{1 - \epsilon}{\epsilon}}$ **(c)** $N = -\sqrt{\frac{1 - \epsilon}{\epsilon}}$

39. $\frac{1}{x^2} < 0.01 \text{ if } |x| > 10, N = 10$

41. $\left|\frac{x}{x + 1} - 1\right| = \left|\frac{1}{x + 1}\right| < 0.001 \text{ if } |x + 1| > 1000, x > 999, N = 999$

43. $\left|\frac{1}{x + 2} - 0\right| < 0.005 \text{ if } |x + 2| > 200, -x - 2 > 200, x < -202, N = -202$

45. $\left|\frac{4x - 1}{2x + 5} - 2\right| = \left|\frac{11}{2x + 5}\right| < 0.1 \text{ if } |2x + 5| > 110, -2x - 5 > 110, 2x < -115, x < -57.5, N = -57.5$

47. $\left|\frac{1}{x^2}\right| < \epsilon \text{ if } |x| > \frac{1}{\sqrt{\epsilon}}, N = \frac{1}{\sqrt{\epsilon}}$

49. $\left|\frac{4x - 1}{2x + 5} - 2\right| = \left|\frac{11}{2x + 5}\right| < \epsilon \text{ if } |2x + 5| > \frac{11}{\epsilon}, -2x - 5 > \frac{11}{\epsilon}, 2x < -\frac{11}{\epsilon} - 5, x < -\frac{11}{2\epsilon} - \frac{5}{2},$
   $N = -\frac{5}{2} - \frac{11}{2\epsilon}$

51. $\left|\frac{2\sqrt{x}}{\sqrt{x} - 1} - 2\right| = \left|\frac{2}{\sqrt{x} - 1}\right| < \epsilon \text{ if } \sqrt{x} - 1 > \frac{2}{\epsilon}, \quad \sqrt{x} > 1 + \frac{2}{\epsilon}, \quad x > \left(1 + \frac{2}{\epsilon}\right)^2, \quad N > \left(1 + \frac{2}{\epsilon}\right)^2$

53. **(a)** $\dfrac{1}{x^2} > 100$ if $|x| < \dfrac{1}{10}$ **(b)** $\dfrac{1}{|x-1|} > 1000$ if $|x-1| < \dfrac{1}{1000}$

  (c) $\dfrac{-1}{(x-3)^2} < -1000$ if $|x-3| < \dfrac{1}{10\sqrt{10}}$ **(d)** $-\dfrac{1}{x^4} < -10000$ if $x^4 < \dfrac{1}{10000}$, $|x| < \dfrac{1}{10}$

55. if $M > 0$ then $\dfrac{1}{(x-3)^2} > M$, $0 < (x-3)^2 < \dfrac{1}{M}$, $0 < |x-3| < \dfrac{1}{\sqrt{M}}$, $\delta = \dfrac{1}{\sqrt{M}}$

57. if $M > 0$ then $\dfrac{1}{|x|} > M$, $0 < |x| < \dfrac{1}{M}$, $\delta = \dfrac{1}{M}$

59. if $M < 0$ then $-\dfrac{1}{x^4} < M$, $0 < x^4 < -\dfrac{1}{M}$, $|x| < \dfrac{1}{(-M)^{1/4}}$, $\delta = \dfrac{1}{(-M)^{1/4}}$

61. if $x > 2$ then $|x+1-3| = |x-2| = x-2 < \epsilon$ if $2 < x < 2+\epsilon$, $\delta = \epsilon$

63. if $x > 4$ then $\sqrt{x-4} < \epsilon$ if $x-4 < \epsilon^2$, $4 < x < 4+\epsilon^2$, $\delta = \epsilon^2$

65. if $x > 2$ then $|f(x) - 2| = |x-2| = x-2 < \epsilon$ if $2 < x < 2+\epsilon$, $\delta = \epsilon$

67. **(a)** if $M < 0$ and $x > 1$ then $\dfrac{1}{1-x} < M$, $x-1 < -\dfrac{1}{M}$, $1 < x < 1 - \dfrac{1}{M}$, $\delta = -\dfrac{1}{M}$

  (b) if $M > 0$ and $x < 1$ then $\dfrac{1}{1-x} > M$, $1-x < \dfrac{1}{M}$, $1 - \dfrac{1}{M} < x < 1$, $\delta = \dfrac{1}{M}$

69. **(a)** Given any $M > 0$ there corresponds $N > 0$ such that if $x > N$ then $f(x) > M$, $x+1 > M$, $x > M - 1$, $N = M - 1$.

  (b) Given any $M < 0$ there corresponds $N < 0$ such that if $x < N$ then $f(x) < M$, $x+1 < M$, $x < M - 1$, $N = M - 1$.

71. if $\delta \le 2$ then $|x-3| < 2$, $-2 < x-3 < 2$, $1 < x < 5$, and $|x^2 - 9| = |x+3||x-3| < 8|x-3| < \epsilon$ if $|x-3| < \frac{1}{8}\epsilon$, $\delta = \min\left(2, \frac{1}{8}\epsilon\right)$

73. **(a)** 0.4 amperes **(b)** $[0.3947, 0.4054]$ **(c)** $\left[\dfrac{3}{7.5+\delta}, \dfrac{3}{7.5-\delta}\right]$

  (d) 0.0187 **(e)** It becomes infinite.

EXERCISE SET 2.5

1. **(a)** no, $x = 2$ **(b)** no, $x = 2$ **(c)** no, $x = 2$ **(d)** yes
  (e) yes **(f)** yes

3. **(a)** no, $x = 1, 3$ **(b)** yes **(c)** no, $x = 1$ **(d)** yes
  (e) no, $x = 3$ **(f)** yes

5. **(a)** 3 **(b)** 3

7. **(a)** 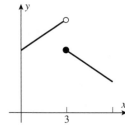 **(b)**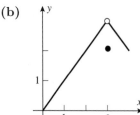

(c)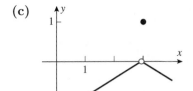

(d)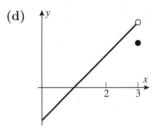

9. (a)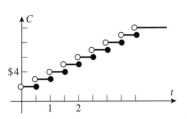

(b) One second could cost you one dollar.

11. none   **13.** none   **15.** $x = 0, -1/2$

17. $x = -1, 0, 1$ **19.** none

21. none; $f(x) = 2x + 3$ is continuous on $x < 4$ and $f(x) = 7 + \dfrac{16}{x}$ is continuous on $4 < x$; $\lim\limits_{x \to 4^-} f(x) = \lim\limits_{x \to 4^+} f(x) = f(4) = 11$ so f is continuous at $x = 4$

23. (a) f is continuous for $x < 1$, and for $x > 1$; $\lim\limits_{x \to 1^-} f(x) = 5$, $\lim\limits_{x \to 1^+} f(x) = k$, so if $k = 5$ then f is continuous for all x

(b) f is continuous for $x < 2$, and for $x > 2$; $\lim\limits_{x \to 2^-} f(x) = 4k$, $\lim\limits_{x \to 2^+} f(x) = 4 + k$, so if $4k = 4 + k$, $k = 4/3$ then f is continuous for all x

25. f is continuous for $x < -1$, $-1 < x < 2$ and $x > 2$; $\lim\limits_{x \to -1^-} f(x) = 4$, $\lim\limits_{x \to -1^+} f(x) = k$, so $k = 4$ is required. Next, $\lim\limits_{x \to 2^-} f(x) = 3m + k = 3m + 4$, $\lim\limits_{x \to 2^+} f(x) = 9$, so $3m + 4 = 9, m = 5/3$ and f is continuous everywhere if $k = 4, m = 5/3$

27. (a) 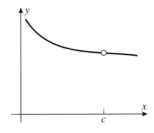   **(b)**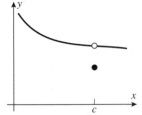

29. (a) $x = 0$, $\lim\limits_{x \to 0^-} f(x) = -1 \neq +1 = \lim\limits_{x \to 0^+} f(x)$ so the discontinuity is not removable

(b) $x = -3$; define $f(-3) = -3 = \lim\limits_{x \to -3} f(x)$, then the discontinuity is removable

(c) f is undefined at $x = \pm 2$; at $x = 2$, $\lim\limits_{x \to 2} f(x) = 1$, so define $f(2) = 1$ and f becomes continuous there; at $x = -2$, $\lim\limits_{x \to -2}$ does not exist, so the discontinuity is not removable

31. **(a)** discontinuity at $x = 1/2$, not removable; **(b)** $2x^2 + 5x - 3 = (2x - 1)(x + 3)$
at $x = -3$, removable

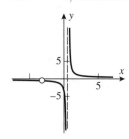

33. For $x > 0$, $f(x) = x^{3/5} = (x^3)^{1/5}$ is the composition (Theorem 2.5.6) of the two continuous functions $g(x) = x^3$ and $h(x) = x^{1/5}$ and is thus continuous. For $x < 0$, $f(x) = f(-x)$ which is the composition of the continuous functions $f(x)$ (for positive x) and the continuous function $y = -x$. Hence $f(-x)$ is continuous for all $x > 0$. At $x = 0$, $f(0) = \lim\limits_{x \to 0} f(x) = 0$.

35. **(a)** Let $f(x) = k$ for $x \neq c$ and $f(c) = 0$; $g(x) = l$ for $x \neq c$ and $g(c) = 0$. If $k = -l$ then $f + g$ is continuous; otherwise it's not.

     (b) $f(x) = k$ for $x \neq c$, $f(c) = 1$; $g(x) = l \neq 0$ for $x \neq c$, $g(c) = 1$. If $kl = 1$, then fg is continuous; otherwise it's not.

37. Since f and g are continuous at $x = c$ we know that $\lim\limits_{x \to c} f(x) = f(c)$ and $\lim\limits_{x \to c} g(x) = g(c)$. In the following we use Theorem 2.2.2.

     (a) $f(c) + g(c) = \lim\limits_{x \to c} f(x) + \lim\limits_{x \to c} g(x) = \lim\limits_{x \to c}(f(x) + g(x))$ so $f + g$ is continuous at $x = c$.

     (b) same as (a) except the $+$ sign becomes a $-$ sign

     (c) $f(c) \cdot g(c) = \left(\lim\limits_{x \to c} f(x)\right)\left(\lim\limits_{x \to c} g(x)\right) = \lim\limits_{x \to c} f(x) \cdot g(x)$ so $f \cdot g$ is continuous at $x = c$

39. Of course such a function must be discontinuous. Let $f(x) = 1$ on $0 \leq x < 1$, and $f(x) = -1$ on $1 \leq x \leq 2$.

41. The cone has volume $\pi r^2 h/3$. The function $V(r) = \pi r^2 h$ (for variable r and fixed h) gives the volume of a right circular cylinder of height h and radius r, and satisfies $V(0) < \pi r^2 h/3 < V(r)$. By the Intermediate Value Theorem there is a value c between 0 and r such that $V(c) = \pi r^2 h/3$, so the cylinder of radius c (and height h) has volume equal to that of the cone.

43. If $f(x) = x^3 + x^2 - 2x$ then $f(-1) = 2$, $f(1) = 0$. Use the Intermediate Value Theorem.

45. For the negative root, use intervals on the x-axis as follows: $[-2, -1]$; since $f(-1.3) < 0$ and $f(-1.2) > 0$, the midpoint $x = -1.25$ of $[-1.3, -1.2]$ is the required approximation of the root. For the positive root use the interval $[0, 1]$; since $f(0.7) < 0$ and $f(0.8) > 0$, the midpoint $x = 0.75$ of $[0.7, 0.8]$ is the required approximation.

47. For the negative root, use intervals on the x-axis as follows: $[-2, -1]$; since $f(-1.7) < 0$ and $f(-1.6) > 0$, use the interval $[-1.7, -1.6]$. Since $f(-1.61) < 0$ and $f(-1.60) > 0$ the midpoint $x = -1.605$ of $[-1.61, -1.60]$ is the required approximation of the root. For the positive root use the interval $[1, 2]$; since $f(1.3) > 0$ and $f(1.4) < 0$, use the interval $[1.3, 1.4]$. Since $f(1.37) > 0$ and $f(1.38) < 0$, the midpoint $x = 1.375$ of $[1.37, 1.38]$ is the required approximation.

49. $x = 2.24$

51. The uncoated sphere has volume $4\pi(x-1)^3/3$ and the coated sphere has volume $4\pi x^3/3$. If the volume of the uncoated sphere and of the coating itself are the same, then the coated sphere has twice the volume of the uncoated sphere. Thus $2(4\pi(x-1)^3/3) = 4\pi x^3/3$, or $x^3 - 6x^2 + 6x - 2 = 0$, with the solution $x = 4.847$ cm.

53. We must show $\lim\limits_{x\to c} f(x) = f(c)$. Let $\epsilon > 0$; then there exists $\delta > 0$ such that if $|x - c| < \delta$ then $|f(x) - f(c)| < \epsilon$. But this certainly satisfies Definition 2.4.1.

EXERCISE SET 2.6

1. none

3. $n\pi, n = 0, \pm 1, \pm 2, \dots$

5. $x = n\pi, n = 0, \pm 1, \pm 2, \dots$

7. $2n\pi + \pi/6, 2n\pi + 5\pi/6, n = 0, \pm 1, \pm 2, \dots$

9. (a) $\sin x, x^3 + 7x + 1$ **(b)** $|x|, \sin x$ **(c)** $x^3, \cos x, x + 1$
    (d) $\sqrt{x}, 3 + x, \sin x, 2x$ **(e)** $\sin x, \sin x$ **(f)** $x^5 - 2x^3 + 1, \cos x$

11. $\cos\left(\lim\limits_{x\to+\infty} \dfrac{1}{x}\right) = \cos 0 = 1$

13. $3\lim\limits_{\theta\to 0} \dfrac{\sin 3\theta}{3\theta} = 3$

15. $\left(\lim\limits_{\theta\to 0^+} \dfrac{1}{\theta}\right) \lim\limits_{\theta\to 0^+} \dfrac{\sin\theta}{\theta} = +\infty$

17. $\dfrac{\tan 7x}{\sin 3x} = \dfrac{7}{3\cos 7x} \dfrac{\sin 7x}{7x} \dfrac{3x}{\sin 3x}$ so $\lim\limits_{x\to 0} \dfrac{\tan 7x}{\sin 3x} = \dfrac{7}{3(1)}(1)(1) = \dfrac{7}{3}$

19. $\dfrac{1}{5} \lim\limits_{x\to 0^+} \sqrt{x} \lim\limits_{x\to 0^+} \dfrac{\sin x}{x} = 0$

21. $\left(\lim\limits_{x\to 0} x\right)\left(\lim\limits_{x\to 0} \dfrac{\sin x^2}{x^2}\right) = 0$

23. $\dfrac{t^2}{1 - \cos^2 t} = \left(\dfrac{t}{\sin t}\right)^2$, so $\lim\limits_{t\to 0} \dfrac{t^2}{1 - \cos^2 t} = 1$

25. $\dfrac{\theta^2}{1 - \cos\theta} \dfrac{1 + \cos\theta}{1 + \cos\theta} = \dfrac{\theta^2(1 + \cos\theta)}{1 - \cos^2\theta} = \left(\dfrac{\theta}{\sin\theta}\right)^2 (1 + \cos\theta)$ so $\lim\limits_{\theta\to 0} \dfrac{\theta^2}{1 - \cos\theta} = (1)^2 2 = 2$

27. $\lim\limits_{x\to 0^+} \sin\left(\dfrac{1}{x}\right) = \lim\limits_{t\to +\infty} \sin t$; limit does not exist

29. $\dfrac{2 - \cos 3x - \cos 4x}{x} = \dfrac{1 - \cos 3x}{x} + \dfrac{1 - \cos 4x}{x}$. Note that
$\dfrac{1 - \cos 3x}{x} = \dfrac{1 - \cos 3x}{x} \dfrac{1 + \cos 3x}{1 + \cos 3x} = \dfrac{\sin^2 3x}{x(1 + \cos 3x)} = \dfrac{\sin 3x}{x} \dfrac{\sin 3x}{1 + \cos 3x}$. Thus
$\lim\limits_{x\to 0} \dfrac{2 - \cos 3x - \cos 4x}{x} = \lim\limits_{x\to 0} \dfrac{\sin 3x}{x} \dfrac{\sin 3x}{1 + \cos 3x} + \lim\limits_{x\to 0} \dfrac{\sin 4x}{x} \dfrac{\sin 4x}{1 + \cos 4x} = 0 + 0 = 0$

31. a/b

33.

5.1	5.01	5.001	5.0001	5.00001	4.9	4.99	4.999	4.9999	4.99999
0.098845	0.099898	0.99990	0.099999	0.100000	0.10084	0.10010	0.10001	0.10000	0.10000

The limit is 0.1.

35.

−1.9	−1.99	−1.999	−1.9999	−1.99999	−2.1	−2.01	−2.001	−2.0001	−2.00001
−0.898785	−0.989984	−0.999000	−0.999900	−0.999990	−1.097783	−1.009983	−1.001000	−1.000100	−1.000010

The limit is −1.

37. Since $\lim\limits_{x\to 0}\sin(1/x)$ does not exist, no conclusions can be drawn.

39. $\lim\limits_{x\to 0^-}f(x)=k\lim\limits_{x\to 0}\dfrac{\sin kx}{kx\cos kx}=k$, $\lim\limits_{x\to 0^+}f(x)=2k^2$, so $k=2k^2$, $k=\dfrac{1}{2}$

41. **(a)** $\lim\limits_{t\to 0^+}\dfrac{\sin t}{t}=1$ **(b)** $\lim\limits_{t\to 0^-}\dfrac{1-\cos t}{t}=0$ (Theorem 2.6.3)

     (c) $\sin(\pi-t)=\sin t$, so $\lim\limits_{x\to\pi}\dfrac{\pi-x}{\sin x}=\lim\limits_{t\to 0}\dfrac{t}{\sin t}=1$

43. $t=x-1$; $\sin(\pi x)=\sin(\pi t+\pi)=-\sin\pi t$; and $\lim\limits_{x\to 1}\dfrac{\sin(\pi x)}{x-1}=-\lim\limits_{t\to 0}\dfrac{\sin\pi t}{t}=-\pi$

45. $-|x|\le x\cos\left(\dfrac{50\pi}{x}\right)\le |x|$

47. $\lim\limits_{x\to 0}f(x)=1$ by the Squeezing Theorem

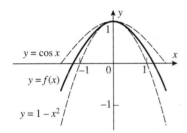

49. Let $g(x)=-\dfrac{1}{x}$ and $h(x)=\dfrac{1}{x}$; thus $\lim\limits_{x\to+\infty}\dfrac{\sin x}{x}=0$ by the Squeezing Theorem.

51. **(a)** $\sin x=\sin t$ where x is measured in degrees, t is measured in radians and $t=\dfrac{\pi x}{180}$. Thus

     $\lim\limits_{x\to 0}\dfrac{\sin x}{x}=\lim\limits_{t\to 0}\dfrac{\sin t}{(180t/\pi)}=\dfrac{\pi}{180}$.

53. **(a)** $\sin 10°=0.17365$ **(b)** $\sin 10°=\sin\dfrac{\pi}{18}\approx\dfrac{\pi}{18}=0.17453$

55. **(a)** 0.08749 **(b)** $\tan 5°\approx\dfrac{\pi}{36}=0.08727$

57. **(a)** Let $f(x)=x-\cos x$; $f(0)=-1$, $f(\pi/2)=\pi/2$. By the IVT there must be a solution of $f(x)=0$.

     (b) 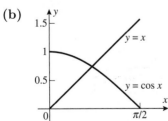   **(c)** 0.739

59. **(a)** Gravity is stronger at the poles than at the equator.

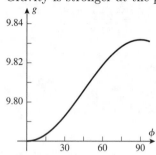

(b) Let $g(\phi)$ be the given function. Then $g(38) < 9.8$ and $g(39) > 9.8$, so by the Intermediate Value Theorem there is a value c between 38 and 39 for which $g(c) = 9.8$ exactly.

REVIEW EXERCISES CHAPTER 2

1. **(a)** 1   **(b)** no limit   **(c)** no limit
     (d) 1   **(e)** 3   **(f)** 0
     (g) 0   **(h)** 2   **(i)** 1/2

3. **(a)**

x	1	0.1	0.01	0.001	0.0001	0.00001	0.000001
$f(x)$	1.000	0.443	0.409	0.406	0.406	0.405	0.405

(b)

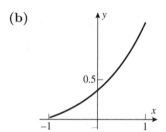

5. 1

7. If $x \neq -3$ then $\dfrac{3x + 9}{x^2 + 4x + 3} = \dfrac{3}{x + 1}$ with limit $-\dfrac{3}{2}$

9. $\dfrac{2^5}{3} = \dfrac{32}{3}$

11. **(a)** $y = 0$ **(b)** none   **(c)** $y = 2$

13. 1   **15.** $3 - k$ **17.** $-\dfrac{1}{2}$

19. **(a)** $f(x) = 2x/(x - 1)$ **(b)**

21. (a) $\lim\limits_{x \to 2} f(x) = 5$ **(b)** 0.0045

23. (a) $|4x - 7 - 1| < 0.01, 4|x - 2| < 0.01, |x - 2| < 0.0025$, let $\delta = 0.0025$

      (b) $\left| \dfrac{4x^2 - 9}{2x - 3} - 6 \right| < 0.05, |2x + 3 - 6| < 0.05, |x - 1.5| < 0.025$, take $\delta = 0.025$

      (c) $|x^2 - 16| < 0.001$; if $\delta < 1$ then $|x + 4| < 9$ if $|x - 4| < 1$;   then
         $\pi|x^2 - 16| = |x - 4||x + 4| \le 9|x - 4| < 0.001$ provided $|x - 4| < 0.001/9$, take $\delta = 0.0001$,
         then $|x^2 - 16| < 9|x - 4| < 9(0.0001) = 0.0009 < 0.001$

25. Let $\epsilon = f(x_0)/2 > 0$; then there corresponds $\delta > 0$ such that if $|x - x_0| < \delta$ then $|f(x) - f(x_0)| < \epsilon$, $-\epsilon < f(x) - f(x_0) < \epsilon, f(x) > f(x_0) - \epsilon = f(x_0)/2 > 0$ for $x_0 - \delta < x < x_0 + \delta$.

27. (a) f is not defined at $x = \pm 1$, continuous elsewhere
      (b) none
      (c) f is not defined at $x = 0, -3$

29. For $x < 2$ f is a polynomial and is continuous; for $x > 2$ f is a polynomial and is continuous. At $x = 2$, $f(2) = -13 \ne 13 = \lim\limits_{x \to 2^+} f(x)$ so f is not continuous there.

31. $f(x) = -1$ for $a \le x < \dfrac{a + b}{2}$ and $f(x) = 1$ for $\dfrac{a + b}{2} \le x \le b$

33. $f(-6) = 185$, $f(0) = -1$, $f(2) = 65$; apply Theorem 2.4.8 twice, once on $[-6, 0]$ and once on $[0, 2]$

CHAPTER 3

The Derivative

EXERCISE SET 3.1

1. (a) $m_{\tan} = (50 - 10)/(15 - 5)$
$= 40/10$
$= 4$ m/s

(b)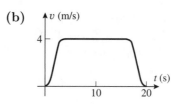

3. (a) $m_{\tan} = (600 - 0)/(20 - 2.2)$
$= 600/17.8$
≈ 33.71 m/s

(b) $m_{\tan} \approx (820 - 600)/(20 - 16)$
$= 220/4$
$= 55$ m/s
The speed is increasing with time.

5. From the figure:

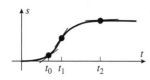

(a) The particle is moving faster at time t_0 because the slope of the tangent to the curve at t_0 is greater than that at t_2.

(b) The initial velocity is 0 because the slope of a horizontal line is 0.

(c) The particle is speeding up because the slope increases as t increases from t_0 to t_1.

(d) The particle is slowing down because the slope decreases as t increases from t_1 to t_2.

7. It is a straight line with slope equal to the velocity.

9. (a) $m_{\sec} = \dfrac{f(1) - f(0)}{1 - 0} = \dfrac{2}{1} = 2$

(b) $m_{\tan} = \lim_{x_1 \to 0} \dfrac{f(x_1) - f(0)}{x_1 - 0} = \lim_{x_1 \to 0} \dfrac{2x_1^2 - 0}{x_1 - 0} = \lim_{x_1 \to 0} 2x_1 = 0$

(c) $m_{\tan} = \lim_{x_1 \to x_0} \dfrac{f(x_1) - f(x_0)}{x_1 - x_0}$

$= \lim_{x_1 \to x_0} \dfrac{2x_1^2 - 2x_0^2}{x_1 - x_0}$

$= \lim_{x_1 \to x_0} (2x_1 + 2x_0)$

$= 4x_0$

(d)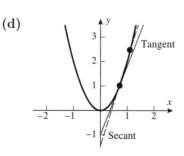

11. (a) $m_{\sec} = \dfrac{f(3) - f(2)}{3 - 2} = \dfrac{1/3 - 1/2}{1} = -\dfrac{1}{6}$

(b) $m_{\tan} = \lim_{x_1 \to 2} \dfrac{f(x_1) - f(2)}{x_1 - 2} = \lim_{x_1 \to 2} \dfrac{1/x_1 - 1/2}{x_1 - 2}$

$= \lim_{x_1 \to 2} \dfrac{2 - x_1}{2x_1(x_1 - 2)} = \lim_{x_1 \to 2} \dfrac{-1}{2x_1} = -\dfrac{1}{4}$

(c) $m_{\tan} = \lim\limits_{x_1 \to x_0} \dfrac{f(x_1) - f(x_0)}{x_1 - x_0}$

$= \lim\limits_{x_1 \to x_0} \dfrac{1/x_1 - 1/x_0}{x_1 - x_0}$

$= \lim\limits_{x_1 \to x_0} \dfrac{x_0 - x_1}{x_0 x_1 (x_1 - x_0)}$

$= \lim\limits_{x_1 \to x_0} \dfrac{-1}{x_0 x_1} = -\dfrac{1}{x_0^2}$

(d)

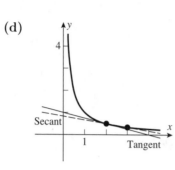

13. (a) $m_{\tan} = \lim\limits_{x_1 \to x_0} \dfrac{f(x_1) - f(x_0)}{x_1 - x_0} = \lim\limits_{x_1 \to x_0} \dfrac{(x_1^2 - 1) - (x_0^2 - 1)}{x_1 - x_0}$

$= \lim\limits_{x_1 \to x_0} \dfrac{(x_1^2 - x_0^2)}{x_1 - x_0} = \lim\limits_{x_1 \to x_0} (x_1 + x_0) = 2x_0$

(b) $m_{\tan} = 2(-1) = -2$

15. (a) $m_{\tan} = \lim\limits_{x_1 \to x_0} \dfrac{f(x_1) - f(x_0)}{x_1 - x_0} = \lim\limits_{x_1 \to x_0} \dfrac{\sqrt{x_1} - \sqrt{x_0}}{x_1 - x_0}$

$= \lim\limits_{x_1 \to x_0} \dfrac{1}{\sqrt{x_1} + \sqrt{x_0}} = \dfrac{1}{2\sqrt{x_0}}$

(b) $m_{\tan} = \dfrac{1}{2\sqrt{1}} = \dfrac{1}{2}$

17. (a) 72°F at about 4:30 P.M.   (b) about $(67 - 43)/6 = 4$°F/h

(c) decreasing most rapidly at about 9 P.M.; rate of change of temperature is about -7°F/h (slope of estimated tangent line to curve at 9 P.M.)

19. (a) during the first year after birth

(b) about 6 cm/year (slope of estimated tangent line at age 5)

(c) the growth rate is greatest at about age 14; about 10 cm/year

(d)

21. (a) $(40)^3/\sqrt{10} = 20{,}238.6$ ft   (b) $v_{\text{ave}} = 20{,}238.6/40 = 505.96$ ft/s

(c) Solve $s = t^3/\sqrt{10} = 135$, $t \approx 7.53$, $v_{\text{ave}} = 135/7.53 = 17.93$ ft/s.

(d) $v_{\text{inst}} = \lim\limits_{t_1 \to 40} \dfrac{t_1^3/\sqrt{10} - (40)^3/\sqrt{10}}{t_1 - 40} = \lim\limits_{t_1 \to 40} \dfrac{(t_1^3 - 40^3)}{(t_1 - 40)\sqrt{10}}$

$= \lim\limits_{t_1 \to 40} \dfrac{1}{\sqrt{10}}(t_1^2 + 40t_1 + 1600) = 1517.89$ ft/s

23. **(a)** $v_{\text{ave}} = \dfrac{6(4)^4 - 6(2)^4}{4 - 2} = 720 \text{ ft/min}$

(b) $v_{\text{inst}} = \lim\limits_{t_1 \to 2} \dfrac{6t_1^4 - 6(2)^4}{t_1 - 2} = \lim\limits_{t_1 \to 2} \dfrac{6(t_1^4 - 16)}{t_1 - 2}$

$\qquad = \lim\limits_{t_1 \to 2} \dfrac{6(t_1^2 + 4)(t_1^2 - 4)}{t_1 - 2} = \lim\limits_{t_1 \to 2} 6(t_1^2 + 4)(t_1 + 2) = 192 \text{ ft/min}$

EXERCISE SET 3.2

1. $f'(1) = 2, \ f'(3) = 0, \ f'(5) = -2, \ f'(6) = -1/2$

3. **(b)** $m = f'(2) = 3$ $\qquad\qquad\qquad\qquad$ **(c)** the same, $f'(2) = 3$

5.

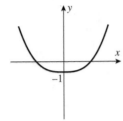

7. $y - (-1) = 5(x - 3), \ y = 5x - 16$

9. $f'(1) = \lim\limits_{h \to 0} \dfrac{f(1 + h) - f(1)}{h} = \lim\limits_{h \to 0} \dfrac{2(1 + h)^2 - 2 \cdot 1^2}{h} = \lim\limits_{h \to 0} \dfrac{4h + 2h^2}{h} = 4;$

$y - 2 = 4(x - 1), y = 4x - 2$

11. $f'(0) = \lim\limits_{h \to 0} \dfrac{f(0 + h) - f(0)}{h} = \lim\limits_{h \to 0} \dfrac{h^3 - 0^3}{h} = \lim\limits_{h \to 0} h^2 = 0;$

$f'(0) = 0$ so $y - 0 = (0)(x - 0), \ y = 0$

13. $f'(8) = \lim\limits_{h \to 0} \dfrac{f(8 + h) - f(8)}{h} = \lim\limits_{h \to 0} \dfrac{\sqrt{9 + h} - 3}{h}$

$\qquad = \lim\limits_{h \to 0} \dfrac{\sqrt{9 + h} - 3}{h} \dfrac{\sqrt{9 + h} + 3}{\sqrt{9 + h} + 3} = \lim\limits_{h \to 0} \dfrac{h}{h(\sqrt{9 + h} + 3)} = \dfrac{1}{6};$

$\qquad f(8) = \sqrt{8 + 1} = 3, \ f'(8) = \dfrac{1}{6}$ so $y - 3 = \dfrac{1}{6}(x - 8), \ y = \dfrac{1}{6}x + \dfrac{5}{3}$

15. $f'(x) = \lim\limits_{\Delta x \to 0} \dfrac{\dfrac{1}{x + \Delta x} - \dfrac{1}{x}}{\Delta x} = \lim\limits_{\Delta x \to 0} \dfrac{\dfrac{x - (x + \Delta x)}{x(x + \Delta x)}}{\Delta x}$

$\qquad = \lim\limits_{\Delta x \to 0} \dfrac{-\Delta x}{x \Delta x(x + \Delta x)} = \lim\limits_{\Delta x \to 0} -\dfrac{1}{x(x + \Delta x)} = -\dfrac{1}{x^2}$

17. $f'(x) = \lim\limits_{\Delta x \to 0} \dfrac{(x + \Delta x)^2 - (x + \Delta x) - (x^2 - x)}{\Delta x} = \lim\limits_{\Delta x \to 0} \dfrac{2x\Delta x + \Delta x^2 - \Delta x}{\Delta x}$

$\qquad = \lim\limits_{\Delta x \to 0} (2x - 1 + \Delta x) = 2x - 1$

19. $f'(x) = \lim\limits_{\Delta x \to 0} \dfrac{\dfrac{1}{\sqrt{x + \Delta x}} - \dfrac{1}{\sqrt{x}}}{\Delta x} = \lim\limits_{\Delta x \to 0} \dfrac{\sqrt{x} - \sqrt{x + \Delta x}}{\Delta x \sqrt{x}\sqrt{x + \Delta x}}$

$= \lim\limits_{\Delta x \to 0} \dfrac{x - (x + \Delta x)}{\Delta x \sqrt{x}\sqrt{x + \Delta x}(\sqrt{x} + \sqrt{x + \Delta x})} = \lim\limits_{\Delta x \to 0} \dfrac{-1}{\sqrt{x}\sqrt{x + \Delta x}(\sqrt{x} + \sqrt{x + \Delta x})} = -\dfrac{1}{2x^{3/2}}$

21. $f'(t) = \lim\limits_{h \to 0} \dfrac{f(t + h) - f(t)}{h} = \lim\limits_{h \to 0} \dfrac{[4(t + h)^2 + (t + h)] - [4t^2 + t]}{h}$

$= \lim\limits_{h \to 0} \dfrac{4t^2 + 8th + 4h^2 + t + h - 4t^2 - t}{h}$

$= \lim\limits_{h \to 0} \dfrac{8th + 4h^2 + h}{h} = \lim\limits_{h \to 0}(8t + 4h + 1) = 8t + 1$

23. **(a)** D   **(b)** F   **(c)** B   **(d)** C   **(e)** A   **(f)** E

25. **(a)** 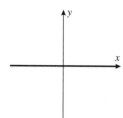   **(b)** 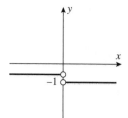   **(c)**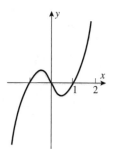

27. **(a)** $f(x) = \sqrt{x}$ and $a = 1$ **(b)** $f(x) = x^2$ and $a = 3$

29. $\dfrac{dy}{dx} = \lim\limits_{h \to 0} \dfrac{(1 - (x + h)^2) - (1 - x^2)}{h} = \lim\limits_{h \to 0} \dfrac{-2xh - h^2}{h} = \lim\limits_{h \to 0} -2(x + h) = -2x,$

and $\left.\dfrac{dy}{dx}\right|_{x=1} = -2$

31. $y = -2x + 1$

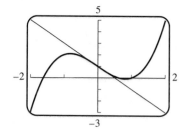

33. **(b)**

h	0.5	0.1	0.01	0.001	0.0001	0.00001
$(f(1 + h) - f(1))/h$	1.6569	1.4355	1.3911	1.3868	1.3863	1.3863

35. **(a)** dollars/ft

(b) As you go deeper the price per foot may increase dramatically, so $f'(x)$ is roughly the price per additional foot.

(c) If each additional foot costs extra money (this is to be expected) then $f'(x)$ remains positive.

(d) From the approximation $1000 = f'(300) \approx \dfrac{f(301) - f(300)}{301 - 300}$

we see that $f(301) \approx f(300) + 1000$, so the extra foot will cost around \$1000.

37. **(a)** $F \approx 200$ lb, $dF/d\theta \approx 50$ lb/rad   **(b)** $\mu = (dF/d\theta)/F \approx 50/200 = 0.25$

39. **(a)** $T \approx 115°F$, $dT/dt \approx -3.35°F/\text{min}$

     (b) $k = (dT/dt)/(T - T_0) \approx (-3.35)/(115 - 75) = -0.084$

41. $\lim\limits_{x \to 0} f(x) = \lim\limits_{x \to 0} \sqrt[3]{x} = 0 = f(0)$, so f is continuous at $x = 0$.

$\lim\limits_{h \to 0} \dfrac{f(0+h) - f(0)}{h} = \lim\limits_{h \to 0} \dfrac{\sqrt[3]{h} - 0}{h} = \lim\limits_{h \to 0} \dfrac{1}{h^{2/3}} = +\infty$, so

$f'(0)$ does not exist.

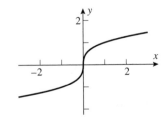

43. $\lim\limits_{x \to 1^-} f(x) = \lim\limits_{x \to 1^+} f(x) = f(1)$, so f is continuous at $x = 1$.

$\lim\limits_{h \to 0^-} \dfrac{f(1+h) - f(1)}{h} = \lim\limits_{h \to 0^-} \dfrac{[(1+h)^2 + 1] - 2}{h} = \lim\limits_{h \to 0^-} (2 + h) = 2$;

$\lim\limits_{h \to 0^+} \dfrac{f(1+h) - f(1)}{h} = \lim\limits_{h \to 0^+} \dfrac{2(1+h) - 2}{h} = \lim\limits_{h \to 0^+} 2 = 2$, so $f'(1) = 2$.

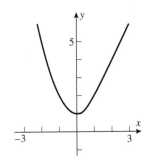

45. Since $-|x| \le x \sin(1/x) \le |x|$ it follows by the Squeezing Theorem

(Theorem 2.6.2) that $\lim\limits_{x \to 0} x \sin(1/x) = 0$. The derivative cannot

exist: consider $\dfrac{f(x) - f(0)}{x} = \sin(1/x)$. This function oscillates

between -1 and $+1$ and does not tend to zero as x tends to zero.

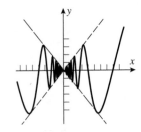

47. Let $\epsilon = |f'(x_0)/2|$. Then there exists $\delta > 0$ such that if $0 < |x - x_0| < \delta$, then

$\left| \dfrac{f(x) - f(x_0)}{x - x_0} - f'(x_0) \right| < \epsilon$. Since $f'(x_0) > 0$ and $\epsilon = f'(x_0)/2$ it follows that

$\dfrac{f(x) - f(x_0)}{x - x_0} > \epsilon > 0$. If $x = x_1 < x_0$ then $f(x_1) < f(x_0)$ and if $x = x_2 > x_0$ then $f(x_2) > f(x_0)$.

49. **(a)** Let $\epsilon = |m|/2$. Since $m \ne 0, \epsilon > 0$. Since $f(0) = f'(0) = 0$ we know there exists $\delta > 0$

     such that $\left| \dfrac{f(0+h) - f(0)}{h} \right| < \epsilon$ whenever $|h| < \delta$. It follows that $|f(h)| < \frac{1}{2}|hm|$ for $|h| < \delta$.

     Replace h with x to get the result.

     (b) For $|x| < \delta, |f(x)| < \frac{1}{2}|mx|$. Moreover $|mx| = |mx - f(x) + f(x)| \le |f(x) - mx| + |f(x)|$,

     which yields $|f(x) - mx| \ge |mx| - |f(x)| > \frac{1}{2}|mx| > |f(x)|$, i.e. $|f(x) - mx| > |f(x)|$.

     (c) If any straight line $y = mx + b$ is to approximate the curve $y = f(x)$ for small values of x,

     then $b = 0$ since $f(0) = 0$. The inequality $|f(x) - mx| > |f(x)|$ can also be interpreted as

     $|f(x) - mx| > |f(x) - 0|$, i.e. the line $y = 0$ is a better approximation than is $y = mx$.

EXERCISE SET 3.3

1. $28x^6$ **3.** $24x^7 + 2$ **5.** 0 **7.** $-\dfrac{1}{3}(7x^6 + 2)$

9. $-3x^{-4} - 7x^{-8}$ **11.** $24x^{-9} + 1/\sqrt{x}$ **13.** $12x(3x^2 + 1)$ **15.** $3ax^2 + 2bx + c$

17. $y' = 10x - 3$, $y'(1) = 7$ **19.** $2t - 1$

21. $dy/dx = 1 + 2x + 3x^2 + 4x^3 + 5x^4$, $dy/dx\big|_{x=1} = 15$

23. $\quad y = (1 - x^2)(1 + x^2)(1 + x^4) = (1 - x^4)(1 + x^4) = 1 - x^8$

$\dfrac{dy}{dx} = -8x^7$, $dy/dx\big|_{x=1} = -8$

25. $f'(1) \approx \dfrac{f(1.01) - f(1)}{0.01} = \dfrac{-0.999699 - (-1)}{0.01} = 0.0301$, and by differentiation, $f'(1) = 3(1)^2 - 3 = 0$

27. From the graph, $f'(1) \approx \dfrac{2.0088 - 2.0111}{1.0981 - 0.9000} \approx -0.0116$,

and by differentiation, $f'(1) = 0$

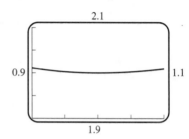

29. $32t$ **31.** $3\pi r^2$

33. **(a)** $\dfrac{dV}{dr} = 4\pi r^2$ **(b)** $\dfrac{dV}{dr}\bigg|_{r=5} = 4\pi(5)^2 = 100\pi$

35. $y - 2 = 5(x + 3)$, $y = 5x + 17$

37. **(a)** $dy/dx = 21x^2 - 10x + 1$, $d^2y/dx^2 = 42x - 10$
(b) $dy/dx = 24x - 2$, $d^2y/dx^2 = 24$ **(c)** $dy/dx = -1/x^2$, $d^2y/dx^2 = 2/x^3$
(d) $y = 35x^5 - 16x^3 - 3x$, $dy/dx = 175x^4 - 48x^2 - 3$, $d^2y/dx^2 = 700x^3 - 96x$

39. **(a)** $y' = -5x^{-6} + 5x^4$, $y'' = 30x^{-7} + 20x^3$, $y''' = -210x^{-8} + 60x^2$
(b) $y = x^{-1}$, $y' = -x^{-2}$, $y'' = 2x^{-3}$, $y''' = -6x^{-4}$
(c) $y' = 3ax^2 + b$, $y'' = 6ax$, $y''' = 6a$

41. **(a)** $f'(x) = 6x$, $f''(x) = 6$, $f'''(x) = 0$, $f'''(2) = 0$

(b) $\dfrac{dy}{dx} = 30x^4 - 8x$, $\dfrac{d^2y}{dx^2} = 120x^3 - 8$, $\dfrac{d^2y}{dx^2}\bigg|_{x=1} = 112$

(c) $\dfrac{d}{dx}\left[x^{-3}\right] = -3x^{-4}$, $\dfrac{d^2}{dx^2}\left[x^{-3}\right] = 12x^{-5}$, $\dfrac{d^3}{dx^3}\left[x^{-3}\right] = -60x^{-6}$, $\dfrac{d^4}{dx^4}\left[x^{-3}\right] = 360x^{-7}$,

$\dfrac{d^4}{dx^4}\left[x^{-3}\right]\bigg|_{x=1} = 360$

43. $y' = 3x^2 + 3$, $y'' = 6x$, and $y''' = 6$ so
$y''' + xy'' - 2y' = 6 + x(6x) - 2(3x^2 + 3) = 6 + 6x^2 - 6x^2 - 6 = 0$

45. The graph has a horizontal tangent at points where $\dfrac{dy}{dx} = 0$, but $\dfrac{dy}{dx} = x^2 - 3x + 2 = (x-1)(x-2) = 0$ if $x = 1, 2$. The corresponding values of y are $5/6$ and $2/3$ so the tangent line is horizontal at $(1, 5/6)$ and $(2, 2/3)$.

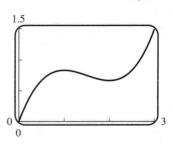

47. The y-intercept is -2 so the point $(0, -2)$ is on the graph; $-2 = a(0)^2 + b(0) + c$, $c = -2$. The x-intercept is 1 so the point $(1,0)$ is on the graph; $0 = a + b - 2$. The slope is $dy/dx = 2ax + b$; at $x = 0$ the slope is b so $b = -1$, thus $a = 3$. The function is $y = 3x^2 - x - 2$.

49. The points $(-1, 1)$ and $(2, 4)$ are on the secant line so its slope is $(4 - 1)/(2 + 1) = 1$. The slope of the tangent line to $y = x^2$ is $y' = 2x$ so $2x = 1$, $x = 1/2$.

51. $y' = -2x$, so at any point (x_0, y_0) on $y = 1 - x^2$ the tangent line is $y - y_0 = -2x_0(x - x_0)$, or $y = -2x_0 x + x_0^2 + 1$. The point $(2, 0)$ is to be on the line, so $0 = -4x_0 + x_0^2 + 1$, $x_0^2 - 4x_0 + 1 = 0$. Use the quadratic formula to get $x_0 = \dfrac{4 \pm \sqrt{16 - 4}}{2} = 2 \pm \sqrt{3}$.

53. $y' = 3ax^2 + b$; the tangent line at $x = x_0$ is $y - y_0 = (3ax_0^2 + b)(x - x_0)$ where $y_0 = ax_0^3 + bx_0$. Solve with $y = ax^3 + bx$ to get

$$(ax^3 + bx) - (ax_0^3 + bx_0) = (3ax_0^2 + b)(x - x_0)$$
$$ax^3 + bx - ax_0^3 - bx_0 = 3ax_0^2 x - 3ax_0^3 + bx - bx_0$$
$$x^3 - 3x_0^2 x + 2x_0^3 = 0$$
$$(x - x_0)(x^2 + xx_0 - 2x_0^2) = 0$$
$$(x - x_0)^2 (x + 2x_0) = 0, \text{ so } x = -2x_0.$$

55. $y' = -\dfrac{1}{x^2}$; the tangent line at $x = x_0$ is $y - y_0 = -\dfrac{1}{x_0^2}(x - x_0)$, or $y = -\dfrac{x}{x_0^2} + \dfrac{2}{x_0}$. The tangent line crosses the x-axis at $2x_0$, the y-axis at $2/x_0$, so that the area of the triangle is $\dfrac{1}{2}(2/x_0)(2x_0) = 2$.

57. $F = GmMr^{-2}$, $\dfrac{dF}{dr} = -2GmMr^{-3} = -\dfrac{2GmM}{r^3}$

59. $f'(x) = 1 + 1/x^2 > 0$ for all $x \neq 0$

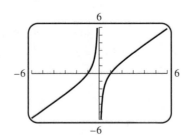

61. f is continuous at 1 because $\lim\limits_{x \to 1^-} f(x) = \lim\limits_{x \to 1^+} f(x) = f(1)$; also $\lim\limits_{x \to 1^-} f'(x) = \lim\limits_{x \to 1^-} (2x + 1) = 3$ and $\lim\limits_{x \to 1^+} f'(x) = \lim\limits_{x \to 1^+} 3 = 3$ so f is differentiable at 1.

63. f is continuous at 1 because $\lim\limits_{x \to 1^-} f(x) = \lim\limits_{x \to 1^+} f(x) = f(1)$, also $\lim\limits_{x \to 1^-} f'(x) = \lim\limits_{x \to 1^-} 2x = 2$ and $\lim\limits_{x \to 1^+} f'(x) = \lim\limits_{x \to 1^+} \dfrac{1}{2\sqrt{x}} = \dfrac{1}{2}$ so f is not differentiable at 1.

65. (a) $f(x) = 3x - 2$ if $x \geq 2/3$, $f(x) = -3x + 2$ if $x < 2/3$ so f is differentiable everywhere except perhaps at $2/3$. f is continuous at $2/3$, also $\lim\limits_{x \to 2/3^-} f'(x) = \lim\limits_{x \to 2/3^-} (-3) = -3$ and $\lim\limits_{x \to 2/3^+} f'(x) = \lim\limits_{x \to 2/3^+} (3) = 3$ so f is not differentiable at $x = 2/3$.

(b) $f(x) = x^2 - 4$ if $|x| \geq 2$, $f(x) = -x^2 + 4$ if $|x| < 2$ so f is differentiable everywhere except perhaps at ± 2. f is continuous at -2 and 2, also $\lim\limits_{x \to 2^-} f'(x) = \lim\limits_{x \to 2^-} (-2x) = -4$ and $\lim\limits_{x \to 2^+} f'(x) = \lim\limits_{x \to 2^+} (2x) = 4$ so f is not differentiable at $x = 2$. Similarly, f is not differentiable at $x = -2$.

67. (a) $\dfrac{d^2}{dx^2}[cf(x)] = \dfrac{d}{dx}\left[\dfrac{d}{dx}[cf(x)]\right] = \dfrac{d}{dx}\left[c\dfrac{d}{dx}[f(x)]\right] = c\dfrac{d}{dx}\left[\dfrac{d}{dx}[f(x)]\right] = c\dfrac{d^2}{dx^2}[f(x)]$

$\dfrac{d^2}{dx^2}[f(x) + g(x)] = \dfrac{d}{dx}\left[\dfrac{d}{dx}[f(x) + g(x)]\right] = \dfrac{d}{dx}\left[\dfrac{d}{dx}[f(x)] + \dfrac{d}{dx}[g(x)]\right]$

$\qquad = \dfrac{d^2}{dx^2}[f(x)] + \dfrac{d^2}{dx^2}[g(x)]$

(b) yes, by repeated application of the procedure illustrated in Part (a)

69. (a) $f'(x) = nx^{n-1}$, $f''(x) = n(n-1)x^{n-2}$, $f'''(x) = n(n-1)(n-2)x^{n-3}, \ldots,$
$f^{(n)}(x) = n(n-1)(n-2)\cdots 1$

(b) from Part (a), $f^{(k)}(x) = k(k-1)(k-2)\cdots 1$ so $f^{(k+1)}(x) = 0$ thus $f^{(n)}(x) = 0$ if $n > k$

(c) from Parts (a) and (b), $f^{(n)}(x) = a_n n(n-1)(n-2)\cdots 1$

71. Let $g(x) = x^n$, $f(x) = (mx + b)^n$. Use Exercise 48 in Section 3.2, but with f and g permuted. If $x_0 = mx_1 + b$ then Exercise 48 says that f is differentiable at x_1 and $f'(x_1) = mg'(x_0)$. Since $g'(x_0) = nx_0^{n-1}$, the result follows.

73. $f'(x) = 3 \cdot 3(3x - 1)^2 = 81x^2 - 54x + 9$

75. $f'(x) = 2 \cdot 3 \cdot (-2)(2x + 1)^{-3} = -12/(2x + 1)^3$

77. $f(x) = \dfrac{2x^2 + 4x + 2 + 1}{(x + 1)^2} = 2 + \dfrac{1}{(x + 1)^2}$, so $f'(x) = -2(x + 1)^{-3} = -2/(x + 1)^3$

EXERCISE SET 3.4

1. (a) $f(x) = 2x^2 + x - 1$, $f'(x) = 4x + 1$
   (b) $f'(x) = (x + 1) \cdot (2) + (2x - 1) \cdot (1) = 4x + 1$

3. (a) $f(x) = x^4 - 1$, $f'(x) = 4x^3$
   (b) $f'(x) = (x^2 + 1) \cdot (2x) + (x^2 - 1) \cdot (2x) = 4x^3$

5. $f'(x) = (3x^2 + 6)\dfrac{d}{dx}\left(2x - \dfrac{1}{4}\right) + \left(2x - \dfrac{1}{4}\right)\dfrac{d}{dx}(3x^2 + 6) = (3x^2 + 6)(2) + \left(2x - \dfrac{1}{4}\right)(6x)$

$\qquad = 18x^2 - \dfrac{3}{2}x + 12$

7. $f'(x) = (x^3 + 7x^2 - 8)\dfrac{d}{dx}(2x^{-3} + x^{-4}) + (2x^{-3} + x^{-4})\dfrac{d}{dx}(x^3 + 7x^2 - 8)$

$\qquad = (x^3 + 7x^2 - 8)(-6x^{-4} - 4x^{-5}) + (2x^{-3} + x^{-4})(3x^2 + 14x)$

$\qquad = -15x^{-2} - 14x^{-3} + 48x^{-4} + 32x^{-5}$

9. $f'(x) = x^2 + 2x + 4 + (2x + 2)(x - 2) = 3x^2$

11. $\dfrac{dy}{dx} = \dfrac{(5x - 3)\dfrac{d}{dx}(1) - (1)\dfrac{d}{dx}(5x - 3)}{(5x - 3)^2} = -\dfrac{5}{(5x - 3)^2}; \quad y'(1) = -5/4$

13. $\dfrac{dy}{dx} = \dfrac{(x + 3)\dfrac{d}{dx}(2x - 1) - (2x - 1)\dfrac{d}{dx}(x + 3)}{(x + 3)^2}$

$\qquad = \dfrac{(x + 3)(2) - (2x - 1)(1)}{(x + 3)^2} = \dfrac{7}{(x + 3)^2}; \left.\dfrac{dy}{dx}\right|_{x=1} = \dfrac{7}{16}$

15. $\dfrac{dy}{dx} = \left(\dfrac{3x + 2}{x}\right)\dfrac{d}{dx}\left(x^{-5} + 1\right) + \left(x^{-5} + 1\right)\dfrac{d}{dx}\left(\dfrac{3x + 2}{x}\right)$

$\qquad = \left(\dfrac{3x + 2}{x}\right)\left(-5x^{-6}\right) + \left(x^{-5} + 1\right)\left[\dfrac{x(3) - (3x + 2)(1)}{x^2}\right]$

$\qquad = \left(\dfrac{3x + 2}{x}\right)\left(-5x^{-6}\right) + \left(x^{-5} + 1\right)\left(-\dfrac{2}{x^2}\right);$

$\left.\dfrac{dy}{dx}\right|_{x=1} = 5(-5) + 2(-2) = -29$

17. $f'(1) = 0$

19. **(a)** $g'(x) = \sqrt{x}f'(x) + \dfrac{1}{2\sqrt{x}}f(x), g'(4) = (2)(-5) + \dfrac{1}{4}(3) = -37/4$

$\qquad$ **(b)** $g'(x) = \dfrac{xf'(x) - f(x)}{x^2}, g'(4) = \dfrac{(4)(-5) - 3}{16} = -23/16$

21. **(a)** $F'(x) = 5f'(x) + 2g'(x), F'(2) = 5(4) + 2(-5) = 10$
$\qquad$ **(b)** $F'(x) = f'(x) - 3g'(x), F'(2) = 4 - 3(-5) = 19$
$\qquad$ **(c)** $F'(x) = f(x)g'(x) + g(x)f'(x), F'(2) = (-1)(-5) + (1)(4) = 9$
$\qquad$ **(d)** $F'(x) = [g(x)f'(x) - f(x)g'(x)]/g^2(x), F'(2) = [(1)(4) - (-1)(-5)]/(1)^2 = -1$

23. $\dfrac{dy}{dx} = \dfrac{2x(x + 2) - (x^2 - 1)}{(x + 2)^2},$

$\dfrac{dy}{dx} = 0$ if $x^2 + 4x + 1 = 0$. By the quadratic formula,

$x = \dfrac{-4 \pm \sqrt{16 - 4}}{2} = -2 \pm \sqrt{3}$. The tangent line is horizontal at $x = -2 \pm \sqrt{3}$.

25. The tangent line is parallel to the line $y = x$ when it has slope 1.
$\dfrac{dy}{dx} = \dfrac{2x(x + 1) - (x^2 + 1)}{(x + 1)^2} = \dfrac{x^2 + 2x - 1}{(x + 1)^2} = 1$ if $x^2 + 2x - 1 = (x + 1)^2$, which reduces to $-1 = +1$,
impossible. Thus the tangent line is never parallel to the line $y = x$.

27. Fix x_0. The slope of the tangent line to the curve $y = \dfrac{1}{x+4}$ at the point $(x_0, 1/(x_0+4))$ is

given by $\dfrac{dy}{dx} = \dfrac{-1}{(x+4)^2}\bigg|_{x=x_0} = \dfrac{-1}{(x_0+4)^2}$. The tangent line to the curve at (x_0, y_0) thus has the

equation $y - y_0 = \dfrac{-1}{(x_0+4)^2}(x - x_0)$, and this line passes through the origin if its constant term

$y_0 - x_0 \dfrac{-1}{(x_0+4)^2}$ is zero. Then $\dfrac{1}{x_0+4} = \dfrac{-x_0}{(x_0+4)^2}$, $x_0 + 4 = -x_0$, $x_0 = -2$.

29. (b) They intersect when $\dfrac{1}{x} = \dfrac{1}{2-x}$, $x = 2 - x$, $x = 1$, $y = 1$. The first curve has derivative

$y = -\dfrac{1}{x^2}$, so the slope when $x = 1$ is m $= -1$. Second curve has derivative $y = \dfrac{1}{(2-x)^2}$ so

the slope when $x = 1$ is m $= 1$. Since the two slopes are negative reciprocals of each other, the tangent lines are perpendicular at the point $(1, 1)$.

31. $F'(x) = xf'(x) + f(x)$, $F''(x) = xf''(x) + f'(x) + f'(x) = xf''(x) + 2f'(x)$

33. $(f \cdot g \cdot h)' = [(f \cdot g) \cdot h]' = (f \cdot g)h' + h(f \cdot g)' = (f \cdot g)h' + h[fg' + f'g] = fgh' + fg'h + f'gh$

35. (a) $2(1 + x^{-1})(x^{-3} + 7) + (2x + 1)(-x^{-2})(x^{-3} + 7) + (2x + 1)(1 + x^{-1})(-3x^{-4})$

(b) $(x^7 + 2x - 3)^3 = (x^7 + 2x - 3)(x^7 + 2x - 3)(x^7 + 2x - 3)$ so

$$\frac{d}{dx}(x^7 + 2x - 3)^3 = (7x^6 + 2)(x^7 + 2x - 3)(x^7 + 2x - 3)$$
$$+ (x^7 + 2x - 3)(7x^6 + 2)(x^7 + 2x - 3)$$
$$+ (x^7 + 2x - 3)(x^7 + 2x - 3)(7x^6 + 2)$$
$$= 3(7x^6 + 2)(x^7 + 2x - 3)^2$$

37. By the product rule, $g'(x)$ is the sum of n terms, each containing n factors of the form $f'(x)f(x)f(x)\ldots f(x)$; the function $f(x)$ occurs $n - 1$ times, and $f'(x)$ occurs once. Each of these n terms is equal to $f'(x)(f(x))^{n-1}$, and so $g'(x) = n(f(x))^{n-1}f'(x)$.

39. $f(x) = \dfrac{1}{x^n}$ so $f'(x) = \dfrac{x^n \cdot (0) - 1 \cdot (nx^{n-1})}{x^{2n}} = -\dfrac{n}{x^{n+1}}$

EXERCISE SET 3.5

1. $f'(x) = -4\sin x + 2\cos x$

3. $f'(x) = 4x^2 \sin x - 8x\cos x$

5. $f'(x) = \dfrac{\sin x(5 + \sin x) - \cos x(5 - \cos x)}{(5 + \sin x)^2} = \dfrac{1 + 5(\sin x - \cos x)}{(5 + \sin x)^2}$

7. $f'(x) = \sec x \tan x - \sqrt{2}\sec^2 x$

9. $f'(x) = -4\csc x \cot x + \csc^2 x$

11. $f'(x) = \sec x(\sec^2 x) + (\tan x)(\sec x \tan x) = \sec^3 x + \sec x \tan^2 x$

13. $f'(x) = \dfrac{(1 + \csc x)(-\csc^2 x) - (\cot x)(0 - \csc x \cot x)}{(1 + \csc x)^2} = \dfrac{\csc x(-\csc x - \csc^2 x + \cot^2 x)}{(1 + \csc x)^2}$ but

$1 + \cot^2 x = \csc^2 x$ (identity) thus $\cot^2 x - \csc^2 x = -1$ so

$f'(x) = \dfrac{\csc x(-\csc x - 1)}{(1 + \csc x)^2} = -\dfrac{\csc x}{1 + \csc x}$

15. $f(x) = \sin^2 x + \cos^2 x = 1$ (identity) so $f'(x) = 0$

17. $f(x) = \dfrac{\tan x}{1 + x \tan x}$ (because $\sin x \sec x = (\sin x)(1/\cos x) = \tan x$),

$$f'(x) = \frac{(1 + x\tan x)(\sec^2 x) - \tan x [x(\sec^2 x) + (\tan x)(1)]}{(1 + x\tan x)^2}$$

$$= \frac{\sec^2 x - \tan^2 x}{(1 + x\tan x)^2} = \frac{1}{(1 + x\tan x)^2} \quad \text{(because } \sec^2 x - \tan^2 x = 1)$$

19. $dy/dx = -x\sin x + \cos x$, $d^2y/dx^2 = -x\cos x - \sin x - \sin x = -x\cos x - 2\sin x$

21. $dy/dx = x(\cos x) + (\sin x)(1) - 3(-\sin x) = x\cos x + 4\sin x$,
$d^2y/dx^2 = x(-\sin x) + (\cos x)(1) + 4\cos x = -x\sin x + 5\cos x$

23. $dy/dx = (\sin x)(-\sin x) + (\cos x)(\cos x) = \cos^2 x - \sin^2 x$,
$d^2y/dx^2 = (\cos x)(-\sin x) + (\cos x)(-\sin x) - [(\sin x)(\cos x) + (\sin x)(\cos x)] = -4\sin x\cos x$

25. Let $f(x) = \tan x$, then $f'(x) = \sec^2 x$.
  (a) $f(0) = 0$ and $f'(0) = 1$ so $y - 0 = (1)(x - 0)$, $y = x$.
  (b) $f\left(\dfrac{\pi}{4}\right) = 1$ and $f'\left(\dfrac{\pi}{4}\right) = 2$ so $y - 1 = 2\left(x - \dfrac{\pi}{4}\right)$, $y = 2x - \dfrac{\pi}{2} + 1$.
  (c) $f\left(-\dfrac{\pi}{4}\right) = -1$ and $f'\left(-\dfrac{\pi}{4}\right) = 2$ so $y + 1 = 2\left(x + \dfrac{\pi}{4}\right)$, $y = 2x + \dfrac{\pi}{2} - 1$.

27. **(a)** If $y = x\sin x$ then $y' = \sin x + x\cos x$ and $y'' = 2\cos x - x\sin x$ so $y'' + y = 2\cos x$.
  (b) If $y = x\sin x$ then $y' = \sin x + x\cos x$ and $y'' = 2\cos x - x\sin x$ so $y'' + y = 2\cos x$; differentiate twice more to get $y^{(4)} + y'' = -2\cos x$.

29. **(a)** $f'(x) = \cos x = 0$ at $x = \pm\pi/2, \pm 3\pi/2$.
  (b) $f'(x) = 1 - \sin x = 0$ at $x = -3\pi/2, \pi/2$.
  (c) $f'(x) = \sec^2 x \geq 1$ always, so no horizontal tangent line.
  (d) $f'(x) = \sec x\tan x = 0$ when $\sin x = 0$, $x = \pm 2\pi, \pm\pi, 0$

31. $x = 10\sin\theta$, $dx/d\theta = 10\cos\theta$; if $\theta = 60°$, then
$dx/d\theta = 10(1/2) = 5$ ft/rad $= \pi/36$ ft/deg ≈ 0.087 ft/deg

33. $D = 50\tan\theta$, $dD/d\theta = 50\sec^2\theta$; if $\theta = 45°$, then
$dD/d\theta = 50(\sqrt{2})^2 = 100$ m/rad $= 5\pi/9$ m/deg ≈ 1.75 m/deg

35. **(a)** $\dfrac{d^4}{dx^4}\sin x = \sin x$, so $\dfrac{d^{4k}}{dx^{4k}}\sin x = \sin x$; $\dfrac{d^{87}}{dx^{87}}\sin x = \dfrac{d^3}{dx^3}\dfrac{d^{4\cdot 21}}{dx^{4\cdot 21}}\sin x = \dfrac{d^3}{dx^3}\sin x = -\cos x$

  (b) $\dfrac{d^{100}}{dx^{100}}\cos x = \dfrac{d^{4k}}{dx^{4k}}\cos x = \cos x$

37. $f'(x) = -\sin x$, $f''(x) = -\cos x$, $f'''(x) = \sin x$, and $f^{(4)}(x) = \cos x$ with higher order derivatives repeating this pattern, so $f^{(n)}(x) = \sin x$ for $n = 3, 7, 11, \ldots$

39. **(a)** all x **(b)** all x
  (c) $x \neq \pi/2 + n\pi$, $n = 0, \pm 1, \pm 2, \ldots$ **(d)** $x \neq n\pi$, $n = 0, \pm 1, \pm 2, \ldots$
  (e) $x \neq \pi/2 + n\pi$, $n = 0, \pm 1, \pm 2, \ldots$ **(f)** $x \neq n\pi$, $n = 0, \pm 1, \pm 2, \ldots$
  (g) $x \neq (2n + 1)\pi$, $n = 0, \pm 1, \pm 2, \ldots$ **(h)** $x \neq n\pi/2$, $n = 0, \pm 1, \pm 2, \ldots$
  (i) all x

41. $\dfrac{d}{dx}\sin x = \displaystyle\lim_{w\to x}\dfrac{\sin w - \sin x}{w - x} = \lim_{w\to x}\dfrac{2\sin\frac{w-x}{2}\cos\frac{w+x}{2}}{w-x}$

$\qquad\qquad = \displaystyle\lim_{w\to x}\dfrac{\sin\frac{w-x}{2}}{\frac{w-x}{2}}\cos\dfrac{w+x}{2} = 1\cdot\cos x = \cos x$

43. (a) $\displaystyle\lim_{h\to 0}\dfrac{\tan h}{h} = \lim_{h\to 0}\dfrac{\left(\frac{\sin h}{\cos h}\right)}{h} = \lim_{h\to 0}\dfrac{\left(\frac{\sin h}{h}\right)}{\cos h} = \dfrac{1}{1} = 1$

(b) $\dfrac{d}{dx}[\tan x] = \displaystyle\lim_{h\to 0}\dfrac{\tan(x+h) - \tan x}{h} = \lim_{h\to 0}\dfrac{\frac{\tan x + \tan h}{1 - \tan x\tan h} - \tan x}{h}$

$\qquad\qquad = \displaystyle\lim_{h\to 0}\dfrac{\tan x + \tan h - \tan x + \tan^2 x\tan h}{h(1 - \tan x\tan h)} = \lim_{h\to 0}\dfrac{\tan h(1 + \tan^2 x)}{h(1 - \tan x\tan h)}$

$\qquad\qquad = \displaystyle\lim_{h\to 0}\dfrac{\tan h\sec^2 x}{h(1 - \tan x\tan h)} = \sec^2 x\lim_{h\to 0}\dfrac{\frac{\tan h}{h}}{1 - \tan x\tan h}$

$\qquad\qquad = \sec^2 x\dfrac{\displaystyle\lim_{h\to 0}\frac{\tan h}{h}}{\displaystyle\lim_{h\to 0}(1 - \tan x\tan h)} = \sec^2 x$

45. $\dfrac{d}{dx}[\cos kx] = \displaystyle\lim_{h\to 0}\dfrac{\cos k(x+h) - \cos kx}{h} = k\lim_{kh\to 0}\dfrac{\cos(kx + kh) - \cos kx}{kh} = -k\sin kx$

47. Let t be the radian measure, then $h = \dfrac{180}{\pi}t$ and $\cos h = \cos t$, $\sin h = \sin t$. Then

$\displaystyle\lim_{h\to 0}\dfrac{\cos h - 1}{h} = \lim_{t\to 0}\dfrac{\cos t - 1}{180t/\pi} = \dfrac{\pi}{180}\lim_{t\to 0}\dfrac{\cos t - 1}{t} = 0$

$\displaystyle\lim_{h\to 0}\dfrac{\sin h}{h} = \lim_{t\to 0}\dfrac{\sin t}{180t/\pi} = \dfrac{\pi}{180}\lim_{t\to 0}\dfrac{\sin t}{t} = \dfrac{\pi}{180}$

(a) $\dfrac{d}{dx}[\sin x] = \displaystyle\lim_{h\to 0}\dfrac{\sin(x+h) - \sin x}{h}$

$\qquad\qquad = \sin x\displaystyle\lim_{h\to 0}\dfrac{\cos h - 1}{h} + \cos x\lim_{h\to 0}\dfrac{\sin h}{h}$

$\qquad\qquad = (\sin x)(0) + (\cos x)(\pi/180) = \dfrac{\pi}{180}\cos x$

(b) $\dfrac{d}{dx}[\cos x] = \displaystyle\lim_{h\to 0}\dfrac{\cos(x+h) - \cos x}{h}$

$\qquad\qquad = \displaystyle\lim_{h\to 0}\dfrac{\cos x\cos h - \sin x\sin h - \cos x}{h}$

$\qquad\qquad = \cos x\displaystyle\lim_{h\to 0}\dfrac{\cos h - 1}{h} - \sin x\lim_{h\to 0}\dfrac{\sin h}{h}$

$\qquad\qquad = 0\cdot\cos x - \dfrac{\pi}{180}\cdot\sin x = -\dfrac{\pi}{180}\sin x$

EXERCISE SET 3.6

1. $(f\circ g)'(x) = f'(g(x))g'(x)$ so $(f\circ g)'(0) = f'(g(0))g'(0) = f'(0)(3) = (2)(3) = 6$

3. (a) $(f\circ g)(x) = f(g(x)) = (2x-3)^5$ and $(f\circ g)'(x) = f'(g(x))g'(x) = 5(2x-3)^4(2) = 10(2x-3)^4$

(b) $(g \circ f)(x) = g(f(x)) = 2x^5 - 3$ and $(g \circ f)'(x) = g'(f(x))f'(x) = 2(5x^4) = 10x^4$

5. (a) $F'(x) = f'(g(x))g'(x) = f'(g(3))g'(3) = -1(7) = -7$

    (b) $G'(x) = g'(f(x))f'(x) = g'(f(3))f'(3) = 4(-2) = -8$

7. $f'(x) = 37(x^3 + 2x)^{36} \dfrac{d}{dx}(x^3 + 2x) = 37(x^3 + 2x)^{36}(3x^2 + 2)$

9. $f'(x) = -2\left(x^3 - \dfrac{7}{x}\right)^{-3} \dfrac{d}{dx}\left(x^3 - \dfrac{7}{x}\right) = -2\left(x^3 - \dfrac{7}{x}\right)^{-3}\left(3x^2 + \dfrac{7}{x^2}\right)$

11. $f(x) = 4(3x^2 - 2x + 1)^{-3}$,

    $f'(x) = -12(3x^2 - 2x + 1)^{-4} \dfrac{d}{dx}(3x^2 - 2x + 1) = -12(3x^2 - 2x + 1)^{-4}(6x - 2) = \dfrac{24(1 - 3x)}{(3x^2 - 2x + 1)^4}$

13. $f'(x) = \dfrac{1}{2\sqrt{4 + 3\sqrt{x}}} \dfrac{d}{dx}(4 + 3\sqrt{x}) = \dfrac{3}{4\sqrt{x}\sqrt{4 + 3\sqrt{x}}}$

15. $f'(x) = \cos(1/x^2) \dfrac{d}{dx}(1/x^2) = -\dfrac{2}{x^3}\cos(1/x^2)$

17. $f'(x) = 20\cos^4 x \dfrac{d}{dx}(\cos x) = 20\cos^4 x(-\sin x) = -20\cos^4 x \sin x$

19. $f'(x) = 2\cos(3\sqrt{x}) \dfrac{d}{dx}[\cos(3\sqrt{x})] = -2\cos(3\sqrt{x})\sin(3\sqrt{x})\dfrac{d}{dx}(3\sqrt{x}) = -\dfrac{3\cos(3\sqrt{x})\sin(3\sqrt{x})}{\sqrt{x}}$

21. $f'(x) = 4\sec(x^7)\dfrac{d}{dx}[\sec(x^7)] = 4\sec(x^7)\sec(x^7)\tan(x^7)\dfrac{d}{dx}(x^7) = 28x^6\sec^2(x^7)\tan(x^7)$

23. $f'(x) = \dfrac{1}{2\sqrt{\cos(5x)}} \dfrac{d}{dx}[\cos(5x)] = -\dfrac{5\sin(5x)}{2\sqrt{\cos(5x)}}$

25. $f'(x) = -3\left[x + \csc(x^3 + 3)\right]^{-4} \dfrac{d}{dx}\left[x + \csc(x^3 + 3)\right]$

    $= -3\left[x + \csc(x^3 + 3)\right]^{-4}\left[1 - \csc(x^3 + 3)\cot(x^3 + 3)\dfrac{d}{dx}(x^3 + 3)\right]$

    $= -3\left[x + \csc(x^3 + 3)\right]^{-4}\left[1 - 3x^2\csc(x^3 + 3)\cot(x^3 + 3)\right]$

27. $\dfrac{dy}{dx} = x^3(2\sin 5x)\dfrac{d}{dx}(\sin 5x) + 3x^2\sin^2 5x = 10x^3\sin 5x\cos 5x + 3x^2\sin^2 5x$

29. $\dfrac{dy}{dx} = x^5\sec\left(\dfrac{1}{x}\right)\tan\left(\dfrac{1}{x}\right)\dfrac{d}{dx}\left(\dfrac{1}{x}\right) + \sec\left(\dfrac{1}{x}\right)(5x^4)$

    $= x^5\sec\left(\dfrac{1}{x}\right)\tan\left(\dfrac{1}{x}\right)\left(-\dfrac{1}{x^2}\right) + 5x^4\sec\left(\dfrac{1}{x}\right)$

    $= -x^3\sec\left(\dfrac{1}{x}\right)\tan\left(\dfrac{1}{x}\right) + 5x^4\sec\left(\dfrac{1}{x}\right)$

31. $\dfrac{dy}{dx} = -\sin(\cos x)\dfrac{d}{dx}(\cos x) = -\sin(\cos x)(-\sin x) = \sin(\cos x)\sin x$

33. $\dfrac{dy}{dx} = 3\cos^2(\sin 2x)\dfrac{d}{dx}[\cos(\sin 2x)] = 3\cos^2(\sin 2x)[-\sin(\sin 2x)]\dfrac{d}{dx}(\sin 2x)$

$\qquad = -6\cos^2(\sin 2x)\sin(\sin 2x)\cos 2x$

35. $\dfrac{dy}{dx} = (5x+8)^7\dfrac{d}{dx}(1-\sqrt{x})^6 + (1-\sqrt{x})^6\dfrac{d}{dx}(5x+8)^7$

$\qquad = 6(5x+8)^7(1-\sqrt{x})^5\dfrac{-1}{2\sqrt{x}} + 7\cdot 5(1-\sqrt{x})^6(5x+8)^6$

$\qquad = \dfrac{-3}{\sqrt{x}}(5x+8)^7(1-\sqrt{x})^5 + 35(1-\sqrt{x})^6(5x+8)^6$

$(5x+8)^6 (1-\sqrt{x})^5$

$\left[\dfrac{-3}{\sqrt{x}}(5x+8) + 35(1-\sqrt{x})\right]$

37. $\dfrac{dy}{dx} = 3\left[\dfrac{x-5}{2x+1}\right]^2\dfrac{d}{dx}\left[\dfrac{x-5}{2x+1}\right] = 3\left[\dfrac{x-5}{2x+1}\right]^2\cdot\dfrac{11}{(2x+1)^2} = \dfrac{33(x-5)^2}{(2x+1)^4}$

39. $\dfrac{dy}{dx} = \dfrac{(4x^2-1)^8(3)(2x+3)^2(2) - (2x+3)^3(8)(4x^2-1)^7(8x)}{(4x^2-1)^{16}}$

$\qquad = \dfrac{2(2x+3)^2(4x^2-1)^7[3(4x^2-1) - 32x(2x+3)]}{(4x^2-1)^{16}} = -\dfrac{2(2x+3)^2(52x^2+96x+3)}{(4x^2-1)^9}$

41. $\dfrac{dy}{dx} = 5\left[x\sin 2x + \tan^4(x^7)\right]^4\dfrac{d}{dx}\left[x\sin 2x\tan^4(x^7)\right]$

$\qquad = 5\left[x\sin 2x + \tan^4(x^7)\right]^4\left[x\cos 2x\dfrac{d}{dx}(2x) + \sin 2x + 4\tan^3(x^7)\dfrac{d}{dx}\tan(x^7)\right]$

$\qquad = 5\left[x\sin 2x + \tan^4(x^7)\right]^4\left[2x\cos 2x + \sin 2x + 28x^6\tan^3(x^7)\sec^2(x^7)\right]$

43. $\dfrac{dy}{dx} = \cos 3x - 3x\sin 3x$; if $x = \pi$ then $\dfrac{dy}{dx} = -1$ and $y = -\pi$, so the equation of the tangent line is $y + \pi = -(x-\pi), y = x$

45. $\dfrac{dy}{dx} = -3\sec^3(\pi/2 - x)\tan(\pi/2 - x)$; if $x = -\pi/2$ then $\dfrac{dy}{dx} = 0, y = -1$ so the equation of the tangent line is $y + 1 = 0, y = -1$

47. $\dfrac{dy}{dx} = \sec^2(4x^2)\dfrac{d}{dx}(4x^2) = 8x\sec^2(4x^2)$, $\dfrac{dy}{dx}\Big|_{x=\sqrt{\pi}} = 8\sqrt{\pi}\sec^2(4\pi) = 8\sqrt{\pi}$. When $x = \sqrt{\pi}$, $y = \tan(4\pi) = 0$, so the equation of the tangent line is $y = 8\sqrt{\pi}(x - \sqrt{\pi}) = 8\sqrt{\pi}x - 8\pi$.

49. $\dfrac{dy}{dx} = 2x\sqrt{5-x^2} + \dfrac{x^2}{2\sqrt{5-x^2}}(-2x)$, $\dfrac{dy}{dx}\Big|_{x=1} = 4 - 1/2 = 7/2$. When $x = 1, y = 2$, so the equation of the tangent line is $y - 2 = (7/2)(x-1)$, or $y = \dfrac{7}{2}x - \dfrac{3}{2}$.

51. $\dfrac{dy}{dx} = x(-\sin(5x))\dfrac{d}{dx}(5x) + \cos(5x) - 2\sin x\dfrac{d}{dx}(\sin x)$

$\qquad = -5x\sin(5x) + \cos(5x) - 2\sin x\cos x = -5x\sin(5x) + \cos(5x) - \sin(2x)$,

$\dfrac{d^2y}{dx^2} = -5x\cos(5x)\dfrac{d}{dx}(5x) - 5\sin(5x) - \sin(5x)\dfrac{d}{dx}(5x) - \cos(2x)\dfrac{d}{dx}(2x)$

$\qquad = -25x\cos(5x) - 10\sin(5x) - 2\cos(2x)$

53. $\dfrac{dy}{dx} = \dfrac{(1-x)+(1+x)}{(1-x)^2} = \dfrac{2}{(1-x)^2} = 2(1-x)^{-2}$ and $\dfrac{d^2y}{dx^2} = -2(2)(-1)(1-x)^{-3} = 4(1-x)^{-3}$

55. $y = \cot^3(\pi - \theta) = -\cot^3\theta$ so $dy/dx = 3\cot^2\theta\csc^2\theta$

57. $\dfrac{d}{d\omega}[a\cos^2\pi\omega + b\sin^2\pi\omega] = -2\pi a\cos\pi\omega\sin\pi\omega + 2\pi b\sin\pi\omega\cos\pi\omega$

$$= \pi(b-a)(2\sin\pi\omega\cos\pi\omega) = \pi(b-a)\sin 2\pi\omega$$

59. **(a)**

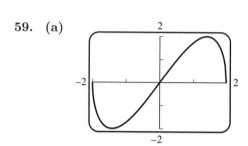

(c) $f'(x) = x\dfrac{-x}{\sqrt{4-x^2}} + \sqrt{4-x^2} = \dfrac{4-2x^2}{\sqrt{4-x^2}}$

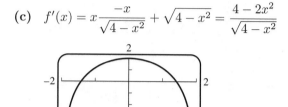

(d) $f(1) = \sqrt{3}$ and $f'(1) = \dfrac{2}{\sqrt{3}}$ so the tangent line has the equation $y - \sqrt{3} = \dfrac{2}{\sqrt{3}}(x-1)$.

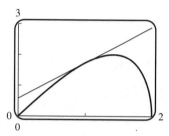

61. **(a)** $dy/dt = -A\omega\sin\omega t, \; d^2y/dt^2 = -A\omega^2\cos\omega t = -\omega^2 y$

(b) one complete oscillation occurs when ωt increases over an interval of length 2π, or if t increases over an interval of length $2\pi/\omega$

(c) $f = 1/T$

(d) amplitude $= 0.6$ cm, $\quad T = 2\pi/15$ s/oscillation, $\quad f = 15/(2\pi)$ oscillations/s

63. $f(x) = \begin{cases} \frac{4}{3}x + 5, & x < 0 \\[2mm] -\frac{5}{2}x + 5, & x > 0 \end{cases}$

$\dfrac{d}{dx}\left[\sqrt{x + f(x)}\right] = \begin{cases} \frac{7}{2\sqrt{21x+45}}, & x < 0 \\[2mm] -\frac{3}{2\sqrt{-6x+20}}, & x > 0 \end{cases} = \dfrac{7\sqrt{6}}{24}$ when $x = -1$

65. **(a)** $p \approx 10$ lb/in^2, $dp/dh \approx -2$ lb/in^2/mi

(b) $\dfrac{dp}{dt} = \dfrac{dp}{dh}\dfrac{dh}{dt} \approx (-2)(0.3) = -0.6$ lb/in^2/s

67. With $u = \sin x$, $\dfrac{d}{dx}(|\sin x|) = \dfrac{d}{dx}(|u|) = \dfrac{d}{du}(|u|)\dfrac{du}{dx} = \dfrac{d}{du}(|u|)\cos x = \begin{cases} \cos x, & u > 0 \\ -\cos x, & u < 0 \end{cases}$

$$= \begin{cases} \cos x, & \sin x > 0 \\ -\cos x, & \sin x < 0 \end{cases} = \begin{cases} \cos x, & 0 < x < \pi \\ -\cos x, & -\pi < x < 0 \end{cases}$$

69. (a) For $x \neq 0, |f(x)| \leq |x|$, and $\lim\limits_{x \to 0} |x| = 0$, so by the Squeezing Theorem, $\lim\limits_{x \to 0} f(x) = 0$.

(b) If $f'(0)$ were to exist, then the limit $\dfrac{f(x) - f(0)}{x - 0} = \sin(1/x)$ would have to exist, but it doesn't.

(c) For $x \neq 0$, $f'(x) = x\left(\cos\dfrac{1}{x}\right)\left(-\dfrac{1}{x^2}\right) + \sin\dfrac{1}{x} = -\dfrac{1}{x}\cos\dfrac{1}{x} + \sin\dfrac{1}{x}$

(d) $\lim\limits_{x \to 0} \dfrac{f(x) - f(0)}{x - 0} = \lim\limits_{x \to 0} \sin(1/x)$, which does not exist, thus $f'(0)$ does not exist.

71. (a) $g'(x) = 3[f(x)]^2 f'(x)$, $g'(2) = 3[f(2)]^2 f'(2) = 3(1)^2(7) = 21$

(b) $h'(x) = f'(x^3)(3x^2)$, $h'(2) = f'(8)(12) = (-3)(12) = -36$

73. $F'(x) = f'(g(x))g'(x) = f'(\sqrt{3x-1})\dfrac{3}{2\sqrt{3x-1}} = \dfrac{\sqrt{3x-1}}{(3x-1)+1}\dfrac{3}{2\sqrt{3x-1}} = \dfrac{1}{2x}$

75. $\dfrac{d}{dx}[f(3x)] = f'(3x)\dfrac{d}{dx}(3x) = 3f'(3x) = 6x$, so $f'(3x) = 2x$. Let $u = 3x$ to get $f'(u) = \dfrac{2}{3}u$;

$\dfrac{d}{dx}[f(x)] = f'(x) = \dfrac{2}{3}x$.

77. For an even function, the graph is symmetric about the y-axis; the slope of the tangent line at $(a, f(a))$ is the negative of the slope of the tangent line at $(-a, f(-a))$. For an odd function, the graph is symmetric about the origin; the slope of the tangent line at $(a, f(a))$ is the same as the slope of the tangent line at $(-a, f(-a))$.

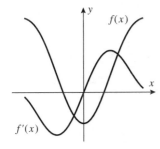 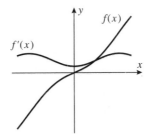

79. $\dfrac{d}{dx}[f(g(h(x)))] = \dfrac{d}{dx}[f(g(u))], \quad u = h(x)$

$$= \dfrac{d}{du}[f(g(u))]\dfrac{du}{dx} = f'(g(u))g'(u)\dfrac{du}{dx} = f'(g(h(x)))g'(h(x))h'(x)$$

EXERCISE SET 3.7

1. $y = (2x - 5)^{1/3}$; $dy/dx = \dfrac{2}{3}(2x - 5)^{-2/3}$

3. $dy/dx = \dfrac{2}{3}\left(\dfrac{x+1}{x-2}\right)^{-1/3}\dfrac{x - 2 - (x+1)}{(x-2)^2} = -\dfrac{2}{(x+1)^{1/3}(x-2)^{5/3}}$

5. $dy/dx = x^3\left(-\dfrac{2}{3}\right)(5x^2 + 1)^{-5/3}(10x) + 3x^2(5x^2 + 1)^{-2/3} = \dfrac{1}{3}x^2(5x^2 + 1)^{-5/3}(25x^2 + 9)$

7. $dy/dx = \dfrac{5}{2}[\sin(3/x)]^{3/2}[\cos(3/x)](-3/x^2) = -\dfrac{15[\sin(3/x)]^{3/2}\cos(3/x)}{2x^2}$

9. **(a)** $1 + y + x\dfrac{dy}{dx} - 6x^2 = 0$, $\dfrac{dy}{dx} = \dfrac{6x^2 - y - 1}{x}$

(b) $y = \dfrac{2 + 2x^3 - x}{x} = \dfrac{2}{x} + 2x^2 - 1$, $\dfrac{dy}{dx} = -\dfrac{2}{x^2} + 4x$

(c) From Part (a), $\dfrac{dy}{dx} = 6x - \dfrac{1}{x} - \dfrac{1}{x}y = 6x - \dfrac{1}{x} - \dfrac{1}{x}\left(\dfrac{2}{x} + 2x^2 - 1\right) = 4x - \dfrac{2}{x^2}$

11. $2x + 2y\dfrac{dy}{dx} = 0$ so $\dfrac{dy}{dx} = -\dfrac{x}{y}$

13. $x^2\dfrac{dy}{dx} + 2xy + 3x(3y^2)\dfrac{dy}{dx} + 3y^3 - 1 = 0$

$(x^2 + 9xy^2)\dfrac{dy}{dx} = 1 - 2xy - 3y^3$ so $\dfrac{dy}{dx} = \dfrac{1 - 2xy - 3y^3}{x^2 + 9xy^2}$

15. $-\dfrac{1}{2x^{3/2}} - \dfrac{\frac{dy}{dx}}{2y^{3/2}} = 0$, $\dfrac{dy}{dx} = -\dfrac{y^{3/2}}{x^{3/2}}$

17. $\cos(x^2y^2)\left[x^2(2y)\dfrac{dy}{dx} + 2xy^2\right] = 1$, $\dfrac{dy}{dx} = \dfrac{1 - 2xy^2\cos(x^2y^2)}{2x^2y\cos(x^2y^2)}$

19. $3\tan^2(xy^2 + y)\sec^2(xy^2 + y)\left(2xy\dfrac{dy}{dx} + y^2 + \dfrac{dy}{dx}\right) = 1$

so $\dfrac{dy}{dx} = \dfrac{1 - 3y^2\tan^2(xy^2 + y)\sec^2(xy^2 + y)}{3(2xy + 1)\tan^2(xy^2 + y)\sec^2(xy^2 + y)}$

21. $4x - 6y\dfrac{dy}{dx} = 0$, $\dfrac{dy}{dx} = \dfrac{2x}{3y}$, $4 - 6\left(\dfrac{dy}{dx}\right)^2 - 6y\dfrac{d^2y}{dx^2} = 0$, $\dfrac{d^2y}{dx^2} = -\dfrac{3\left(\frac{dy}{dx}\right)^2 - 2}{3y} = \dfrac{2(3y^2 - 2x^2)}{9y^3}$

23. $\dfrac{dy}{dx} = -\dfrac{y}{x}$, $\dfrac{d^2y}{dx^2} = -\dfrac{x(dy/dx) - y(1)}{x^2} = -\dfrac{x(-y/x) - y}{x^2} = \dfrac{2y}{x^2}$

25. $\dfrac{dy}{dx} = (1 + \cos y)^{-1}$, $\dfrac{d^2y}{dx^2} = -(1 + \cos y)^{-2}(-\sin y)\dfrac{dy}{dx} = \dfrac{\sin y}{(1 + \cos y)^3}$

27. By implicit differentiation, $2x + 2y(dy/dx) = 0$, $\dfrac{dy}{dx} = -\dfrac{x}{y}$; at $(1/2, \sqrt{3}/2)$, $\dfrac{dy}{dx} = -\sqrt{3}/3$; at $(1/2, -\sqrt{3}/2)$, $\dfrac{dy}{dx} = +\sqrt{3}/3$. Directly, at the upper point $y = \sqrt{1 - x^2}$, $\dfrac{dy}{dx} = \dfrac{-x}{\sqrt{1 - x^2}} = -\dfrac{1/2}{\sqrt{3/4}} = -1/\sqrt{3}$ and at the lower point $y = -\sqrt{1 - x^2}$, $\dfrac{dy}{dx} = \dfrac{x}{\sqrt{1 - x^2}} = +1/\sqrt{3}$.

29. $4x^3 + 4y^3\dfrac{dy}{dx} = 0$, so $\dfrac{dy}{dx} = -\dfrac{x^3}{y^3} = -\dfrac{1}{15^{3/4}} \approx -0.1312$.

31. $4(x^2 + y^2)\left(2x + 2y\dfrac{dy}{dx}\right) = 25\left(2x - 2y\dfrac{dy}{dx}\right)$,

$\dfrac{dy}{dx} = \dfrac{x[25 - 4(x^2 + y^2)]}{y[25 + 4(x^2 + y^2)]}$; at $(3, 1)$ $\dfrac{dy}{dx} = -9/13$

33. $4a^3\dfrac{da}{dt} - 4t^3 = 6\left(a^2 + 2at\dfrac{da}{dt}\right)$, solve for $\dfrac{da}{dt}$ to get $\dfrac{da}{dt} = \dfrac{2t^3 + 3a^2}{2a^3 - 6at}$

35. $2a^2\omega\dfrac{d\omega}{d\lambda} + 2b^2\lambda = 0$ so $\dfrac{d\omega}{d\lambda} = -\dfrac{b^2\lambda}{a^2\omega}$

37. **(a)**

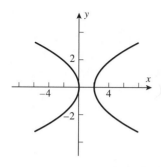

(b) Implicit differentiation of the equation of the curve yields $(4y^3 + 2y)\dfrac{dy}{dx} = 2x - 1$ so $\dfrac{dy}{dx} = 0$ only if $x = 1/2$ but $y^4 + y^2 \geq 0$, so $x = 1/2$ is impossible.

(c) $x^2 - x - (y^4 + y^2) = 0$, so by the Quadratic Formula $x = \dfrac{1 \pm \sqrt{1 + 4y^2 + 4y^4}}{2} = 1 + y^2, -y^2$ which gives the parabolas $x = 1 + y^2, x = -y^2$.

39. **(a)**

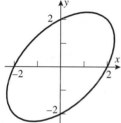

(b) $x \approx \pm 1.1547$

(c) Implicit differentiation yields $2x - x\dfrac{dy}{dx} - y + 2y\dfrac{dy}{dx} = 0$. Solve for $\dfrac{dy}{dx} = \dfrac{y - 2x}{2y - x}$. If $\dfrac{dy}{dx} = 0$ then $y - 2x = 0$ or $y = 2x$. Thus $4 = x^2 - xy + y^2 = x^2 - 2x^2 + 4x^2 = 3x^2$, $x = \pm\dfrac{2}{\sqrt{3}}$.

41. Solve the simultaneous equations $y = x, x^2 - xy + y^2 = 4$ to get $x^2 - x^2 + x^2 = 4, x = \pm 2, y = x = \pm 2$, so the points of intersection are $(2, 2)$ and $(-2, -2)$.

From Exercise 39 part (c), $\dfrac{dy}{dx} = \dfrac{y - 2x}{2y - x}$. When $x = y = 2, \dfrac{dy}{dx} = -1$; when $x = y = -2, \dfrac{dy}{dx} = -1$; the slopes are equal.

43. The point $(1,1)$ is on the graph, so $1 + a = b$. The slope of the tangent line at $(1,1)$ is $-4/3$; use implicit differentiation to get $\dfrac{dy}{dx} = -\dfrac{2xy}{x^2 + 2ay}$ so at $(1,1)$, $-\dfrac{2}{1 + 2a} = -\dfrac{4}{3}, 1 + 2a = 3/2, a = 1/4$ and hence $b = 1 + 1/4 = 5/4$.

45. **(a)**

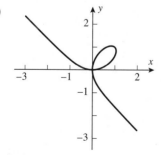

(b) $x \approx 0.84$

(c) Use implicit differentiation to get $dy/dx = (2y - 3x^2)/(3y^2 - 2x)$, so $dy/dx = 0$ if $y = (3/2)x^2$. Substitute this into $x^3 - 2xy + y^3 = 0$ to obtain $27x^6 - 16x^3 = 0, x^3 = 16/27, x = 2^{4/3}/3$ and hence $y = 2^{5/3}/3$.

47. (a) The curve is the circle $(x - 2)^2 + y^2 = 1$ about the point $(2, 0)$ of radius 1. One tangent line is tangent at a point $P(x, y)$ in the first quadrant. Let $Q(2, 0)$ be the center of the circle. Then OPQ is a right angle, with sides $|PQ| = r = 1$ and $|OP| = \sqrt{x^2 + y^2}$. By the Pythagorean Theorem $x^2 + y^2 + 1^2 = 2^2$. Substitute this into $(x - 2)^2 + y^2 = 1$ to obtain $3 - 4x + 4 = 1, x = 3/2, y = \sqrt{3 - x^2} = \sqrt{3}/2$. So the required tangent lines are $y = \pm(\sqrt{3}/3)x$.

(b) Let $P(x_0, y_0)$ be a point where a line through the origin is tangent to the curve $x^2 - 4x + y^2 + 3 = 0$. Implicit differentiation applied to the equation of the curve gives $dy/dx = (2 - x)/y$. At P the slope of the curve must equal the slope of the line so $(2 - x_0)/y_0 = y_0/x_0$, or $y_0^2 = 2x_0 - x_0^2$. But $x_0^2 - 4x_0 + y_0^2 + 3 = 0$ because (x_0, y_0) is on the curve, and elimination of y_0^2 in the latter two equations gives $x_0^2 - 4x_0 + (2x_0 - x_0^2) + 3 = 0$, $x_0 = 3/2$ which when substituted into $y_0^2 = 2x_0 - x_0^2$ yields $y_0^2 = 3/4$, so $y_0 = \pm\sqrt{3}/2$. The slopes of the lines are $(\pm\sqrt{3}/2)/(3/2) = \pm\sqrt{3}/3$ and their equations are $y = (\sqrt{3}/3)x$ and $y = -(\sqrt{3}/3)x$.

49. The linear equation $ax_0^{r-1}x + by_0^{r-1}y = c$ is the equation of a line ℓ. Implicit differentiation of the equation of the curve yields $rax^{r-1} + rby^{r-1}\dfrac{dy}{dx} = 0$, $\dfrac{dy}{dx} = -\dfrac{ax^{r-1}}{by^{r-1}}$. At the point (x_0, y_0) the slope of the line must be $-\dfrac{ax_0^{r-1}}{by_0^{r-1}}$, which is the slope of ℓ. Moreover, the equation of ℓ is satisfied by the point (x_0, y_0), so this point lies on ℓ. By the point-slope formula, ℓ must be the line tangent to the curve at (x_0, y_0).

51. By the chain rule, $\dfrac{dy}{dx} = \dfrac{dy}{dt}\dfrac{dt}{dx}$. Use implicit differentiation on $2y^3t + t^3y = 1$ to get

$$\frac{dy}{dt} = -\frac{2y^3 + 3t^2y}{6ty^2 + t^3}, \text{ but } \frac{dt}{dx} = \frac{1}{\cos t} \text{ so } \frac{dy}{dx} = -\frac{2y^3 + 3t^2y}{(6ty^2 + t^3)\cos t}.$$

53. $y' = rx^{r-1}, y'' = r(r-1)x^{r-2}$ so $3x^2\left[r(r-1)x^{r-2}\right] + 4x\left(rx^{r-1}\right) - 2x^r = 0$, $3r(r-1)x^r + 4rx^r - 2x^r = 0, (3r^2 + r - 2)x^r = 0$, $3r^2 + r - 2 = 0, (3r - 2)(r + 1) = 0; r = -1, 2/3$

55. We shall find when the curves intersect and check that the slopes are negative reciprocals. For the intersection solve the simultaneous equations $x^2 + (y - c)^2 = c^2$ and $(x - k)^2 + y^2 = k^2$ to obtain $cy = kx = \dfrac{1}{2}(x^2 + y^2)$. Thus $x^2 + y^2 = cy + kx$, or $y^2 - cy = -x^2 + kx$, and $\dfrac{y - c}{x} = -\dfrac{x - k}{y}$. Differentiating the two families yields (black) $\dfrac{dy}{dx} = -\dfrac{x}{y - c}$, and (gray) $\dfrac{dy}{dx} = -\dfrac{x - k}{y}$. But it was proven that these quantities are negative reciprocals of each other.

EXERCISE SET 3.8

1. $\dfrac{dy}{dt} = 3\dfrac{dx}{dt}$

    (a) $\dfrac{dy}{dt} = 3(2) = 6$

    (b) $-1 = 3\dfrac{dx}{dt}, \dfrac{dx}{dt} = -\dfrac{1}{3}$

3. $8x\dfrac{dx}{dt} + 18y\dfrac{dy}{dt} = 0$

    (a) $8\dfrac{1}{2\sqrt{2}} \cdot 3 + 18\dfrac{1}{3\sqrt{2}}\dfrac{dy}{dt} = 0,\ \dfrac{dy}{dt} = -2$
           (b) $8\left(\dfrac{1}{3}\right)\dfrac{dx}{dt} - 18\dfrac{\sqrt{5}}{9} \cdot 8 = 0,\ \dfrac{dx}{dt} = 6\sqrt{5}$

5. **(b)** $A = x^2$ **(c)** $\dfrac{dA}{dt} = 2x\dfrac{dx}{dt}$

    (d) Find $\dfrac{dA}{dt}\Big|_{x=3}$ given that $\dfrac{dx}{dt}\Big|_{x=3} = 2$. From Part (c), $\dfrac{dA}{dt}\Big|_{x=3} = 2(3)(2) = 12\ \text{ft}^2/\text{min}$.

7. **(a)** $V = \pi r^2 h$, so $\dfrac{dV}{dt} = \pi\left(r^2\dfrac{dh}{dt} + 2rh\dfrac{dr}{dt}\right)$.

    (b) Find $\dfrac{dV}{dt}\Big|_{\substack{h=6,\\ r=10}}$ given that $\dfrac{dh}{dt}\Big|_{\substack{h=6,\\ r=10}} = 1$ and $\dfrac{dr}{dt}\Big|_{\substack{h=6,\\ r=10}} = -1$. From Part (a),

      $\dfrac{dV}{dt}\Big|_{\substack{h=6,\\ r=10}} = \pi[10^2(1) + 2(10)(6)(-1)] = -20\pi\ \text{in}^3/\text{s}$; the volume is decreasing.

9. **(a)** $\tan\theta = \dfrac{y}{x}$, so $\sec^2\theta\,\dfrac{d\theta}{dt} = \dfrac{x\dfrac{dy}{dt} - y\dfrac{dx}{dt}}{x^2}$, $\dfrac{d\theta}{dt} = \dfrac{\cos^2\theta}{x^2}\left(x\dfrac{dy}{dt} - y\dfrac{dx}{dt}\right)$

    (b) Find $\dfrac{d\theta}{dt}\Big|_{\substack{x=2,\\ y=2}}$ given that $\dfrac{dx}{dt}\Big|_{\substack{x=2,\\ y=2}} = 1$ and $\dfrac{dy}{dt}\Big|_{\substack{x=2,\\ y=2}} = -\dfrac{1}{4}$.

      When $x = 2$ and $y = 2$, $\tan\theta = 2/2 = 1$ so $\theta = \dfrac{\pi}{4}$ and $\cos\theta = \cos\dfrac{\pi}{4} = \dfrac{1}{\sqrt{2}}$. Thus

      from Part (a), $\dfrac{d\theta}{dt}\Big|_{\substack{x=2,\\ y=2}} = \dfrac{(1/\sqrt{2})^2}{2^2}\left[2\left(-\dfrac{1}{4}\right) - 2(1)\right] = -\dfrac{5}{16}\ \text{rad/s}$; θ is decreasing.

11. Let A be the area swept out, and θ the angle through which the minute hand has rotated.
    Find $\dfrac{dA}{dt}$ given that $\dfrac{d\theta}{dt} = \dfrac{\pi}{30}$ rad/min; $A = \dfrac{1}{2}r^2\theta = 8\theta$, so $\dfrac{dA}{dt} = 8\dfrac{d\theta}{dt} = \dfrac{4\pi}{15}\ \text{in}^2/\text{min}$.

13. Find $\dfrac{dr}{dt}\Big|_{A=9}$ given that $\dfrac{dA}{dt} = 6$. From $A = \pi r^2$ we get $\dfrac{dA}{dt} = 2\pi r\dfrac{dr}{dt}$ so $\dfrac{dr}{dt} = \dfrac{1}{2\pi r}\dfrac{dA}{dt}$. If $A = 9$
    then $\pi r^2 = 9$, $r = 3/\sqrt{\pi}$ so $\dfrac{dr}{dt}\Big|_{A=9} = \dfrac{1}{2\pi(3/\sqrt{\pi})}(6) = 1/\sqrt{\pi}\ \text{mi/h}$.

15. Find $\dfrac{dV}{dt}\Big|_{r=9}$ given that $\dfrac{dr}{dt} = -15$. From $V = \dfrac{4}{3}\pi r^3$ we get $\dfrac{dV}{dt} = 4\pi r^2\dfrac{dr}{dt}$ so

    $\dfrac{dV}{dt}\Big|_{r=9} = 4\pi(9)^2(-15) = -4860\pi$. Air must be removed at the rate of $4860\pi\ \text{cm}^3/\text{min}$.

17. Find $\dfrac{dx}{dt}\Big|_{y=5}$ given that $\dfrac{dy}{dt} = -2$. From $x^2 + y^2 = 13^2$

    we get $2x\dfrac{dx}{dt} + 2y\dfrac{dy}{dt} = 0$ so $\dfrac{dx}{dt} = -\dfrac{y}{x}\dfrac{dy}{dt}$. Use

    $x^2 + y^2 = 169$ to find that $x = 12$ when $y = 5$ so

    $\dfrac{dx}{dt}\Big|_{y=5} = -\dfrac{5}{12}(-2) = \dfrac{5}{6}\ \text{ft/s}$.

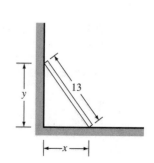

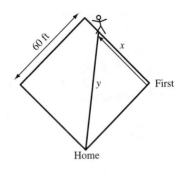

19. Let x denote the distance from first base and y the distance from home plate. Then $x^2 + 60^2 = y^2$ and $2x\dfrac{dx}{dt} = 2y\dfrac{dy}{dt}$. When $x = 50$ then $y = 10\sqrt{61}$ so

$$\frac{dy}{dt} = \frac{x}{y}\frac{dx}{dt} = \frac{50}{10\sqrt{61}}(25) = \frac{125}{\sqrt{61}} \text{ ft/s.}$$

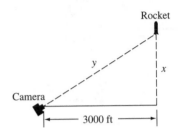

21. Find $\left.\dfrac{dy}{dt}\right|_{x=4000}$ given that $\left.\dfrac{dx}{dt}\right|_{x=4000} = 880$. From

$y^2 = x^2 + 3000^2$ we get $2y\dfrac{dy}{dt} = 2x\dfrac{dx}{dt}$ so $\dfrac{dy}{dt} = \dfrac{x}{y}\dfrac{dx}{dt}$.
If $x = 4000$, then $y = 5000$ so

$$\left.\frac{dy}{dt}\right|_{x=4000} = \frac{4000}{5000}(880) = 704 \text{ ft/s.}$$

23. (a) If x denotes the altitude, then $r - x = 3960$, the radius of the Earth. $\theta = 0$ at perigee, so $r = 4995/1.12 \approx 4460$; the altitude is $x = 4460 - 3960 = 500$ miles. $\theta = \pi$ at apogee, so $r = 4995/0.88 \approx 5676$; the altitude is $x = 5676 - 3960 = 1716$ miles.

(b) If $\theta = 120°$, then $r = 4995/0.94 \approx 5314$; the altitude is $5314 - 3960 = 1354$ miles. The rate of change of the altitude is given by

$$\frac{dx}{dt} = \frac{dr}{dt} = \frac{dr}{d\theta}\frac{d\theta}{dt} = \frac{4995(0.12\sin\theta)}{(1 + 0.12\cos\theta)^2}\frac{d\theta}{dt}.$$

Use $\theta = 120°$ and $d\theta/dt = 2.7°/\text{min} = (2.7)(\pi/180)$ rad/min to get $dr/dt \approx 27.7$ mi/min.

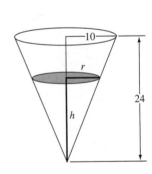

25. Find $\left.\dfrac{dh}{dt}\right|_{h=16}$ given that $\dfrac{dV}{dt} = 20$. The volume of

water in the tank at a depth h is $V = \dfrac{1}{3}\pi r^2 h$. Use

similar triangles (see figure) to get $\dfrac{r}{h} = \dfrac{10}{24}$ so $r = \dfrac{5}{12}h$

thus $V = \dfrac{1}{3}\pi\left(\dfrac{5}{12}h\right)^2 h = \dfrac{25}{432}\pi h^3$, $\dfrac{dV}{dt} = \dfrac{25}{144}\pi h^2\dfrac{dh}{dt}$;

$\dfrac{dh}{dt} = \dfrac{144}{25\pi h^2}\dfrac{dV}{dt}$, $\left.\dfrac{dh}{dt}\right|_{h=16} = \dfrac{144}{25\pi(16)^2}(20) = \dfrac{9}{20\pi}$
ft/min.

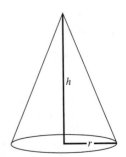

27. Find $\left.\dfrac{dV}{dt}\right|_{h=10}$ given that $\dfrac{dh}{dt} = 5$. $V = \dfrac{1}{3}\pi r^2 h$, but

$r = \dfrac{1}{2}h$ so $V = \dfrac{1}{3}\pi\left(\dfrac{h}{2}\right)^2 h = \dfrac{1}{12}\pi h^3$, $\dfrac{dV}{dt} = \dfrac{1}{4}\pi h^2\dfrac{dh}{dt}$,

$$\left.\frac{dV}{dt}\right|_{h=10} = \frac{1}{4}\pi(10)^2(5) = 125\pi \text{ ft}^3/\text{min.}$$

29. With s and h as shown in the figure, we want to find $\dfrac{dh}{dt}$ given that $\dfrac{ds}{dt} = 500$. From the figure,

$$h = s \sin 30° = \frac{1}{2}s \text{ so } \frac{dh}{dt} = \frac{1}{2}\frac{ds}{dt} = \frac{1}{2}(500) = 250 \text{ mi/h.}$$

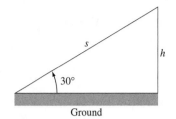

Ground

31. Find $\dfrac{dy}{dt}$ given that $\dfrac{dx}{dt}\bigg|_{y=125} = -12$. From $x^2 + 10^2 = y^2$

we get $2x\dfrac{dx}{dt} = 2y\dfrac{dy}{dt}$ so $\dfrac{dy}{dt} = \dfrac{x}{y}\dfrac{dx}{dt}$. Use $x^2 + 100 = y^2$

to find that $x = \sqrt{15,525} = 15\sqrt{69}$ when $y = 125$ so

$\dfrac{dy}{dt} = \dfrac{15\sqrt{69}}{125}(-12) = -\dfrac{36\sqrt{69}}{25}$. The rope must be pulled

at the rate of $\dfrac{36\sqrt{69}}{25}$ ft/min.

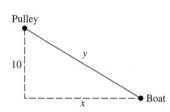

Pulley

10

y

x

Boat

33. Find $\dfrac{dx}{dt}\bigg|_{\theta=\pi/4}$ given that $\dfrac{d\theta}{dt} = \dfrac{2\pi}{10} = \dfrac{\pi}{5}$ rad/s.

Then $x = 4\tan\theta$ (see figure) so $\dfrac{dx}{dt} = 4\sec^2\theta\dfrac{d\theta}{dt}$,

$$\frac{dx}{dt}\bigg|_{\theta=\pi/4} = 4\left(\sec^2\frac{\pi}{4}\right)\left(\frac{\pi}{5}\right) = 8\pi/5 \text{ km/s.}$$

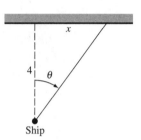

x

4

θ

Ship

35. We wish to find $\dfrac{dz}{dt}\bigg|_{\substack{x=2,\\y=4}}$ given $\dfrac{dx}{dt} = -600$ and

$\dfrac{dy}{dt}\bigg|_{\substack{x=2,\\y=4}} = -1200$ (see figure). From the law of cosines,

$$z^2 = x^2 + y^2 - 2xy\cos 120° = x^2 + y^2 - 2xy(-1/2)$$

$$= x^2 + y^2 + xy, \text{ so } 2z\frac{dz}{dt} = 2x\frac{dx}{dt} + 2y\frac{dy}{dt} + x\frac{dy}{dt} + y\frac{dx}{dt},$$

$$\frac{dz}{dt} = \frac{1}{2z}\left[(2x+y)\frac{dx}{dt} + (2y+x)\frac{dy}{dt}\right].$$

When $x = 2$ and $y = 4$, $z^2 = 2^2 + 4^2 + (2)(4) = 28$, so $z = \sqrt{28} = 2\sqrt{7}$, thus

$$\frac{dz}{dt}\bigg|_{\substack{x=2,\\y=4}} = \frac{1}{2(2\sqrt{7})}[(2(2)+4)(-600) + (2(4)+2)(-1200)] = -\frac{4200}{\sqrt{7}} = -600\sqrt{7} \text{ mi/h;}$$

the distance between missile and aircraft is decreasing at the rate of $600\sqrt{7}$ mi/h.

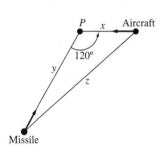

P x Aircraft

120°

y

z

Missile

37. (a) We want $\dfrac{dy}{dt}\Big|_{\substack{x=1,\\y=2}}$ given that $\dfrac{dx}{dt}\Big|_{\substack{x=1,\\y=2}} = 6$. For convenience, first rewrite the equation as

$$xy^3 = \frac{8}{5} + \frac{8}{5}y^2 \text{ then } 3xy^2\frac{dy}{dt} + y^3\frac{dx}{dt} = \frac{16}{5}y\frac{dy}{dt}, \ \frac{dy}{dt} = \frac{y^3}{\frac{16}{5}y - 3xy^2}\frac{dx}{dt} \text{ so}$$

$$\frac{dy}{dt}\Big|_{\substack{x=1,\\y=2}} = \frac{2^3}{\frac{16}{5}(2) - 3(1)2^2}(6) = -60/7 \text{ units/s}.$$

(b) falling, because $\dfrac{dy}{dt} < 0$

39. The coordinates of P are $(x, 2x)$, so the distance between P and the point $(3, 0)$ is

$$D = \sqrt{(x-3)^2 + (2x-0)^2} = \sqrt{5x^2 - 6x + 9}. \text{ Find } \frac{dD}{dt}\Big|_{x=3} \text{ given that } \frac{dx}{dt}\Big|_{x=3} = -2.$$

$$\frac{dD}{dt} = \frac{5x-3}{\sqrt{5x^2 - 6x + 9}}\frac{dx}{dt}, \text{ so } \frac{dD}{dt}\Big|_{x=3} = \frac{12}{\sqrt{36}}(-2) = -4 \text{ units/s}.$$

41. Solve $\dfrac{dx}{dt} = 3\dfrac{dy}{dt}$ given $y = x/(x^2 + 1)$. Then $y(x^2 + 1) = x$. Differentiating with respect to x,

$(x^2 + 1)\dfrac{dy}{dx} + y(2x) = 1$. But $\dfrac{dy}{dx} = \dfrac{dy/dt}{dx/dt} = \dfrac{1}{3}$ so $(x^2 + 1)\dfrac{1}{3} + 2xy = 1$, $x^2 + 1 + 6xy = 3$,

$x^2 + 1 + 6x^2/(x^2 + 1) = 3$, $(x^2 + 1)^2 + 6x^2 - 3x^2 - 3 = 0$, $x^4 + 5x^2 - 3 = 0$. By the binomial theorem applied to x^2 we obtain $x^2 = (-5 \pm \sqrt{25 + 12})/2$. The minus sign is spurious since x^2 cannot be negative, so $x^2 = (\sqrt{37} - 5)/2$, $x \approx \pm 0.7357861545$, $y = \pm -0.4773550654$.

43. Find $\dfrac{dS}{dt}\Big|_{s=10}$ given that $\dfrac{ds}{dt}\Big|_{s=10} = -2$. From $\dfrac{1}{s} + \dfrac{1}{S} = \dfrac{1}{6}$ we get $-\dfrac{1}{s^2}\dfrac{ds}{dt} - \dfrac{1}{S^2}\dfrac{dS}{dt} = 0$, so

$\dfrac{dS}{dt} = -\dfrac{S^2}{s^2}\dfrac{ds}{dt}$. If $s = 10$, then $\dfrac{1}{10} + \dfrac{1}{S} = \dfrac{1}{6}$ which gives $S = 15$. So $\dfrac{dS}{dt}\Big|_{s=10} = -\dfrac{225}{100}(-2) = 4.5$ cm/s. The image is moving away from the lens.

45. Let r be the radius, V the volume, and A the surface area of a sphere. Show that $\dfrac{dr}{dt}$ is a constant given that $\dfrac{dV}{dt} = -kA$, where k is a positive constant. Because $V = \dfrac{4}{3}\pi r^3$,

$$\frac{dV}{dt} = 4\pi r^2\frac{dr}{dt} \tag{1}$$

But it is given that $\dfrac{dV}{dt} = -kA$ or, because $A = 4\pi r^2$, $\dfrac{dV}{dt} = -4\pi r^2 k$ which when substituted into equation (1) gives $-4\pi r^2 k = 4\pi r^2\dfrac{dr}{dt}$, $\dfrac{dr}{dt} = -k$.

47. Extend sides of cup to complete the cone and let V_0 be the volume of the portion added, then (see figure)

$$V = \frac{1}{3}\pi r^2 h - V_0 \text{ where } \frac{r}{h} = \frac{4}{12} = \frac{1}{3} \text{ so } r = \frac{1}{3}h \text{ and}$$

$$V = \frac{1}{3}\pi\left(\frac{h}{3}\right)^2 h - V_0 = \frac{1}{27}\pi h^3 - V_0, \ \frac{dV}{dt} = \frac{1}{9}\pi h^2\frac{dh}{dt},$$

$$\frac{dh}{dt} = \frac{9}{\pi h^2}\frac{dV}{dt}, \ \frac{dh}{dt}\Big|_{h=9} = \frac{9}{\pi(9)^2}(20) = \frac{20}{9\pi} \text{ cm/s}.$$

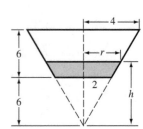

EXERCISE SET 3.9

1. **(a)** $f(x) \approx f(1) + f'(1)(x-1) = 1 + 3(x-1)$
(b) $f(1 + \Delta x) \approx f(1) + f'(1)\Delta x = 1 + 3\Delta x$
(c) From Part (a), $(1.02)^3 \approx 1 + 3(0.02) = 1.06$. From Part (b), $(1.02)^3 \approx 1 + 3(0.02) = 1.06$.

3. **(a)** $f(x) \approx f(x_0) + f'(x_0)(x - x_0) = 1 + (1/(2\sqrt{1})(x - 0) = 1 + (1/2)x$, so with $x_0 = 0$ and $x = -0.1$, we have $\sqrt{0.9} = f(-0.1) \approx 1 + (1/2)(-0.1) = 1 - 0.05 = 0.95$. With $x = 0.1$ we have $\sqrt{1.1} = f(0.1) \approx 1 + (1/2)(0.1) = 1.05$.

(b)

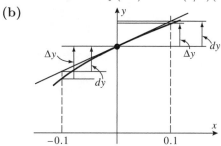

5. $f(x) = (1+x)^{15}$ and $x_0 = 0$. Thus $(1+x)^{15} \approx f(x_0) + f'(x_0)(x - x_0) = 1 + 15(1)^{14}(x - 0) = 1 + 15x$.

7. $\tan x \approx \tan(0) + \sec^2(0)(x - 0) = x$

9. $x^4 \approx (1)^4 + 4(1)^3(x - 1)$. Set $\Delta x = x - 1$; then $x = \Delta x + 1$ and $(1 + \Delta x)^4 = 1 + 4\Delta x$.

11. $\dfrac{1}{2 + x} \approx \dfrac{1}{2 + 1} - \dfrac{1}{(2+1)^2}(x - 1)$, and $2 + x = 3 + \Delta x$, so $\dfrac{1}{3 + \Delta x} \approx \dfrac{1}{3} - \dfrac{1}{9}\Delta x$

13. $f(x) = \sqrt{x + 3}$ and $x_0 = 0$, so
$$\sqrt{x + 3} \approx \sqrt{3} + \frac{1}{2\sqrt{3}}(x - 0) = \sqrt{3} + \frac{1}{2\sqrt{3}}x, \text{ and}$$
$$\left| f(x) - \left(\sqrt{3} + \frac{1}{2\sqrt{3}}x \right) \right| < 0.1 \text{ if } |x| < 1.692.$$

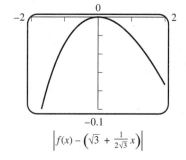

$\left| f(x) - \left(\sqrt{3} + \frac{1}{2\sqrt{3}}x \right) \right|$

15. $\tan 2x \approx \tan 0 + (\sec^2 0)(2x - 0) = 2x$,
and $|\tan x - x| < 0.1$ if $|x| < 0.6316$

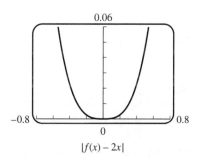

$|f(x) - 2x|$

17. **(a)** The local linear approximation $\sin x \approx x$ gives $\sin 1° = \sin(\pi/180) \approx \pi/180 = 0.0174533$ and a calculator gives $\sin 1° = 0.0174524$. The relative error $|\sin(\pi/180) - (\pi/180)|/(\sin \pi/180) = 0.000051$ is very small, so for such a small value of x the approximation is very good.

(b) Use $x_0 = 45°$ (this assumes you know, or can approximate, $\sqrt{2}/2$).

(c) $44° = \dfrac{44\pi}{180}$ radians, and $45° = \dfrac{45\pi}{180} = \dfrac{\pi}{4}$ radians. With $x = \dfrac{44\pi}{180}$ and $x_0 = \dfrac{\pi}{4}$ we obtain

$$\sin 44° = \sin \frac{44\pi}{180} \approx \sin \frac{\pi}{4} + \left(\cos \frac{\pi}{4}\right)\left(\frac{44\pi}{180} - \frac{\pi}{4}\right) = \frac{\sqrt{2}}{2} + \frac{\sqrt{2}}{2}\left(\frac{-\pi}{180}\right) = 0.694765. \text{ With a}$$

calculator, $\sin 44° = 0.694658$.

19. $f(x) = x^4$, $f'(x) = 4x^3$, $x_0 = 3$, $\Delta x = 0.02$; $(3.02)^4 \approx 3^4 + (108)(0.02) = 81 + 2.16 = 83.16$

21. $f(x) = \sqrt{x}$, $f'(x) = \dfrac{1}{2\sqrt{x}}$, $x_0 = 64$, $\Delta x = 1$; $\sqrt{65} \approx \sqrt{64} + \dfrac{1}{16}(1) = 8 + \dfrac{1}{16} = 8.0625$

23. $f(x) = \sqrt{x}$, $f'(x) = \dfrac{1}{2\sqrt{x}}$, $x_0 = 81$, $\Delta x = -0.1$; $\sqrt{80.9} \approx \sqrt{81} + \dfrac{1}{18}(-0.1) \approx 8.9944$

25. $f(x) = \sin x$, $f'(x) = \cos x$, $x_0 = 0$, $\Delta x = 0.1$; $\sin 0.1 \approx \sin 0 + (\cos 0)(0.1) = 0.1$

27. $f(x) = \cos x$, $f'(x) = -\sin x$, $x_0 = \pi/6$, $\Delta x = \pi/180$;

$$\cos 31° \approx \cos 30° + \left(-\frac{1}{2}\right)\left(\frac{\pi}{180}\right) = \frac{\sqrt{3}}{2} - \frac{\pi}{360} \approx 0.8573$$

29. **(a)** $dy = f'(x)dx = 2xdx = 4(1) = 4$ and
$\Delta y = (x + \Delta x)^2 - x^2 = (2+1)^2 - 2^2 = 5$

(b)

31. **(a)** $dy = (-1/x^2)dx = (-1)(-0.5) = 0.5$ and
$\Delta y = 1/(x + \Delta x) - 1/x$
$= 1/(1 - 0.5) - 1/1 = 2 - 1 = 1$

(b)

33. $dy = 3x^2 dx$;
$\Delta y = (x + \Delta x)^3 - x^3 = x^3 + 3x^2\Delta x + 3x(\Delta x)^2 + (\Delta x)^3 - x^3 = 3x^2\Delta x + 3x(\Delta x)^2 + (\Delta x)^3$

35. $dy = (2x - 2)dx$;
$\Delta y = [(x + \Delta x)^2 - 2(x + \Delta x) + 1] - [x^2 - 2x + 1]$
$= x^2 + 2x\,\Delta x + (\Delta x)^2 - 2x - 2\Delta x + 1 - x^2 + 2x - 1 = 2x\,\Delta x + (\Delta x)^2 - 2\Delta x$

37. **(a)** $dy = (12x^2 - 14x)dx$
(b) $dy = x\,d(\cos x) + \cos x\,dx = x(-\sin x)dx + \cos x dx = (-x\sin x + \cos x)dx$

39. **(a)** $dy = \left(\sqrt{1 - x} - \dfrac{x}{2\sqrt{1 - x}}\right)dx = \dfrac{2 - 3x}{2\sqrt{1 - x}}dx$

(b) $dy = -17(1 + x)^{-18}dx$

41. $dy = \dfrac{3}{2\sqrt{3x - 2}}dx$, $x = 2$, $dx = 0.03$; $\Delta y \approx dy = \dfrac{3}{4}(0.03) = 0.0225$

43. $dy = \dfrac{1 - x^2}{(x^2 + 1)^2}dx$, $x = 2$, $dx = -0.04$; $\Delta y \approx dy = \left(-\dfrac{3}{25}\right)(-0.04) = 0.0048$

45. **(a)** $A = x^2$ where x is the length of a side; $dA = 2x\,dx = 2(10)(\pm 0.1) = \pm 2$ ft^2.

(b) relative error in x is $\approx \dfrac{dx}{x} = \dfrac{\pm 0.1}{10} = \pm 0.01$ so percentage error in x is $\approx \pm 1\%$; relative error

in A is $\approx \dfrac{dA}{A} = \dfrac{2x\,dx}{x^2} = 2\dfrac{dx}{x} = 2(\pm 0.01) = \pm 0.02$ so percentage error in A is $\approx \pm 2\%$

47. **(a)** $x = 10\sin\theta$, $y = 10\cos\theta$ (see figure),

$$dx = 10\cos\theta\,d\theta = 10\left(\cos\dfrac{\pi}{6}\right)\left(\pm\dfrac{\pi}{180}\right) = 10\left(\dfrac{\sqrt{3}}{2}\right)\left(\pm\dfrac{\pi}{180}\right)$$

$$\approx \pm 0.151 \text{ in},$$

$$dy = -10(\sin\theta)d\theta = -10\left(\sin\dfrac{\pi}{6}\right)\left(\pm\dfrac{\pi}{180}\right) = -10\left(\dfrac{1}{2}\right)\left(\pm\dfrac{\pi}{180}\right)$$

$$\approx \pm 0.087 \text{ in}$$

(b) relative error in x is $\approx \dfrac{dx}{x} = (\cot\theta)d\theta = \left(\cot\dfrac{\pi}{6}\right)\left(\pm\dfrac{\pi}{180}\right) = \sqrt{3}\left(\pm\dfrac{\pi}{180}\right) \approx \pm 0.030$
so percentage error in x is $\approx \pm 3.0\%$;
relative error in y is $\approx \dfrac{dy}{y} = -\tan\theta\,d\theta = -\left(\tan\dfrac{\pi}{6}\right)\left(\pm\dfrac{\pi}{180}\right) = -\dfrac{1}{\sqrt{3}}\left(\pm\dfrac{\pi}{180}\right) \approx \pm 0.010$
so percentage error in y is $\approx \pm 1.0\%$

49. $\dfrac{dR}{R} = \dfrac{(-2k/r^3)dr}{(k/r^2)} = -2\dfrac{dr}{r}$, but $\dfrac{dr}{r} \approx \pm 0.05$ so $\dfrac{dR}{R} \approx -2(\pm 0.05) = \pm 0.10$; percentage error in R
is $\approx \pm 10\%$

51. $A = \dfrac{1}{4}(4)^2\sin 2\theta = 4\sin 2\theta$ thus $dA = 8\cos 2\theta\,d\theta$ so, with $\theta = 30° = \pi/6$ radians and
$d\theta = \pm 15' = \pm 1/4° = \pm\pi/720$ radians, $dA = 8\cos(\pi/3)(\pm\pi/720) = \pm\pi/180 \approx \pm 0.017$ cm^2

53. $V = x^3$ where x is the length of a side; $\dfrac{dV}{V} = \dfrac{3x^2\,dx}{x^3} = 3\dfrac{dx}{x}$, but $\dfrac{dx}{x} \approx \pm 0.02$
so $\dfrac{dV}{V} \approx 3(\pm 0.02) = \pm 0.06$; percentage error in V is $\approx \pm 6\%$.

55. $A = \dfrac{1}{4}\pi D^2$ where D is the diameter of the circle; $\dfrac{dA}{A} = \dfrac{(\pi D/2)dD}{\pi D^2/4} = 2\dfrac{dD}{D}$, but $\dfrac{dA}{A} \approx \pm 0.01$ so
$2\dfrac{dD}{D} \approx \pm 0.01$, $\dfrac{dD}{D} \approx \pm 0.005$; maximum permissible percentage error in D is $\approx \pm 0.5\%$.

57. $V = $ volume of cylindrical rod $= \pi r^2 h = \pi r^2(15) = 15\pi r^2$; approximate ΔV by dV if $r = 2.5$ and
$dr = \Delta r = 0.001$. $dV = 30\pi r\,dr = 30\pi(2.5)(0.001) \approx 235.62$ cm^3.

59. **(a)** $\alpha = \Delta L/(L\Delta T) = 0.006/(40 \times 10) = 1.5 \times 10^{-5}/°C$
(b) $\Delta L = 2.3 \times 10^{-5}(180)(25) \approx 0.1$ cm, so the pole is about 180.1 cm long.

REVIEW EXERCISES CHAPTER 3

3. (a) $m_{\tan} = \lim\limits_{x_1 \to x_0} \dfrac{f(x_1) - f(x_0)}{x_1 - x_0} = \lim\limits_{x_1 \to x_0} \dfrac{(x_1^2 + 1) - (x_0^2 + 1)}{x_1 - x_0}$

$= \lim\limits_{x_1 \to x_0} \dfrac{x_1^2 - x_0^2}{x_1 - x_0} = \lim\limits_{x_1 \to x_0} (x_1 + x_0) = 2x_0$

(b) $m_{\tan} = 2(2) = 4$

5. $v_{\text{inst}} = \lim\limits_{h \to 0} \dfrac{3(h + 1)^{2.5} + 580h - 3}{10h} = 58 + \dfrac{1}{10} \dfrac{d}{dx} 3x^{2.5} \Big|_{x=1} = 58 + \dfrac{1}{10}(2.5)(3)(1)^{1.5} = 58.75$ ft/s

7. (a) $v_{\text{ave}} = \dfrac{[3(3)^2 + 3] - [3(1)^2 + 1]}{3 - 1} = 13$ mi/h

(b) $v_{\text{inst}} = \lim\limits_{t_1 \to 1} \dfrac{(3t_1^2 + t_1) - 4}{t_1 - 1} = \lim\limits_{t_1 \to 1} \dfrac{(3t_1 + 4)(t_1 - 1)}{t_1 - 1} = \lim\limits_{t_1 \to 1} (3t_1 + 4) = 7$ mi/h

9. (a) $\dfrac{dy}{dx} = \lim\limits_{h \to 0} \dfrac{\sqrt{9 - 4(x + h)} - \sqrt{9 - 4x}}{h} = \lim\limits_{h \to 0} \dfrac{9 - 4(x + h) - (9 - 4x)}{h(\sqrt{9 - 4(x + h)} + \sqrt{9 - 4x})}$

$= \lim\limits_{h \to 0} \dfrac{-4h}{h(\sqrt{9 - 4(x + h)} + \sqrt{9 - 4x})} = \dfrac{-4}{2\sqrt{9 - 4x}} = \dfrac{-2}{\sqrt{9 - 4x}}$

(b) $\dfrac{dy}{dx} = \lim\limits_{h \to 0} \dfrac{\dfrac{x + h}{x + h + 1} - \dfrac{x}{x + 1}}{h} = \lim\limits_{h \to 0} \dfrac{(x + h)(x + 1) - x(x + h + 1)}{h(x + h + 1)(x + 1)}$

$= \lim\limits_{h \to 0} \dfrac{h}{h(x + h + 1)(x + 1)} = \dfrac{1}{(x + 1)^2}$

11. (a) $x = -2, -1, 1, 3$

(b) $(-\infty, -2), (-1, 1), (3, +\infty)$

(c) $(-2, -1), (1, 3)$

(d) $g''(x) = f''(x) \sin x + 2f'(x) \cos x - f(x) \sin x;\ g''(0) = 2f'(0) \cos 0 = 2(2)(1) = 4$

13. (a) The slope of the tangent line $\approx \dfrac{10 - 2.2}{2050 - 1950} = 0.078$ billion, or in 2050 the world population was increasing at the rate of about 78 million per year.

(b) $\dfrac{dN/dt}{N} \approx \dfrac{0.078}{6} = 0.013 = 1.3$ %/year

15. $f'(x) = 2x \sin x + x^2 \cos x$

17. $f'(x) = \dfrac{6x^2 + 8x - 17}{(3x + 2)^2}$

19. (a) $\dfrac{dW}{dt} = 200(t - 15)$; at $t = 5$, $\dfrac{dW}{dt} = -2000$; the water is running out at the rate of 2000 gal/min.

(b) $\dfrac{W(5) - W(0)}{5 - 0} = \dfrac{10000 - 22500}{5} = -2500$; the average rate of flow out is 2500 gal/min.

21. (a) $f'(x) = 2x, f'(1.8) = 3.6$

(b) $f'(x) = (x^2 - 4x)/(x - 2)^2, f'(3.5) \approx -0.7777778$

23. f is continuous at $x = 1$ because it is differentiable there, thus $\lim\limits_{h \to 0} f(1+h) = f(1)$ and so $f(1) = 0$ because $\lim\limits_{h \to 0} \dfrac{f(1+h)}{h}$ exists; $f'(1) = \lim\limits_{h \to 0} \dfrac{f(1+h) - f(1)}{h} = \lim\limits_{h \to 0} \dfrac{f(1+h)}{h} = 5.$

25. The equation of such a line has the form $y = mx$. The points (x_0, y_0) which lie on both the line and the parabola and for which the slopes of both curves are equal satisfy $y_0 = mx_0 = x_0^3 - 9x_0^2 - 16x_0$, so that $m = x_0^2 - 9x_0 - 16$. By differentiating, the slope is also given by $m = 3x_0^2 - 18x_0 - 16$. Equating, we have $x_0^2 - 9x_0 - 16 = 3x_0^2 - 18x_0 - 16$, or $2x_0^2 - 9x_0 = 0$. The root $x_0 = 0$ corresponds to $m = -16, y_0 = 0$ and the root $x_0 = 9/2$ corresponds to $m = -145/4, y_0 = -1305/8$. So the line $y = -16x$ is tangent to the curve at the point $(0,0)$, and the line $y = -145x/4$ is tangent to the curve at the point $(9/2, -1305/8)$.

27. The slope of the tangent line is the derivative

$$y' = 2x\Big|_{x = \frac{1}{2}(a+b)} = a + b. \text{ The slope of the secant is}$$

$$\frac{a^2 - b^2}{a - b} = a + b, \text{ so they are equal.}$$

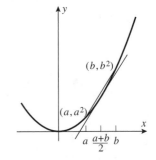

29. **(a)** $8x^7 - \dfrac{3}{2\sqrt{x}} - 15x^{-4}$

    (b) $2 \cdot 101(2x+1)^{100}(5x^2 - 7) + 10x(2x+1)^{101}$

    (c) $2(x-1)\sqrt{3x+1} + \dfrac{3}{2\sqrt{3x+1}}(x-1)^2$

    (d) $3\left(\dfrac{3x+1}{x^2}\right)^2 \dfrac{x^2(3) - (3x+1)(2x)}{x^4} = -\dfrac{3(3x+1)^2(3x+2)}{x^7}$

31. Set $f'(x) = 0$: $f'(x) = 6(2)(2x+7)^5(x-2)^5 + 5(2x+7)^6(x-2)^4 = 0$, so $2x + 7 = 0$ or $x - 2 = 0$ or, factoring out $(2x+7)^5(x-2)^4$, $12(x-2) + 5(2x+7) = 0$. This reduces to $x = -7/2$, $x = 2$, or $22x + 11 = 0$, so the tangent line is horizontal at $x = -7/2, 2, -1/2$.

33. Let $y = mx + b$ be a line tangent to $y = x^2 + 1$ at the point $(x_0, x_0^2 + 1)$ with slope $m = 2x_0$. By inspection if $x_1 = -x_0$ then the same line is also tangent to the curve $y = -x^2 - 1$ at $(-x_1, -y_1)$, since $y_1 = -y_0 = -x_0^2 - 1 = -(x_0^2 + 1) = -x_1^2 - 1$. Thus the tangent line passes through the points (x_0, y_0) and $(x_1, y_1) = (-x_0, -y_0)$, so its slope $m = \dfrac{y_0 - y_1}{x_0 - x_1} = \dfrac{2y_0}{2x_0} = \dfrac{x_0^2 + 1}{x_0}$. But, from the above, $m = 2x_0$; equate and get $\dfrac{x_0^2 + 1}{x_0} = 2x_0$, with solution $x_0 = \pm 1$. Thus the only possible such tangent lines are $y = 2x$ and $y = -2x$.

35. The line $y - x = 2$ has slope $m_1 = 1$ so we set $m_2 = \dfrac{d}{dx}(3x - \tan x) = 3 - \sec^2 x = 1$, or $\sec^2 x = 2$, $\sec x = \pm\sqrt{2}$ so $x = n\pi \pm \pi/4$ where $n = 0, \pm 1, \pm 2, \ldots$.

37. $3 = f(\pi/4) = (M + N)\sqrt{2}/2$ and $1 = f'(\pi/4) = (M - N)\sqrt{2}/2$. Add these two equations to get $4 = \sqrt{2}M, M = 2^{3/2}$. Subtract to obtain $2 = \sqrt{2}N, N = \sqrt{2}$. Thus $f(x) = 2\sqrt{2}\sin x + \sqrt{2}\cos x$.

39. $f'(x) = 2xf(x), f(2) = 5$

    (a) $g(x) = f(\sec x), g'(\pi/3) = f'(\sec \pi/3)\sec \pi/3 \tan \pi/3 = 2\sec \pi/3 f(\sec \pi/3)\sec \pi/3 \tan \pi/3 = 2 \cdot 2f(2) \cdot 2 \cdot \sqrt{3} = 40\sqrt{3}.$

    (b) $h'(x) = 4\left[\dfrac{f(x)}{x-1}\right]^3 \dfrac{(x-1)f'(x) - f(x)}{(x-1)^2},$

    $h'(2) = 4\dfrac{5^3}{1}\dfrac{f'(2) - f(2)}{1} = 4 \cdot 5^3 \dfrac{2 \cdot 2f(2) - f(2)}{1} = 4 \cdot 5^3 \cdot 3 \cdot 5 = 7500$

41. $y = (x^2 + x)^{1/3}, \quad y' = (1/3)(2x+1)(x^2+x)^{-2/3} = \dfrac{2x+1}{3(x^2+x)^{2/3}}$

43. **(a)** $3x^2 + y + xy' - 2 = 0, \quad y' = \dfrac{2 - 3x^2 - y}{x}$

   (b) $y = 2 - x^2 + 1/x, \quad y' = -2x - 1/x^2$

   (c) From (b), $y' = \dfrac{1}{x} - \dfrac{3x}{2} + \dfrac{2}{x} - x + \dfrac{1}{x^2} = \dfrac{3}{x} - \dfrac{5}{2}x + \dfrac{1}{x^2}$

45. $-\dfrac{y'}{y^2} - \dfrac{1}{x^2} = 0, \quad y' = -\dfrac{y^2}{x^2}$

47. $(xy' + y)\sec(xy)\tan(xy) = y', \, y' = \dfrac{y\sec(xy)\tan(xy)}{1 - x\sec(xy)\tan(xy)}$

49. $3x^2 - 4y^2 = 7, 6x - 8yy' = 0, y' = \dfrac{3x}{4y}$

   $y'' = \dfrac{12y - 12xy'}{16y^2} = \dfrac{3}{4y} - \dfrac{3x}{4y^2}\dfrac{3x}{4y} = \dfrac{3}{4y} - \dfrac{9x^2}{16y^3}$

51. $A = \pi r^2$ and $\dfrac{dr}{dt} = -5$, so $\dfrac{dA}{dt} = \dfrac{dA}{dr}\dfrac{dr}{dt} = 2\pi r(-5) = -500\pi$, so the area is shrinking at a rate of $500\pi \, \mathrm{m}^2/\mathrm{min}$.

53. **(a)** $\Delta x = 1.5 - 2 = -0.5; \, dy = \dfrac{-1}{(x-1)^2}\Delta x = \dfrac{-1}{(2-1)^2}(-0.5) = 0.5$; and

   $\Delta y = \dfrac{1}{(1.5-1)} - \dfrac{1}{(2-1)} = 2 - 1 = 1.$

   (b) $\Delta x = 0 - (-\pi/4) = \pi/4; \, dy = \left(\sec^2(-\pi/4)\right)(\pi/4) = \pi/2$; and $\Delta y = \tan 0 - \tan(-\pi/4) = 1.$

   (c) $\Delta x = 3 - 0 = 3; \, dy = \dfrac{-x}{\sqrt{25 - x^2}} = \dfrac{-0}{\sqrt{25 - (0)^2}}(3) = 0$; and

   $\Delta y = \sqrt{25 - 3^2} - \sqrt{25 - 0^2} = 4 - 5 = -1.$

55. **(a)** $h = 115\tan\phi, \, dh = 115\sec^2\phi \, d\phi$; with $\phi = 51° = \dfrac{51}{180}\pi$ radians and

   $d\phi = \pm 0.5° = \pm 0.5\left(\dfrac{\pi}{180}\right)$ radians, $h \pm dh = 115(1.2349) \pm 2.5340 = 142.0135 \pm 2.5340$, so the height lies between 139.48 m and 144.55 m.

   (b) If $|dh| \le 5$ then $|d\phi| \le \dfrac{5}{115}\cos^2\dfrac{51}{180}\pi \approx 0.017$ radians, or $|d\phi| \le 0.98°.$

CHAPTER 4
The Derivative in Graphing and Applications

EXERCISE SET 4.1

1. (a) $f' > 0$ and $f'' > 0$

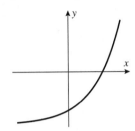

(b) $f' > 0$ and $f'' < 0$

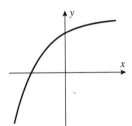

(c) $f' < 0$ and $f'' > 0$

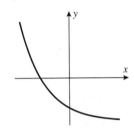

(d) $f' < 0$ and $f'' < 0$

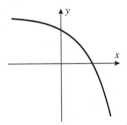

3. A: $dy/dx < 0$, $d^2y/dx^2 > 0$
      B: $dy/dx > 0$, $d^2y/dx^2 < 0$
      C: $dy/dx < 0$, $d^2y/dx^2 < 0$

5. An inflection point occurs when f'' changes sign: at $x = -1, 0, 1$ and 2.

7. (a) $[4, 6]$ **(b)** $[1, 4]$ and $[6, 7]$ **(c)** $(1, 2)$ and $(3, 5)$
    (d) $(2, 3)$ and $(5, 7)$ **(e)** $x = 2, 3, 5$

9. (a) f is increasing on $[1, 3]$ **(b)** f is decreasing on $(-\infty, 1], [3, +\infty]$
    (c) f is concave up on $(-\infty, 2), (4, +\infty)$ **(d)** f is concave down on $(2, 4)$
    (e) points of inflection at $x = 2, 4$

11. $f'(x) = 2(x - 3/2)$ **(a)** $[3/2, +\infty)$ **(b)** $(-\infty, 3/2]$
      $f''(x) = 2$ **(c)** $(-\infty, +\infty)$ **(d)** nowhere
                              (e) none

13. $f'(x) = 6(2x + 1)^2$ **(a)** $(-\infty, +\infty)$ **(b)** nowhere
      $f''(x) = 24(2x + 1)$ **(c)** $(-1/2, +\infty)$ **(d)** $(-\infty, -1/2)$
                          (e) $-1/2$

15. $f'(x) = 12x^2(x - 1)$ **(a)** $[1, +\infty)$ **(b)** $(-\infty, 1]$
      $f''(x) = 36x(x - 2/3)$ **(c)** $(-\infty, 0), (2/3, +\infty)$ **(d)** $(0, 2/3)$
                        (e) $0, 2/3$

17. $f'(x) = -\dfrac{3(x^2 - 3x + 1)}{(x^2 - x + 1)^3}$ **(a)** $[\frac{3-\sqrt{5}}{2}, \frac{3+\sqrt{5}}{2}]$ **(b)** $(-\infty, \frac{3-\sqrt{5}}{2}], [\frac{3+\sqrt{5}}{2}, +\infty)$

      $f''(x) = \dfrac{6x(2x^2 - 8x + 5)}{(x^2 - x + 1)^4}$ **(c)** $(0, 2-\frac{\sqrt{6}}{2}), (2+\frac{\sqrt{6}}{2}, +\infty)$ **(d)** $(-\infty, 0), (2-\frac{\sqrt{6}}{2}, 2+\frac{\sqrt{6}}{2})$

                          (e) $0, 2 - \frac{\sqrt{6}}{2}, 2 + \frac{\sqrt{6}}{2}$

$3 \quad 4x^3 - 16x^2 + 10x$

19. $f'(x) = \dfrac{2x+1}{3(x^2+x+1)^{2/3}}$

(a) $[-1/2, +\infty)$

(b) $(-\infty, -1/2]$

$f''(x) = -\dfrac{2(x+2)(x-1)}{9(x^2+x+1)^{5/3}}$

(c) $(-2, 1)$

(d) $(-\infty, -2), (1, +\infty)$

(e) $-2, 1$

21. $f'(x) = \dfrac{4(x^{2/3} - 1)}{3x^{1/3}}$

(a) $[-1, 0], [1, +\infty)$

(b) $(-\infty, -1], [0, 1]$

$f''(x) = \dfrac{4(x^{2/3} + 1)}{9x^{4/3}}$

(c) $(-\infty, 0), (0, +\infty)$

(d) nowhere

(e) none

23. $f'(x) = \cos x + \sin x$

$f''(x) = -\sin x + \cos x$

(a) $[-\pi/4, 3\pi/4]$

(b) $(-\pi, -\pi/4], [3\pi/4, \pi)$

(c) $(-3\pi/4, \pi/4)$

(d) $(-\pi, -3\pi/4), (\pi/4, \pi)$

(e) $-3\pi/4, \pi/4$

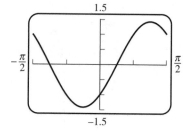

25. $f'(x) = -\dfrac{1}{2}\sec^2(x/2)$

$f''(x) = -\dfrac{1}{2}\tan(x/2)\sec^2(x/2))$

(a) nowhere

(b) $(-\pi, \pi)$

(c) $(-\pi, 0)$

(d) $(0, \pi)$

(e) 0

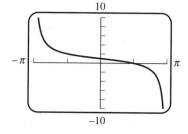

27. $f(x) = 1 + \sin 2x$
$f'(x) = 2\cos 2x$
$f''(x) = -4\sin 2x$

(a) $[-\pi, -3\pi/4], [-\pi/4, \pi/4], [3\pi/4, \pi]$
(b) $[-3\pi/4, -\pi/4], [\pi/4, 3\pi/4]$
(c) $(-\pi/2, 0), (\pi/2, \pi)$
(d) $(-\pi, -\pi/2), (0, \pi/2)$
(e) $-\pi/2, 0, \pi/2$

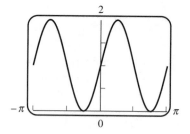

29. (a)

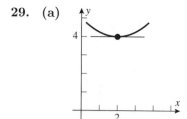

(b)

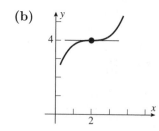

(c)

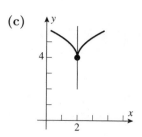

31. (a) $g(x)$ has no zeros:

There can be no zero of $g(x)$ on the interval $-\infty < x < 0$ because if there were, say $g(x_0) = 0$ where $x_0 < 0$, then $g'(x)$ would have to be positive between $x = x_0$ and $x = 0$, say $g'(x_1) > 0$ where $x_0 < x_1 < 0$. But then $g'(x)$ cannot be concave up on the interval $(x_1, 0)$, a contradiction.

There can be no zero of $g(x)$ on $0 < x < 4$ because $g(x)$ is concave up for $0 < x < 4$ and thus the graph of $g(x)$, for $0 < x < 4$, must lie above the line $y = -\frac{2}{3}x + 2$, which is the tangent line to the curve at $(0, 2)$, and above the line $y = 3(x - 4) + 3 = 3x - 9$ also for $0 < x < 4$ (see figure). The first condition says that $g(x)$ could only be zero for $x > 3$ and the second condition says that $g(x)$ could only be zero for $x < 3$, thus $g(x)$ has no zeros for $0 < x < 4$.

Finally, if $4 < x < +\infty$, $g(x)$ could only have a zero if $g'(x)$ were negative somewhere for $x > 4$, and since $g'(x)$ is decreasing there we would ultimately have $g(x) < -10$, a contradiction.

(b) one, between 0 and 4

(c) We must have $\lim\limits_{x \to +\infty} g'(x) = 0$; if the limit were -5 then $g(x)$ would at some time cross the line $g(x) = -10$; if the limit were 5 then, since g is concave down for $x > 4$ and $g'(4) = 3$, g' must decrease for $x > 4$ and thus the limit would be ≤ 4.

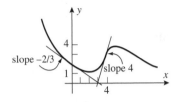

33. For $n \geq 2$, $f''(x) = n(n-1)(x-a)^{n-2}$; there is a sign change of f'' (point of inflection) at $(a, 0)$ if and only if n is odd. For $n = 1$, $y = x - a$, so there is no point of inflection.

35. $f'(x) = 1/3 - 1/[3(1+x)^{2/3}]$ so f is increasing on $[0, +\infty)$ thus if $x > 0$, then $f(x) > f(0) = 0$, $1 + x/3 - \sqrt[3]{1+x} > 0$, $\sqrt[3]{1+x} < 1 + x/3$.

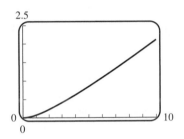

37. $x \geq \sin x$ on $[0, +\infty)$: let $f(x) = x - \sin x$. Then $f(0) = 0$ and $f'(x) = 1 - \cos x \geq 0$, so $f(x)$ is increasing on $[0, +\infty)$.

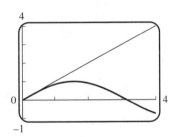

39. Points of inflection at $x = -2, +2$. Concave up on $(-5, -2)$ and $(2, 5)$; concave down on $(-2, 2)$. Increasing on $[-3.5829, 0.2513]$ and $[3.3316, 5]$, and decreasing on $[-5, -3.5829]$ and $[0.2513, 3.3316]$.

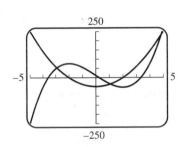

41. $f''(x) = 2\dfrac{90x^3 - 81x^2 - 585x + 397}{(3x^2 - 5x + 8)^3}$. The denominator has complex roots, so is always positive; hence the x-coordinates of the points of inflection of $f(x)$ are the roots of the numerator (if it changes sign). A plot of the numerator over $[-5, 5]$ shows roots lying in $[-3, -2]$, $[0, 1]$, and $[2, 3]$. To six decimal places the roots are $x = -2.464202, 0.662597, 2.701605$.

43. $f(x_1) - f(x_2) = x_1^2 - x_2^2 = (x_1 + x_2)(x_1 - x_2) < 0$ if $x_1 < x_2$ for x_1, x_2 in $[0, +\infty)$, so $f(x_1) < f(x_2)$ and f is thus increasing.

45. (a) If $x_1 < x_2$ where x_1 and x_2 are in I, then $f(x_1) < f(x_2)$ and $g(x_1) < g(x_2)$, so $f(x_1) + g(x_1) < f(x_2) + g(x_2)$, $(f + g)(x_1) < (f + g)(x_2)$. Thus $f + g$ is increasing on I.

(b) Case I: If f and g are ≥ 0 on I, and if $x_1 < x_2$ where x_1 and x_2 are in I, then $0 < f(x_1) < f(x_2)$ and $0 < g(x_1) < g(x_2)$, so $f(x_1)g(x_1) < f(x_2)g(x_2)$, $(f \cdot g)(x_1) < (f \cdot g)(x_2)$. Thus $f \cdot g$ is increasing on I.

Case II: If f and g are not necessarily positive on I then no conclusion can be drawn: for example, $f(x) = g(x) = x$ are both increasing on $(-\infty, 0)$, but $(f \cdot g)(x) = x^2$ is decreasing there.

47. (a) $f(x) = x$, $g(x) = 2x$ **(b)** $f(x) = x$, $g(x) = x + 6$ **(c)** $f(x) = 2x$, $g(x) = x$

49. (a) $f''(x) = 6ax + 2b = 6a\left(x + \dfrac{b}{3a}\right)$, $f''(x) = 0$ when $x = -\dfrac{b}{3a}$. f changes its direction of concavity at $x = -\dfrac{b}{3a}$ so $-\dfrac{b}{3a}$ is an inflection point.

(b) If $f(x) = ax^3 + bx^2 + cx + d$ has three x-intercepts, then it has three roots, say x_1, x_2 and x_3, so we can write $f(x) = a(x - x_1)(x - x_2)(x - x_3) = ax^3 + bx^2 + cx + d$, from which it follows that $b = -a(x_1 + x_2 + x_3)$. Thus $-\dfrac{b}{3a} = \dfrac{1}{3}(x_1 + x_2 + x_3)$, which is the average.

(c) $f(x) = x(x^2 - 3x + 2) = x(x - 1)(x - 2)$ so the intercepts are 0, 1, and 2 and the average is 1. $f''(x) = 6x - 6 = 6(x - 1)$ changes sign at $x = 1$.

51. (a) Let $x_1 < x_2$ belong to (a, b). If both belong to $(a, c]$ or both belong to $[c, b)$ then we have $f(x_1) < f(x_2)$ by hypothesis. So assume $x_1 < c < x_2$. We know by hypothesis that $f(x_1) < f(c)$, and $f(c) < f(x_2)$. We conclude that $f(x_1) < f(x_2)$.

(b) Use the same argument as in Part (a), but with inequalities reversed.

53. By Theorem 4.1.2, f is decreasing on any interval $[(2n\pi + \pi/2, 2(n + 1)\pi + \pi/2]$ $(n = 0, \pm 1, \pm 2, \cdots)$, because $f'(x) = -\sin x + 1 < 0$ on $(2n\pi + \pi/2, 2(n + 1)\pi + \pi/2)$. By Exercise 51 (b) we can piece these intervals together to show that $f(x)$ is decreasing on $(-\infty, +\infty)$.

55.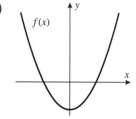

57.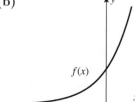

EXERCISE SET 4.2

1. (a)

(b)

(c)

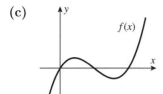

(d)

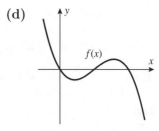

3. (a) $f'(x) = 6x - 6$ and $f''(x) = 6$, with $f'(1) = 0$. For the first derivative test, $f' < 0$ for $x < 1$ and $f' > 0$ for $x > 1$. For the second derivative test, $f''(1) > 0$.

(b) $f'(x) = 3x^2 - 3$ and $f''(x) = 6x$. $f'(x) = 0$ at $x = \pm 1$. First derivative test: $f' > 0$ for $x < -1$ and $x > 1$, and $f' < 0$ for $-1 < x < 1$, so there is a relative maximum at $x = -1$, and a relative minimum at $x = 1$. Second derivative test: $f'' < 0$ at $x = -1$, a relative maximum; and $f'' > 0$ at $x = 1$, a relative minimum.

5. (a) $f'(x) = 4(x - 1)^3$, $g'(x) = 3x^2 - 6x + 3$ so $f'(1) = g'(1) = 0$.

(b) $f''(x) = 12(x - 1)^2$, $g''(x) = 6x - 6$, so $f''(1) = g''(1) = 0$, which yields no information.

(c) $f' < 0$ for $x < 1$ and $f' > 0$ for $x > 1$, so there is a relative minimum at $x = 1$; $g'(x) = 3(x - 1)^2 > 0$ on both sides of $x = 1$, so there is no relative extremum at $x = 1$.

7. $f'(x) = 16x^3 - 32x = 16x(x^2 - 2)$, so $x = 0, \pm\sqrt{2}$ are stationary points.

9. $f'(x) = \dfrac{-x^2 - 2x + 3}{(x^2 + 3)^2}$, so $x = -3, 1$ are the stationary points.

11. $f'(x) = \dfrac{2x}{3(x^2 - 25)^{2/3}}$; so $x = 0$ is the stationary point.

13. $f(x) = |\sin x| = \begin{cases} \sin x, & \sin x \geq 0 \\ -\sin x, & \sin x < 0 \end{cases}$ so $f'(x) = \begin{cases} \cos x, & \sin x > 0 \\ -\cos x, & \sin x < 0 \end{cases}$ and $f'(x)$ does not exist

when $x = n\pi$, $n = 0, \pm 1, \pm 2, \cdots$ (the points where $\sin x = 0$) because $\lim\limits_{x \to n\pi^-} f'(x) \neq \lim\limits_{x \to n\pi^+} f'(x)$ (see Theorem preceding Exercise 61, Section 3.3). Now $f'(x) = 0$ when $\pm \cos x = 0$ provided $\sin x \neq 0$ so $x = \pi/2 + n\pi$, $n = 0, \pm 1, \pm 2, \cdots$ are stationary points.

15. (a) none

(b) $x = 1$ because f' changes sign from $+$ to $-$ there

(c) none because $f'' = 0$ (never changes sign)

17. (a) $x = 2$ because $f'(x)$ changes sign from $-$ to $+$ there.

(b) $x = 0$ because $f'(x)$ changes sign from $+$ to $-$ there.

(c) $x = 1, 3$ because $f''(x)$ changes sign at these points.

19. critical points $x = 0, 5^{1/3}$: f':

$x = 0$: neither
$x = 5^{1/3}$: relative minimum

$$- - - 0 - - - 0 + + +$$
$$\;\;\;\;\;\;\; 0 \;\;\;\;\; 5^{1/3}$$

21. critical points $x = 2/3$: f':
$x = 2/3$: relative maximum;

$$+ + + \; 0 \; - - -$$
$$\;\;\;\;\;\; 2/3$$

23. $f'(x) = 8 - 6x$: critical point $x = 4/3$
$f''(4/3) = -6$: f has a maximum of $19/3$ at $x = 4/3$

25. $f'(x) = 4x^3 - 36x^2$: critical points at $x = 0, 9$
$f''(0) = 0$: Theorem 4.2.5 with $m = 3$: f has an inflection point at $x = 0$
$f''(9) > 0$: f has a minimum of -2187 at $x = 9$

27. $f'(x) = 4x^3 - 12x^2 + 8x$: critical points at $x = 0, 1, 2$
relative minimum of 0 at $x = 0$
relative maximum of 1 at $x = 1$
relative minimum of 0 at $x = 2$

29. $f'(x) = 5x^4 + 8x^3 + 3x^2$: critical points at $x = -3/5, -1, 0$
$f''(-3/5) = 18/25$: f has a relative minimum of $-108/3125$ at $x = -3/5$
$f''(-1) = -2$: f has a relative maximum of 0 at $x = -1$
$f''(0) = 0$: Theorem 4.2.5 with $m = 3$: f has an inflection point at $x = 0$

31. $f'(x) = \dfrac{2(x^{1/3} + 1)}{x^{1/3}}$: critical point at $x = -1, 0$
$f''(-1) = -\dfrac{2}{3}$: f has a relative maximum of 1 at $x = -1$
f' does not exist at $x = 0$. By inspection it is a relative minimum of 0.

33. $f'(x) = -\dfrac{5}{(x-2)^2}$; no extrema

35. $f'(x) = \pm(3 - 2x)$: critical points at $x = 3/2, 0, 3$
$f''(3/2) = -2$, relative maximum of 9/4 at $x = 3/2$
$x = 0, 3$ are not stationary points:
by first derivative test, $x = 0$ and $x = 3$ are each a minimum of 0

37.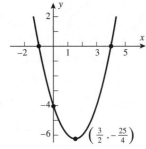

39.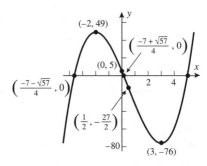

41.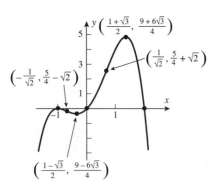

43.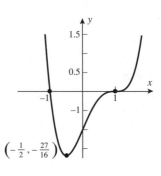

45.

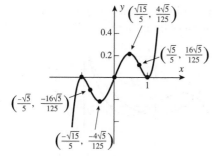

47. (a) $\lim\limits_{x\to-\infty} y = -\infty,\ \lim\limits_{x\to+\infty} y = +\infty;$
curve crosses x-axis at $x = 0, 1, -1$

(b) $\lim\limits_{x\to\pm\infty} y = +\infty;$
curve never crosses x-axis

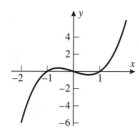

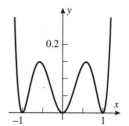

(c) $\lim\limits_{x\to-\infty} y = -\infty,\ \lim\limits_{x\to+\infty} y = +\infty;$
curve crosses x-axis at $x = -1$

(d) $\lim\limits_{x\to\pm\infty} y = +\infty;$
curve crosses x-axis at $x = 0, 1$

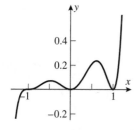

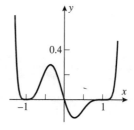

49. $f'(x) = 2\cos 2x$ if $\sin 2x > 0$,
$f'(x) = -2\cos 2x$ if $\sin 2x < 0$,
$f'(x)$ does not exist when $x = \pi/2, \pi, 3\pi/2$;
critical numbers $x = \pi/4, 3\pi/4, 5\pi/4, 7\pi/4, \pi/2, \pi, 3\pi/2$
relative minimum of 0 at $x = \pi/2, \pi, 3\pi/2$;
relative maximum of 1 at $x = \pi/4, 3\pi/4, 5\pi/4, 7\pi/4$

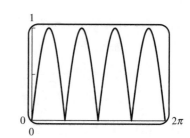

51. $f'(x) = -\sin 2x$;
critical numbers $x = \pi/2, \pi, 3\pi/2$
relative minimum of 0 at $x = \pi/2, 3\pi/2$;
relative maximum of 1 at $x = \pi$

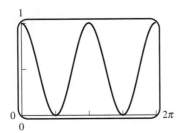

53. Relative minima at $x = -3.58, 3.33$;
relative maximum at $x = 0.25$

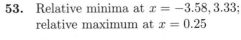

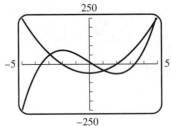

55. Relative maximum at $x = -0.27$, relative
minimum at $x = 0.22$

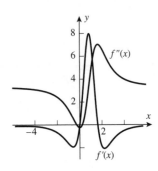

57. $f'(x) = \dfrac{4x^3 - \sin 2x}{2\sqrt{x^4 + \cos^2 x}}$,

$f''(x) = \dfrac{6x^2 - \cos 2x}{\sqrt{x^4 + \cos^2 x}} 54 - \dfrac{(4x^3 - \sin 2x)(4x^3 - \sin 2x)}{4(x^4 + \cos^2 x)^{3/2}}$

Relative minima at $x \approx \pm 0.62$, relative maximum at $x = 0$

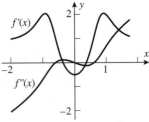

59. (a) Let $f(x) = x^2 + \dfrac{k}{x}$, then $f'(x) = 2x - \dfrac{k}{x^2} = \dfrac{2x^3 - k}{x^2}$. f has a relative extremum when
$2x^3 - k = 0$, so $k = 2x^3 = 2(3)^3 = 54$.

(b) Let $f(x) = \dfrac{x}{x^2 + k}$, then $f'(x) = \dfrac{k - x^2}{(x^2 + k)^2}$. f has a relative extremum when $k - x^2 = 0$, so
$k = x^2 = 3^2 = 9$.

61. (a) $(-2.2, 4), (2, 1.2), (4.2, 3)$

(b) f' exists everywhere, so the critical numbers are when $f' = 0$, i.e. when $x = \pm 2$ or $r(x) = 0$,
so $x \approx -5.1, -2, 0.2, 2$. At $x = -5.1$ f' changes sign from $-$ to $+$, so minimum; at $x = -2$
f' changes sign from $+$ to $-$, so maximum; at $x = 0.2$ f' doesn't change sign, so neither; at
$x = 2$ f' changes sign from $-$ to $+$, so minimum.
Finally, $f''(1) = (1^2 - 4)r'(1) + 2r(1) \approx -3(0.6) + 2(0.3) = -1.2$.

63. $f'(x) = 3ax^2 + 2bx + c$ and $f'(x)$ has roots at $x = 0, 1$, so $f'(x)$ must be of the form
$f'(x) = 3ax(x - 1)$; thus $c = 0$ and $2b = -3a$, $b = -3a/2$. $f''(x) = 6ax + 2b = 6ax - 3a$, so
$f''(0) > 0$ and $f''(1) < 0$ provided $a < 0$. Finally $f(0) = d$, so $d = 0$; and
$f(1) = a + b + c + d = a + b = -a/2$ so $a = -2$. Thus $f(x) = -2x^3 + 3x^2$.

65. (a)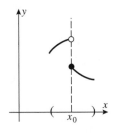

$f(x_0)$ is not an
extreme value.

(b)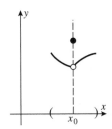

$f(x_0)$ is a relative
maximum.

(c)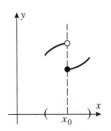

$f(x_0)$ is a relative
minimum.

EXERCISE SET 4.3

1. Vertical asymptote $x = 4$
   horizontal asymptote $y = -2$

   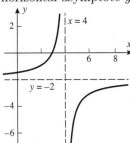

3. Vertical asymptotes $x = \pm 2$
   horizontal asymptote $y = 0$

   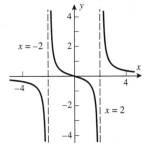

5. No vertical asymptotes
   horizontal asymptote $y = 1$

   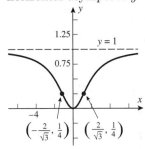

7. Vertical asymptote $x = 1$
   horizontal asymptote $y = 1$

   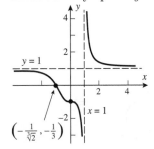

9. Vertical asymptote $x = 0$
   horizontal asymptote $y = 3$

   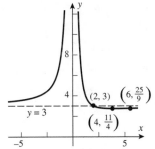

11. Vertical asymptote $x = 1$
    horizontal asymptote $y = 9$

    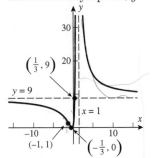

13. Vertical asymptote $x = 1$
horizontal asymptote $y = -1$

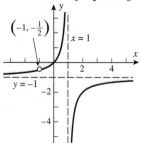

15. (a) horizontal asymptote $y = 3$ as
$x \to \pm\infty$, vertical asymptotes of $x = \pm 2$

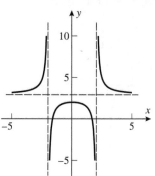

(b) horizontal asymptote of $y = 1$ as
$x \to \pm\infty$, vertical asymptotes at $x = \pm 1$

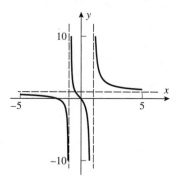

(c) horizontal asymptote of $y = -1$ as
$x \to \pm\infty$, vertical asymptotes at $x = -2, 1$

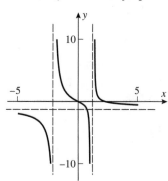

(d) horizontal asymptote of $y = 1$ as
$x \to \pm\infty$, vertical asymptote at $x = -1, 2$

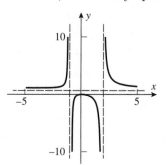

17. $\displaystyle\lim_{x\to\pm\infty}\left|\frac{x^2}{x-3}-(x+3)\right| = \lim_{x\to\pm\infty}\left|\frac{9}{x-3}\right| = 0$

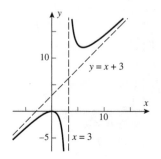

19. $y = x^2 - \dfrac{1}{x} = \dfrac{x^3 - 1}{x}$;

$y' = \dfrac{2x^3 + 1}{x^2}$,

$y' = 0$ when

$x = -\sqrt[3]{\dfrac{1}{2}} \approx -0.8$;

$y'' = \dfrac{2(x^3 - 1)}{x^3}$

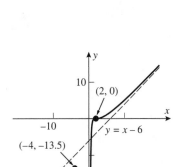

21. $y = \dfrac{(x-2)^3}{x^2} = x - 6 + \dfrac{12x - 8}{x^2}$ so

$y = x - 6$ is an oblique asymptote;

$y' = \dfrac{(x-2)^2(x+4)}{x^3}$,

$y'' = \dfrac{24(x-2)}{x^4}$

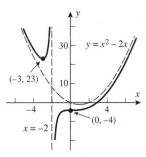

23. $y = \dfrac{x^3 - 4x - 8}{x + 2} = x^2 - 2x - \dfrac{8}{x + 2}$ so

$y = x^2 - 2x$ is a curvilinear asymptote as $x \to \pm\infty$

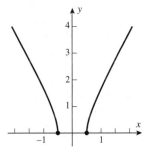

25. **(a)** VI   **(b)** I   **(c)** III   **(d)** V   **(e)** IV   **(f)** II

27. $y = \sqrt{4x^2 - 1}$

$y' = \dfrac{4x}{\sqrt{4x^2 - 1}}$

$y'' = -\dfrac{4}{(4x^2 - 1)^{3/2}}$ so

extrema when $x = \pm\dfrac{1}{2}$, no inflection points

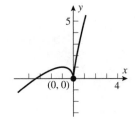

29. $y = 2x + 3x^{2/3}$;

$y' = 2 + 2x^{-1/3}$;

$y'' = -\dfrac{2}{3}x^{-4/3}$

31. $y = x^{1/3}(4 - x)$;

$y' = \dfrac{4(1 - x)}{3x^{2/3}}$;

$y'' = -\dfrac{4(x + 2)}{9x^{5/3}}$

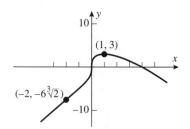

33. $y = x^{2/3} - 2x^{1/3} + 4$

$y' = \dfrac{2(x^{1/3} - 1)}{3x^{2/3}}$

$y'' = -\dfrac{2(x^{1/3} - 2)}{9x^{5/3}}$

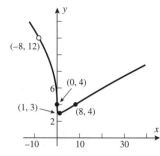

35. $y = x + \sin x$;

$y' = 1 + \cos x$, $y' = 0$ when $x = \pi + 2n\pi$;

$y'' = -\sin x$; $y'' = 0$ when $x = n\pi$

$n = 0, \pm 1, \pm 2, \ldots$

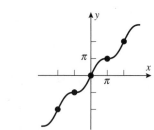

37. $y = \sqrt{3}\cos x + \sin x$;

$y' = -\sqrt{3}\sin x + \cos x$;

$y' = 0$ when $x = \pi/6 + n\pi$;

$y'' = -\sqrt{3}\cos x - \sin x$;

$y'' = 0$ when $x = 2\pi/3 + n\pi$

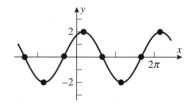

39. $y = \sin^2 x - \cos x$;

$y' = \sin x(2\cos x + 1)$;

$y' = 0$ when $x = \pm\pi, 2\pi, 3\pi$ and when

$x = -\dfrac{2}{3}\pi, \dfrac{2}{3}\pi, \dfrac{4}{3}\pi, \dfrac{8}{3}\pi$;

$y'' = 4\cos^2 x + \cos x - 2$;

$y'' = 0$ when $x \approx \pm 2.57, \pm 0.94, 3.71, 5.35, 7.22, 8.86$

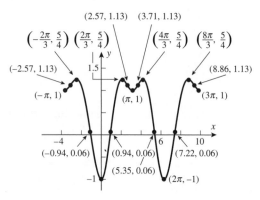

41. **(a)** $x = 1, 2.5, 4$ and $x = 3$, the latter being a cusp

(b) $(-\infty, 1], [2.5, 3)$

(c) relative maxima for $x = 1, 3$; relative minima for $x = 2.5$

(d) $x = 0.8, 1.9, 4$

43. Let y be the length of the other side of the rectangle, then $L = 2x + 2y$ and $xy = 400$ so $y = 400/x$ and hence $L = 2x + 800/x$. $L = 2x$ is an oblique asymptote

$$L = 2x + \frac{800}{x} = \frac{2(x^2 + 400)}{x},$$

$$L' = 2 - \frac{800}{x^2} = \frac{2(x^2 - 400)}{x^2},$$

$$L'' = \frac{1600}{x^3},$$

$L' = 0$ when $x = 20, L = 80$

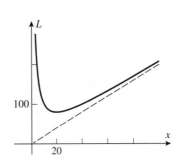

45. $y' = 0.1x^4(6x - 5)$;
critical numbers: $x = 0$, $x = 5/6$;
relative minimum at $x = 5/6$,
$y \approx -6.7 \times 10^{-3}$

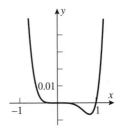

EXERCISE SET 4.4

1. relative maxima at $x = 2, 6$; absolute maximum at $x = 6$; relative and absolute minima at $x = 0, 4$

3. **(a)**

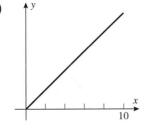

(b)

(c)

5. $x = 1$ is a point of discontinuity of f.

7. $f'(x) = 8x - 12$, $f'(x) = 0$ when $x = 3/2$; $f(1) = 2$, $f(3/2) = 1$, $f(2) = 2$ so the maximum value is 2 at $x = 1, 2$ and the minimum value is 1 at $x = 3/2$.

9. $f'(x) = 3(x - 2)^2, f'(x) = 0$ when $x = 2$; $f(1) = -1, f(2) = 0, f(4) = 8$ so the minimum is -1 at $x = 1$ and the maximum is 8 at $x = 4$.

11. $f'(x) = 3/(4x^2 + 1)^{3/2}$, no critical points; $f(-1) = -3/\sqrt{5}$, $f(1) = 3/\sqrt{5}$ so the maximum value is $3/\sqrt{5}$ at $x = 1$ and the minimum value is $-3/\sqrt{5}$ at $x = -1$.

13. $f'(x) = 1 - 2\cos x, f'(x) = 0$ when $x = \pi/3$; then $f(-\pi/4) = -\pi/4 + \sqrt{2}$; $f(\pi/3) = \pi/3 - \sqrt{3}$; $f(\pi/2) = \pi/2 - 2$, so f has a minimum of $\pi/3 - \sqrt{3}$ at $x = \pi/3$ and a maximum of $-\pi/4 + \sqrt{2}$ at $x = -\pi/4$.

15. $f(x) = 1 + |9 - x^2| = \begin{cases} 10 - x^2, & |x| \le 3 \\ -8 + x^2, & |x| > 3 \end{cases}$, $f'(x) = \begin{cases} -2x, & |x| < 3 \\ 2x, & |x| > 3 \end{cases}$ thus $f'(x) = 0$ when

$x = 0$, $f'(x)$ does not exist for x in $(-5, 1)$ when $x = -3$ because $\lim\limits_{x \to -3^-} f'(x) \ne \lim\limits_{x \to -3^+} f'(x)$ (see Theorem preceding Exercise 61, Section 3.3); $f(-5) = 17$, $f(-3) = 1$, $f(0) = 10$, $f(1) = 9$ so the maximum value is 17 at $x = -5$ and the minimum value is 1 at $x = -3$.

17. $f'(x) = 2x - 1$, $f'(x) = 0$ when $x = 1/2$; $f(1/2) = -9/4$ and $\lim\limits_{x \to \pm\infty} f(x) = +\infty$. Thus f has a minimum of $-9/4$ at $x = 1/2$ and no maximum.

19. $f'(x) = 12x^2(1 - x)$; critical points $x = 0, 1$. Maximum value $f(1) = 1$, no minimum because $\lim\limits_{x \to +\infty} f(x) = -\infty$.

21. No maximum or minimum because $\lim\limits_{x \to +\infty} f(x) = +\infty$ and $\lim\limits_{x \to -\infty} f(x) = -\infty$.

23. $\lim\limits_{x \to -1^-} f(x) = -\infty$, so there is no absolute minimum on the interval; $f'(x) = 0$ at $x \approx -2.414213562$, for which $y \approx -4.828427125$. Also $f(-5) = -13/2$, so the absolute maximum of f on the interval is $y \approx -4.828427125$ taken at $x \approx -2.414213562$.

25. $\lim\limits_{x \to \pm\infty} = +\infty$ so there is no absolute maximum. $f'(x) = 4x(x - 2)(x - 1)$, $f'(x) = 0$ when $x = 0, 1, 2$, and $f(0) = 0$, $f(1) = 1$, $f(2) = 0$ so f has an absolute minimum of 0 at $x = 0, 2$.

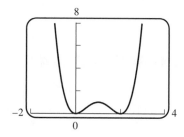

27. $f'(x) = \dfrac{5(8 - x)}{3x^{1/3}}$, $f'(x) = 0$ when $x = 8$ and $f'(x)$ does not exist when $x = 0$; $f(-1) = 21$, $f(0) = 0$, $f(8) = 48$, $f(20) = 0$ so the maximum value is 48 at $x = 8$ and the minimum value is 0 at $x = 0, 20$.

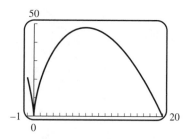

29. $f'(x) = -1/x^2$; no maximum or minimum because there are no critical points in $(0, +\infty)$.

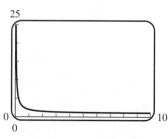

31. $f'(x) = \dfrac{1 - 2\cos x}{\sin^2 x}$; $f'(x) = 0$ on $[\pi/4, 3\pi/4]$ only when $x = \pi/3$. Then $f(\pi/4) = 2\sqrt{2} - 1$, $f(\pi/3) = \sqrt{3}$ and $f(3\pi/4) = 2\sqrt{2} + 1$, so f has an absolute maximum value of $2\sqrt{2} + 1$ at $x = 3\pi/4$ and an absolute minimum value of $\sqrt{3}$ at $x = \pi/3$.

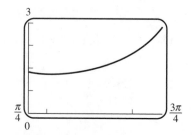

33. $f'(x) = -[\cos(\cos x)] \sin x$; $f'(x) = 0$ if $\sin x = 0$ or if $\cos(\cos x) = 0$. If $\sin x = 0$, then $x = \pi$ is the critical point in $(0, 2\pi)$; $\cos(\cos x) = 0$ has no solutions because $-1 \le \cos x \le 1$. Thus $f(0) = \sin(1)$, $f(\pi) = \sin(-1) = -\sin(1)$, and $f(2\pi) = \sin(1)$ so the maximum value is $\sin(1) \approx 0.84147$ and the minimum value is $-\sin(1) \approx -0.84147$.

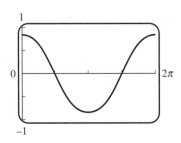

35. $f'(x) = \begin{cases} 4, & x < 1 \\ 2x - 5, & x > 1 \end{cases}$ so $f'(x) = 0$ when $x = 5/2$, and $f'(x)$ does not exist when $x = 1$

because $\lim\limits_{x \to 1^-} f'(x) \ne \lim\limits_{x \to 1^+} f'(x)$ (see Theorem preceding Exercise 61, Section 3.3); $f(1/2) = 0$, $f(1) = 2$, $f(5/2) = -1/4$, $f(7/2) = 3/4$ so the maximum value is 2 and the minimum value is $-1/4$.

37. The period of $f(x)$ is 2π, so check $f(0) = 3$, $f(2\pi) = 3$ and the critical points. $f'(x) = -2\sin x - 2\sin 2x = -2\sin x(1 + 2\cos x) = 0$ on $[0, 2\pi]$ at $x = 0, \pi, 2\pi$ and $x = 2\pi/3, 4\pi/3$. Check $f(\pi) = -1$, $f(2\pi/3) = -3/2$, $f(4\pi/3) = -3/2$. Thus f has an absolute maximum on $(-\infty, +\infty)$ of 3 at $x = 2k\pi, k = 0, \pm 1, \pm 2, \ldots$ and an absolute minimum of $-3/2$ at $x = 2k\pi \pm 2\pi/3, k = 0, \pm 1, \pm 2, \ldots$.

39. Let $f(x) = x - \sin x$, then $f'(x) = 1 - \cos x$ and so $f'(x) = 0$ when $\cos x = 1$ which has no solution for $0 < x < 2\pi$ thus the minimum value of f must occur at 0 or 2π. $f(0) = 0$, $f(2\pi) = 2\pi$ so 0 is the minimum value on $[0, 2\pi]$ thus $x - \sin x \ge 0$, $\sin x \le x$ for all x in $[0, 2\pi]$.

41. Let $m =$ slope at x, then $m = f'(x) = 3x^2 - 6x + 5$, $dm/dx = 6x - 6$; critical point for m is $x = 1$, minimum value of m is $f'(1) = 2$

43. $\lim\limits_{x \to +\infty} f(x) = +\infty$, $\lim\limits_{x \to 8^-} f(x) = +\infty$, so there is no absolute maximum value of f for $x > 8$. By table 4.4.3 there must be a minimum. Since $f'(x) = \dfrac{2x(-520 + 192x - 24x^2 + x^3)}{(x - 8)^3}$, we must solve a quartic equation to find the critical points. But it is easy to see that $x = 0$ and $x = 10$ are real roots, and the other two are complex. Since $x = 0$ is not in the interval in question, we must have an absolute minimum of f on $(8, +\infty)$ of 125 at $x = 10$.

45. The slope of the line is -1, and the slope of the tangent to $y = -x^2$ is $-2x$ so $-2x = -1$, $x = 1/2$. The line lies above the curve so the vertical distance is given by $F(x) = 2 - x + x^2$; $F(-1) = 4$, $F(1/2) = 7/4$, $F(3/2) = 11/4$. The point $(1/2, -1/4)$ is closest, the point $(-1, -1)$ farthest.

47. The absolute extrema of $y(t)$ can occur at the endpoints $t = 0, 12$ or when $dy/dt = 2\sin t = 0$, i.e. $t = 0, 12, k\pi$, $k = 1, 2, 3$; the absolute maximum is $y = 4$ at $t = \pi, 3\pi$; the absolute minimum is $y = 0$ at $t = 0, 2\pi$.

49. $f'(x) = 2ax + b$; critical point is $x = -\dfrac{b}{2a}$

$f''(x) = 2a > 0$ so $f\left(-\dfrac{b}{2a}\right)$ is the minimum value of f, but

$f\left(-\dfrac{b}{2a}\right) = a\left(-\dfrac{b}{2a}\right)^2 + b\left(-\dfrac{b}{2a}\right) + c = \dfrac{-b^2 + 4ac}{4a}$ thus $f(x) \ge 0$ if and only if

$f\left(-\dfrac{b}{2a}\right) \ge 0$, $\dfrac{-b^2 + 4ac}{4a} \ge 0$, $-b^2 + 4ac \ge 0$, $b^2 - 4ac \le 0$

EXERCISE SET 4.5

1. If $y = x + 1/x$ for $1/2 \le x \le 3/2$ then $dy/dx = 1 - 1/x^2 = (x^2 - 1)/x^2$, $dy/dx = 0$ when $x = 1$. If $x = 1/2, 1, 3/2$ then $y = 5/2, 2, 13/6$ so
   (a) y is as small as possible when $x = 1$.
   (b) y is as large as possible when $x = 1/2$.

3. $A = xy$ where $x + 2y = 1000$ so $y = 500 - x/2$ and $A = 500x - x^2/2$ for x in $[0, 1000]$; $dA/dx = 500 - x$, $dA/dx = 0$ when $x = 500$. If $x = 0$ or 1000 then $A = 0$, if $x = 500$ then $A = 125,000$ so the area is maximum when $x = 500$ ft and $y = 500 - 500/2 = 250$ ft.

5. Let x and y be the dimensions shown in the figure and A the area, then $A = xy$ subject to the cost condition $3(2x) + 2(2y) = 6000$, or $y = 1500 - 3x/2$. Thus $A = x(1500 - 3x/2) = 1500x - 3x^2/2$ for x in $[0, 1000]$. $dA/dx = 1500 - 3x$, $dA/dx = 0$ when $x = 500$. If $x = 0$ or 1000 then $A = 0$, if $x = 500$ then $A = 375,000$ so the area is greatest when $x = 500$ ft and (from $y = 1500 - 3x/2$) when $y = 750$ ft.

7. Let x, y, and z be as shown in the figure and A the area of the rectangle, then $A = xy$ and, by similar triangles, $z/10 = y/6$, $z = 5y/3$; also $x/10 = (8 - z)/8 = (8 - 5y/3)/8$ thus $y = 24/5 - 12x/25$ so $A = x(24/5 - 12x/25) = 24x/5 - 12x^2/25$ for x in $[0, 10]$. $dA/dx = 24/5 - 24x/25$, $dA/dx = 0$ when $x = 5$. If $x = 0, 5, 10$ then $A = 0, 12, 0$ so the area is greatest when $x = 5$ in. and $y = 12/5$ in.

9. $A = xy$ where $x^2 + y^2 = 20^2 = 400$ so $y = \sqrt{400 - x^2}$ and $A = x\sqrt{400 - x^2}$ for $0 \le x \le 20$; $dA/dx = 2(200 - x^2)/\sqrt{400 - x^2}$, $dA/dx = 0$ when $x = \sqrt{200} = 10\sqrt{2}$. If $x = 0, 10\sqrt{2}, 20$ then $A = 0, 200, 0$ so the area is maximum when $x = 10\sqrt{2}$ and $y = \sqrt{400 - 200} = 10\sqrt{2}$.

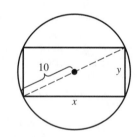

11. Let $x =$ length of each side that uses the \$1 per foot fencing,
    $y =$ length of each side that uses the \$2 per foot fencing.
    The cost is $C = (1)(2x) + (2)(2y) = 2x + 4y$, but $A = xy = 3200$ thus $y = 3200/x$ so

$$C = 2x + 12800/x \text{ for } x > 0,$$
$$dC/dx = 2 - 12800/x^2, \ dC/dx = 0 \text{ when } x = 80, \ d^2C/dx^2 > 0 \text{ so}$$

   C is least when $x = 80$, $y = 40$.

13. Let x and y be the dimensions of a rectangle; the perimeter is $p = 2x + 2y$. But $A = xy$ thus $y = A/x$ so $p = 2x + 2A/x$ for $x > 0$, $dp/dx = 2 - 2A/x^2 = 2(x^2 - A)/x^2$, $dp/dx = 0$ when $x = \sqrt{A}$, $d^2p/dx^2 = 4A/x^3 > 0$ if $x > 0$ so p is a minimum when $x = \sqrt{A}$ and $y = \sqrt{A}$ and thus the rectangle is a square.

15. Let the two sides with fencing have length x, and let the included angle be 2θ. Then the area $A = 2(x\cos\theta)(x\sin\theta) = x^2\sin 2\theta$. But $x + x = 300, x = 150$, so $A = 150^2\sin 2\theta, \dfrac{dA}{d\theta} = 150 \cdot 300\cos 2\theta = 0$ when $2\theta = \frac{1}{2}\pi, \theta = \frac{1}{4}\pi$, and the dimensions are 150 yd $\times$150 yd $\times 150\sqrt{2}$ yd.

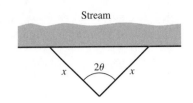

17. Let the box have dimensions x, x, y, with $y \geq x$. The constraint is $4x + y \leq 108$, and the volume $V = x^2 y$. If we take $y = 108 - 4x$ then $V = x^2(108 - 4x)$ and $dV/dx = 12x(-x + 18)$ with roots $x = 0, 18$. The maximum value of V occurs at $x = 18, y = 36$ with $V = 11,664$ in^2.

19. Let x be the length of each side of a square, then $V = x(3 - 2x)(8 - 2x) = 4x^3 - 22x^2 + 24x$ for $0 \leq x \leq 3/2$; $dV/dx = 12x^2 - 44x + 24 = 4(3x - 2)(x - 3)$, $dV/dx = 0$ when $x = 2/3$ for $0 < x < 3/2$. If $x = 0, 2/3, 3/2$ then $V = 0, 200/27, 0$ so the maximum volume is $200/27$ ft^3.

21. Let $x = $ length of each edge of base, $y = $ height, $k = \$/$cm^2 for the sides. The cost is $C = (2k)(2x^2) + (k)(4xy) = 4k(x^2 + xy)$, but $V = x^2 y = 2000$ thus $y = 2000/x^2$ so $C = 4k(x^2 + 2000/x)$ for $x > 0$, $dC/dx = 4k(2x - 2000/x^2)$, $dC/dx = 0$ when $x = \sqrt[3]{1000} = 10$, $d^2C/dx^2 > 0$ so C is least when $x = 10$, $y = 20$.

23. Let $x = $ height and width, $y = $ length. The surface area is $S = 2x^2 + 3xy$ where $x^2 y = V$, so $y = V/x^2$ and $S = 2x^2 + 3V/x$ for $x > 0$; $dS/dx = 4x - 3V/x^2$, $dS/dx = 0$ when $x = \sqrt[3]{3V/4}$, $d^2S/dx^2 > 0$ so S is minimum when $x = \sqrt[3]{\dfrac{3V}{4}}$, $y = \dfrac{4}{3}\sqrt[3]{\dfrac{3V}{4}}$.

25. Let r and h be the dimensions shown in the figure, then the surface area is $S = 2\pi rh + 2\pi r^2$.

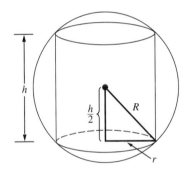

But $r^2 + \left(\dfrac{h}{2}\right)^2 = R^2$ thus $h = 2\sqrt{R^2 - r^2}$ so

$S = 4\pi r\sqrt{R^2 - r^2} + 2\pi r^2$ for $0 \leq r \leq R$,

$\dfrac{dS}{dr} = \dfrac{4\pi(R^2 - 2r^2)}{\sqrt{R^2 - r^2}} + 4\pi r$; $\dfrac{dS}{dr} = 0$ when

$\dfrac{R^2 - 2r^2}{\sqrt{R^2 - r^2}} = -r$ \hfill (1)

$R^2 - 2r^2 = -r\sqrt{R^2 - r^2}$

$R^4 - 4R^2 r^2 + 4r^4 = r^2(R^2 - r^2)$

$5r^4 - 5R^2 r^2 + R^4 = 0$

and using the quadratic formula $r^2 = \dfrac{5R^2 \pm \sqrt{25R^4 - 20R^4}}{10} = \dfrac{5 \pm \sqrt{5}}{10}R^2$, $r = \sqrt{\dfrac{5 \pm \sqrt{5}}{10}}R$, of

which only $r = \sqrt{\dfrac{5 + \sqrt{5}}{10}}R$ satisfies (1). If $r = 0, \sqrt{\dfrac{5 + \sqrt{5}}{10}}R, 0$ then $S = 0, (5 + \sqrt{5})\pi R^2, 2\pi R^2$ so

the surface area is greatest when $r = \sqrt{\dfrac{5 + \sqrt{5}}{10}}R$ and, from $h = 2\sqrt{R^2 - r^2}$, $h = 2\sqrt{\dfrac{5 - \sqrt{5}}{10}}R$.

27. From (13), $S = 2\pi r^2 + 2\pi rh$. But $V = \pi r^2 h$ thus $h = V/(\pi r^2)$ and so $S = 2\pi r^2 + 2V/r$ for $r > 0$. $dS/dr = 4\pi r - 2V/r^2$, $dS/dr = 0$ if $r = \sqrt[3]{V/(2\pi)}$. Since $d^2S/dr^2 = 4\pi + 4V/r^3 > 0$, the minimum surface area is achieved when $r = \sqrt[3]{V/2\pi}$ and so $h = V/(\pi r^2) = [V/(\pi r^3)]r = 2r$.

29. The surface area is $S = \pi r^2 + 2\pi rh$
where $V = \pi r^2 h = 500$ so $h = 500/(\pi r^2)$
and $S = \pi r^2 + 1000/r$ for $r > 0$;
$dS/dr = 2\pi r - 1000/r^2 = (2\pi r^3 - 1000)/r^2$,
$dS/dr = 0$ when $r = \sqrt[3]{500/\pi}$, $d^2 S/dr^2 > 0$
for $r > 0$ so S is minimum when $r = \sqrt[3]{500/\pi}$ cm and
$$h = \frac{500}{\pi r^2} = \frac{500}{\pi} \left(\frac{\pi}{500} \right)^{2/3}$$
$$= \sqrt[3]{500/\pi} \text{ cm}$$

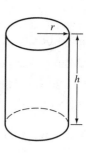

31. Let x be the length of each side of the squares and y the height of the frame, then the volume is
$V = x^2 y$. The total length of the wire is L thus $8x + 4y = L$, $y = (L - 8x)/4$ so
$V = x^2(L - 8x)/4 = (Lx^2 - 8x^3)/4$ for $0 \le x \le L/8$. $dV/dx = (2Lx - 24x^2)/4$, $dV/dx = 0$ for
$0 < x < L/8$ when $x = L/12$. If $x = 0, L/12, L/8$ then $V = 0, L^3/1728, 0$ so the volume is greatest
when $x = L/12$ and $y = L/12$.

33. Let h and r be the dimensions shown in the figure,
then the volume is $V = \frac{1}{3}\pi r^2 h$. But $r^2 + h^2 = L^2$ thus
$r^2 = L^2 - h^2$ so $V = \frac{1}{3}\pi(L^2 - h^2)h = \frac{1}{3}\pi(L^2 h - h^3)$
for $0 \le h \le L$. $\frac{dV}{dh} = \frac{1}{3}\pi(L^2 - 3h^2)$. $\frac{dV}{dh} = 0$ when
$h = L/\sqrt{3}$. If $h = 0, L/\sqrt{3}, 0$ then $V = 0, \frac{2\pi}{9\sqrt{3}}L^3, 0$ so
the volume is as large as possible when $h = L/\sqrt{3}$ and
$r = \sqrt{2/3}L$.

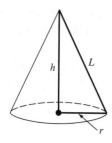

35. The area of the paper is $A = \pi rL = \pi r\sqrt{r^2 + h^2}$, but
$V = \frac{1}{3}\pi r^2 h = 10$ thus $h = 30/(\pi r^2)$
so $A = \pi r\sqrt{r^2 + 900/(\pi^2 r^4)}$.
To simplify the computations let $S = A^2$,
$$S = \pi^2 r^2 \left(r^2 + \frac{900}{\pi^2 r^4} \right) = \pi^2 r^4 + \frac{900}{r^2} \text{ for } r > 0,$$
$$\frac{dS}{dr} = 4\pi^2 r^3 - \frac{1800}{r^3} = \frac{4(\pi^2 r^6 - 450)}{r^3}, \, dS/dr = 0 \text{ when}$$
$r = \sqrt[6]{450/\pi^2}$, $d^2 S/dr^2 > 0$, so S and hence A is least
when $r = \sqrt[6]{450/\pi^2}$ cm, $h = \frac{30}{\pi}\sqrt[3]{\pi^2/450}$ cm.

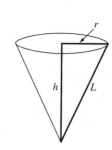

37. The volume of the cone is $V = \dfrac{1}{3}\pi r^2 h$. By similar tri-

angles (see figure) $\dfrac{r}{h} = \dfrac{R}{\sqrt{h^2 - 2Rh}}$, $r = \dfrac{Rh}{\sqrt{h^2 - 2Rh}}$

so $V = \dfrac{1}{3}\pi R^2 \dfrac{h^3}{h^2 - 2Rh} = \dfrac{1}{3}\pi R^2 \dfrac{h^2}{h - 2R}$ for $h > 2R$,

$\dfrac{dV}{dh} = \dfrac{1}{3}\pi R^2 \dfrac{h(h - 4R)}{(h - 2R)^2}$, $\dfrac{dV}{dh} = 0$ for $h > 2R$ when

$h = 4R$, by the first derivative test V is minimum when $h = 4R$. If $h = 4R$ then $r = \sqrt{2}R$.

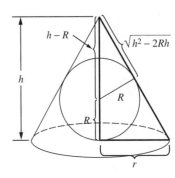

39. Let b and h be the dimensions shown in the figure, then the cross-sectional area is $A = \dfrac{1}{2}h(5 + b)$. But $h = 5\sin\theta$ and $b = 5 + 2(5\cos\theta) = 5 + 10\cos\theta$ so $A = \dfrac{5}{2}\sin\theta(10 + 10\cos\theta) = 25\sin\theta(1 + \cos\theta)$ for $0 \le \theta \le \pi/2$.

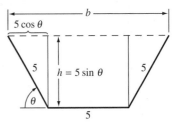

$\begin{aligned} dA/d\theta &= -25\sin^2\theta + 25\cos\theta(1 + \cos\theta) \\ &= 25(-\sin^2\theta + \cos\theta + \cos^2\theta) \\ &= 25(-1 + \cos^2\theta + \cos\theta + \cos^2\theta) \\ &= 25(2\cos^2\theta + \cos\theta - 1) = 25(2\cos\theta - 1)(\cos\theta + 1). \end{aligned}$

$dA/d\theta = 0$ for $0 < \theta < \pi/2$ when $\cos\theta = 1/2$, $\theta = \pi/3$. If $\theta = 0, \pi/3, \pi/2$ then $A = 0, 75\sqrt{3}/4, 25$ so the cross-sectional area is greatest when $\theta = \pi/3$.

41. Let L, L_1, and L_2 be as shown in the figure, then $L = L_1 + L_2 = 8\csc\theta + \sec\theta$,

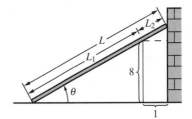

$\dfrac{dL}{d\theta} = -8\csc\theta\cot\theta + \sec\theta\tan\theta, \ 0 < \theta < \pi/2$

$= -\dfrac{8\cos\theta}{\sin^2\theta} + \dfrac{\sin\theta}{\cos^2\theta} = \dfrac{-8\cos^3\theta + \sin^3\theta}{\sin^2\theta\cos^2\theta};$

$\dfrac{dL}{d\theta} = 0$ if $\sin^3\theta = 8\cos^3\theta$, $\tan^3\theta = 8$, $\tan\theta = 2$ which gives the absolute minimum for L because $\lim\limits_{\theta\to 0^+} L = \lim\limits_{\theta\to \pi/2^-} L = +\infty$. If $\tan\theta = 2$, then $\csc\theta = \sqrt{5}/2$ and $\sec\theta = \sqrt{5}$ so $L = 8(\sqrt{5}/2) + \sqrt{5} = 5\sqrt{5}$ ft.

43. (a) The daily profit is

$\begin{aligned} P &= \text{(revenue)} - \text{(production cost)} = 100x - (100,000 + 50x + 0.0025x^2) \\ &= -100,000 + 50x - 0.0025x^2 \end{aligned}$

for $0 \le x \le 7000$, so $dP/dx = 50 - 0.005x$ and $dP/dx = 0$ when $x = 10,000$. Because 10,000 is not in the interval $[0, 7000]$, the maximum profit must occur at an endpoint. When $x = 0$, $P = -100,000$; when $x = 7000$, $P = 127,500$ so 7000 units should be manufactured and sold daily.

(b) Yes, because $dP/dx > 0$ when $x = 7000$ so profit is increasing at this production level.

(c) $dP/dx = 15$ when $x = 7000$, so $P(7001) - P(7000) \approx 15$, and the marginal profit is \$15.

45. The profit is

$$P = \text{(profit on nondefective)} - \text{(loss on defective)} = 100(x - y) - 20y = 100x - 120y$$

but $y = 0.01x + 0.00003x^2$ so $P = 100x - 120(0.01x + 0.00003x^2) = 98.8x - 0.0036x^2$ for $x > 0$, $dP/dx = 98.8 - 0.0072x$, $dP/dx = 0$ when $x = 98.8/0.0072 \approx 13,722$, $d^2P/dx^2 < 0$ so the profit is maximum at a production level of about 13,722 pounds.

47. The distance between the particles is $D = \sqrt{(1 - t - t)^2 + (t - 2t)^2} = \sqrt{5t^2 - 4t + 1}$ for $t \geq 0$. For convenience, we minimize D^2 instead, so $D^2 = 5t^2 - 4t + 1$, $dD^2/dt = 10t - 4$, which is 0 when $t = 2/5$. $d^2D^2/dt^2 > 0$ so D^2 and hence D is minimum when $t = 2/5$. The minimum distance is $D = 1/\sqrt{5}$.

49. Let $P(x, y)$ be a point on the curve $x^2 + y^2 = 1$. The distance between $P(x, y)$ and $P_0(2, 0)$ is $D = \sqrt{(x - 2)^2 + y^2}$, but $y^2 = 1 - x^2$ so $D = \sqrt{(x - 2)^2 + 1 - x^2} = \sqrt{5 - 4x}$ for $-1 \leq x \leq 1$, $\dfrac{dD}{dx} = -\dfrac{2}{\sqrt{5 - 4x}}$ which has no critical points for $-1 < x < 1$. If $x = -1, 1$ then $D = 3, 1$ so the closest point occurs when $x = 1$ and $y = 0$.

51. **(a)** Draw the line perpendicular to the given line that passes through Q.

(b) Let $Q : (x_0, y_0)$. If $(x, mx + b)$ is the point on the graph closest to Q, then the distance squared $d^2 = D = (x - x_0)^2 + (mx + b - y_0)^2$ is minimized. Then

$$dD/dx = 2(x - x_0) + 2m(mx + b - y_0) = 0$$

at the point $(x, mx + b)$ which minimizes the distance.

On the other hand, the line connecting the point Q with the line $y = mx + b$ is perpendicular to this line provided the connecting line has slope $-1/m$. But this condition is $(y - y_0)/(x - x_0) = -1/m$, or $m(y - y_0) = -(x - x_0)$. Since $y = mx + b$ the condition becomes $m(mx + b - y_0) = -(x - x_0)$, which is the equivalent to the expression above which minimizes the distance.

53. **(a)** The line through (x, y) and the origin has slope y/x, and the negative reciprocal is $-x/y$. If (x, y) is a point on the ellipse, then $x^2 - xy + y^2 = 4$ and, differentiating, $2x - x(dy/dx) - y + 2y(dy/dx) = 0$.

For the desired points we have $dy/dx = -x/y$, and inserting that into the previous equation results in $2x + x^2/y - y - 2x = 0$, $2xy + x^2 - y^2 - 2xy = 0$, $x^2 - y^2 = 0$, $y = \pm x$. Inserting this into the equation of the ellipse we have $x^2 \mp x^2 + x^2 = 4$ with solutions $y = x = \pm 2$ or $y = -x = \pm 2/\sqrt{3}$. Thus there are four solutions, $(2, 2), (-2, -2), (2/\sqrt{3}, -2/\sqrt{3})$ and $(-2/\sqrt{3}, 2/\sqrt{3})$.

(b) In general, the shortest/longest distance from a point to a curve is taken on the line connecting the point to the curve which is perpendicular to the tangent line at the point in question.

55. If $P(x_0, y_0)$ is on the curve $y = 1/x^2$, then $y_0 = 1/x_0^2$. At P the slope of the tangent line is $-2/x_0^3$ so its equation is $y - \dfrac{1}{x_0^2} = -\dfrac{2}{x_0^3}(x - x_0)$, or $y = -\dfrac{2}{x_0^3}x + \dfrac{3}{x_0^2}$. The tangent line crosses the y-axis at $\dfrac{3}{x_0^2}$, and the x-axis at $\dfrac{3}{2}x_0$. The length of the segment then is $L = \sqrt{\dfrac{9}{x_0^4} + \dfrac{9}{4}x_0^2}$ for $x_0 > 0$. For convenience, we minimize L^2 instead, so $L^2 = \dfrac{9}{x_0^4} + \dfrac{9}{4}x_0^2$, $\dfrac{dL^2}{dx_0} = -\dfrac{36}{x_0^5} + \dfrac{9}{2}x_0 = \dfrac{9(x_0^6 - 8)}{2x_0^5}$, which is 0 when $x_0^6 = 8$, $x_0 = \sqrt{2}$. $\dfrac{d^2L^2}{dx_0^2} > 0$ so L^2 and hence L is minimum when $x_0 = \sqrt{2}$, $y_0 = 1/2$.

57. At each point (x, y) on the curve the slope of the tangent line is $m = \dfrac{dy}{dx} = -\dfrac{2x}{(1+x^2)^2}$ for any x, $\dfrac{dm}{dx} = \dfrac{2(3x^2-1)}{(1+x^2)^3}$, $\dfrac{dm}{dx} = 0$ when $x = \pm 1/\sqrt{3}$, by the first derivative test the only relative maximum occurs at $x = -1/\sqrt{3}$, which is the absolute maximum because $\lim\limits_{x \to \pm\infty} m = 0$. The tangent line has greatest slope at the point $(-1/\sqrt{3}, 3/4)$.

59. With x and y as shown in the figure, the maximum length of pipe will be the smallest value of $L = x + y$. By similar triangles

$$\frac{y}{8} = \frac{x}{\sqrt{x^2-16}}, \quad y = \frac{8x}{\sqrt{x^2-16}} \text{ so}$$

$$L = x + \frac{8x}{\sqrt{x^2-16}} \text{ for } x > 4, \quad \frac{dL}{dx} = 1 - \frac{128}{(x^2-16)^{3/2}},$$

$$\frac{dL}{dx} = 0 \text{ when}$$

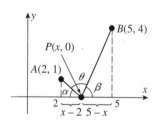

$$(x^2-16)^{3/2} = 128$$
$$x^2 - 16 = 128^{2/3} = 16(2^{2/3})$$
$$x^2 = 16(1 + 2^{2/3})$$
$$x = 4(1 + 2^{2/3})^{1/2},$$

$d^2L/dx^2 = 384x/(x^2-16)^{5/2} > 0$ if $x > 4$ so L is smallest when $x = 4(1 + 2^{2/3})^{1/2}$. For this value of x, $L = 4(1 + 2^{2/3})^{3/2}$ ft.

61. Let $x =$ distance from the weaker light source, $I =$ the intensity at that point, and k the constant of proportionality. Then

$$I = \frac{kS}{x^2} + \frac{8kS}{(90-x)^2} \text{ if } 0 < x < 90;$$

$$\frac{dI}{dx} = -\frac{2kS}{x^3} + \frac{16kS}{(90-x)^3} = \frac{2kS[8x^3 - (90-x)^3]}{x^3(90-x)^3} = 18\frac{kS(x-30)(x^2+2700)}{x^3(x-90)^3},$$

which is 0 when $x = 30$; $\dfrac{dI}{dx} < 0$ if $x < 30$, and $\dfrac{dI}{dx} > 0$ if $x > 30$, so the intensity is minimum at a distance of 30 cm from the weaker source.

63. $\theta = \pi - (\alpha + \beta)$
$$= \pi - \cot^{-1}(x-2) - \cot^{-1}\frac{5-x}{4},$$

$$\frac{d\theta}{dx} = \frac{1}{1+(x-2)^2} + \frac{-1/4}{1+(5-x)^2/16}$$

$$= -\frac{3(x^2-2x-7)}{[1+(x-2)^2][16+(5-x)^2]},$$

$d\theta/dx = 0$ when $x = \dfrac{2 \pm \sqrt{4+28}}{2} = 1 \pm 2\sqrt{2}$,

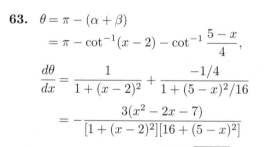

only $1 + 2\sqrt{2}$ is in $[2, 5]$; $d\theta/dx > 0$ for x in $[2, 1+2\sqrt{2})$, $d\theta/dx < 0$ for x in $(1+2\sqrt{2}, 5]$, θ is maximum when $x = 1 + 2\sqrt{2}$.

65. Let $v =$ speed of light in the medium. The total time required for the light to travel from A to P to B is

$$t = \text{(total distance from } A \text{ to } P \text{ to } B)/v = \frac{1}{v}(\sqrt{(c-x)^2 + a^2} + \sqrt{x^2 + b^2}),$$

$$\frac{dt}{dx} = \frac{1}{v}\left[-\frac{c-x}{\sqrt{(c-x)^2 + a^2}} + \frac{x}{\sqrt{x^2 + b^2}}\right]$$

and $\dfrac{dt}{dx} = 0$ when $\dfrac{x}{\sqrt{x^2 + b^2}} = \dfrac{c-x}{\sqrt{(c-x)^2 + a^2}}$. But $x/\sqrt{x^2 + b^2} = \sin\theta_2$ and

$(c-x)/\sqrt{(c-x)^2 + a^2} = \sin\theta_1$ thus $dt/dx = 0$ when $\sin\theta_2 = \sin\theta_1$ so $\theta_2 = \theta_1$.

67. **(a)** The rate at which the farmer walks is analogous to the speed of light in Fermat's principle.

(b) the best path occurs when $\theta_1 = \theta_2$ (see figure).

(c) by similar triangles,
$$x/(1/4) = (1-x)/(3/4)$$
$$3x = 1 - x$$
$$4x = 1$$
$$x = 1/4 \text{ mi.}$$

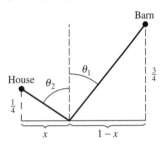

EXERCISE SET 4.6

1. $f(x) = x^2 - 2$, $f'(x) = 2x$, $x_{n+1} = x_n - \dfrac{x_n^2 - 2}{2x_n}$

$x_1 = 1$, $x_2 = 1.5$, $x_3 = 1.416666667, \ldots$, $x_5 = x_6 = 1.414213562$

3. $f(x) = x^3 - 6$, $f'(x) = 3x^2$, $x_{n+1} = x_n - \dfrac{x_n^3 - 6}{3x_n^2}$

$x_1 = 2$, $x_2 = 1.833333333$, $x_3 = 1.817263545, \ldots$, $x_5 = x_6 = 1.817120593$

5. $f(x) = x^3 - 2x - 2$, $f'(x) = 3x^2 - 2$, $x_{n+1} = x_n - \dfrac{x_n^3 - 2x_n - 2}{3x_n^2 - 2}$

$x_1 = 2$, $x_2 = 1.8$, $x_3 = 1.7699481865$, $x_4 = 1.7692926629$, $x_5 = x_6 = 1.7692923542$

7. $f(x) = x^5 + x^4 - 5$, $f'(x) = 5x^4 + 4x^3$, $x_{n+1} = x_n - \dfrac{x_n^5 + x_n^4 - 5}{5x_n^4 + 4x_n^3}$

$x_1 = 1$, $x_2 = 1.333333333$, $x_3 = 1.239420573, \ldots$, $x_6 = x_7 = 1.224439550$

9. $f(x) = x^4 + x^2 - 4$, $f'(x) = 4x^3 + 2x$,

$$x_{n+1} = x_n - \frac{x_n^4 + x_n^2 - 4}{4x_n^3 + 2x_n}$$

$x_1 = 1$, $x_2 = 1.3333$, $x_3 = 1.2561$, $x_4 = 1.24966, \ldots$,
$x_7 = x_8 = 1.249621068$;
by symmetry, $x = -1.249621068$ is the other solution.

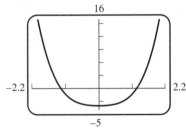

11. $f(x) = 2\cos x - x$, $f'(x) = -2\sin x - 1$

$$x_{n+1} = x_n - \frac{2\cos x - x}{-2\sin x - 1}$$

$x_1 = 1$, $x_2 = 1.03004337$, $x_3 = 1.02986654$,
$x_4 = x_5 = 1.02986653$

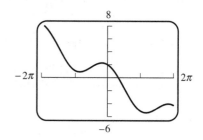

13. $f(x) = x - \tan x$,

$$f'(x) = 1 - \sec^2 x = -\tan^2 x, \quad x_{n+1} = x_n + \frac{x_n - \tan x_n}{\tan^2 x_n}$$

$x_1 = 4.5$, $x_2 = 4.493613903$, $x_3 = 4.493409655$,
$x_4 = x_5 = 4.493409458$

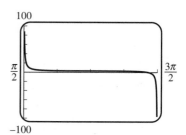

15. The graphs of $y = x^3$ and $y = 1 - x$ intersect near the point $x = 0.7$. Let $f(x) = x^3 + x - 1$, so that $f'(x) = 3x^2 + 1$, and

$$x_{n+1} = x_n - \frac{x^3 + x - 1}{3x^2 + 1}.$$

If $x_1 = 0.7$ then $x_2 = 0.68259109$, $x_3 = 0.68232786$, $x_4 = x_5 = 0.68232780$.

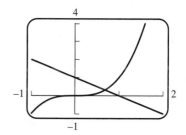

17. The graphs of $y = x^2$ and $y = \sqrt{2x + 1}$ intersect at points near $x = -0.5$ and $x = 1$; $x^2 = \sqrt{2x + 1}$, $x^4 - 2x - 1 = 0$. Let $f(x) = x^4 - 2x - 1$, then $f'(x) = 4x^3 - 2$ so

$$x_{n+1} = x_n - \frac{x_n^4 - 2x_n - 1}{4x_n^3 - 2}.$$

If $x_1 = -0.5$, then $x_2 = -0.475$,
$x_3 = -0.474626695$,
$x_4 = x_5 = -0.474626618$; if
$x_1 = 1$, then $x_2 = 2$,
$x_3 = 1.633333333, \ldots, x_8 = x_9 = 1.395336994$.

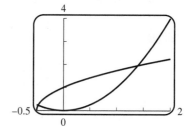

19. (a) $f(x) = x^2 - a$, $f'(x) = 2x$, $x_{n+1} = x_n - \dfrac{x_n^2 - a}{2x_n} = \dfrac{1}{2}\left(x_n + \dfrac{a}{x_n}\right)$

(b) $a = 10$; $x_1 = 3$, $x_2 = 3.166666667$, $x_3 = 3.162280702$, $x_4 = x_5 = 3.162277660$

21. $f'(x) = x^3 + 2x - 5$; solve $f'(x) = 0$ to find the critical points. Graph $y = x^3$ and $y = -2x + 5$ to see that they intersect at a point near $x = 1.25$; $f''(x) = 3x^2 + 2$ so $x_{n+1} = x_n - \dfrac{x_n^3 + 2x_n - 5}{3x_n^2 + 2}$.

$x_1 = 1.25$, $x_2 = 1.3317757009$, $x_3 = 1.3282755613$, $x_4 = 1.3282688557$, $x_5 = 1.3282688557$ so the minimum value of $f(x)$ occurs at $x \approx 1.3282688557$ because $f''(x) > 0$; its value is approximately -4.098859123.

23. Let $f(x)$ be the square of the distance between $(1, 0)$ and any point (x, x^2) on the parabola, then $f(x) = (x - 1)^2 + (x^2 - 0)^2 = x^4 + x^2 - 2x + 1$ and $f'(x) = 4x^3 + 2x - 2$. Solve $f'(x) = 0$ to find the critical points; $f''(x) = 12x^2 + 2$ so $x_{n+1} = x_n - \dfrac{4x_n^3 + 2x_n - 2}{12x_n^2 + 2} = x_n - \dfrac{2x_n^3 + x_n - 1}{6x_n^2 + 1}$.

$x_1 = 1$, $x_2 = 0.714285714$, $x_3 = 0.605168701, \ldots, x_6 = x_7 = 0.589754512$; the coordinates are approximately $(0.589754512, 0.347810385)$.

25. (a) Let s be the arc length, and L the length of the chord, then $s = 1.5L$. But $s = r\theta$ and $L = 2r\sin(\theta/2)$ so $r\theta = 3r\sin(\theta/2)$, $\theta - 3\sin(\theta/2) = 0$.

(b) Let $f(\theta) = \theta - 3\sin(\theta/2)$, then $f'(\theta) = 1 - 1.5\cos(\theta/2)$ so $\theta_{n+1} = \theta_n - \dfrac{\theta_n - 3\sin(\theta_n/2)}{1 - 1.5\cos(\theta_n/2)}$.

$\theta_1 = 3$, $\theta_2 = 2.991592920$, $\theta_3 = 2.991563137$, $\theta_4 = \theta_5 = 2.991563136$ rad so $\theta \approx 171°$.

27. If $x = 1$, then $y^4 + y = 1$, $y^4 + y - 1 = 0$. Graph $z = y^4$ and $z = 1 - y$ to see that they intersect near $y = -1$ and $y = 1$. Let $f(y) = y^4 + y - 1$, then $f'(y) = 4y^3 + 1$ so $y_{n+1} = y_n - \dfrac{y_n^4 + y_n - 1}{4y_n^3 + 1}$.

If $y_1 = -1$, then $y_2 = -1.333333333$, $y_3 = -1.235807860, \ldots, y_6 = y_7 = -1.220744085$; if $y_1 = 1$, then $y_2 = 0.8$, $y_3 = 0.731233596, \ldots, y_6 = y_7 = 0.724491959$.

29. $S(25) = 250{,}000 = \dfrac{5000}{i}\left[(1 + i)^{25} - 1\right]$; set $f(i) = 50i - (1+i)^{25} + 1$, $f'(i) = 50 - 25(1+i)^{24}$; solve $f(i) = 0$. Set $i_0 = .06$ and $i_{k+1} = i_k - \left[50i - (1+i)^{25} + 1\right] / \left[50 - 25(1+i)^{24}\right]$. Then $i_1 = 0.05430$, $i_2 = 0.05338$, $i_3 = 0.05336, \ldots, i = 0.053362$.

31. (a)

x_1	x_2	x_3	x_4	x_5	x_6	x_7	x_8	x_9	x_{10}
0.5000	-0.7500	0.2917	-1.5685	-0.4654	0.8415	-0.1734	2.7970	1.2197	0.1999

(b) The sequence x_n must diverge, since if it did converge then $f(x) = x^2 + 1 = 0$ would have a solution. It seems the x_n are oscillating back and forth in a quasi-cyclical fashion.

33. (a) $|x_{n+1} - x_n| \le |x_{n+1} - c| + |c - x_n| < 1/n + 1/n = 2/n$

(b) The closed interval $[c - 1, c + 1]$ contains all of the x_n, since $|x_n - c| < 1/n$. Let M be an upper bound for $|f'(x)|$ on $[c - 1, c + 1]$. Since $x_{n+1} = x_n - f(x_n)/f'(x_n)$ it follows that $|f(x_n)| \le |f'(x_n)||x_{n+1} - x_n| < M|x_{n+1} - x_n| < 2M/n$.

(c) Assume that $f(c) \ne 0$. The sequence x_n converges to c, since $|x_n - c| < 1/n$. By the continuity of f, $f(c) = f(\lim_{n \to +\infty} x_n) = \lim_{n \to +\infty} f(x_n)$.

Let $\epsilon = |f(c)|/2$. Choose N such that $|f(x_n) - f(c)| < \epsilon/2$ for $n > N$. Then $|f(x_n) - f(c)| < |f(c)|/2$ for $n > N$, so $-|f(c)|/2 < f(x_n) - f(c) < |f(c)/2|$.

If $f(c) > 0$ then $f(x_n) > f(c) - |f(c)|/2 = f(c)/2$.

If $f(c) < 0$, then $f(x_n) < f(c) + |f(c)|/2 = -|f(c)|/2$, or $|f(x_n)| > |f(c)|/2$.

(d) From (b) it follows that $\lim_{n \to +\infty} f(x_n) = 0$. From (c) it follows that if $f(c) \ne 0$ then $\lim_{n \to +\infty} f(x_n) \ne 0$, a contradiction. The conclusion, then, is that $f(c) = 0$.

EXERCISE SET 4.7

1. $f(3) = f(5) = 0$; $f'(x) = 2x - 8$, $2c - 8 = 0$, $c = 4$, $f'(4) = 0$

3. $f(\pi/2) = f(3\pi/2) = 0$, $f'(x) = -\sin x$, $-\sin c = 0$, $c = \pi$

5. $f(0) = f(4) = 0$, $f'(x) = \dfrac{1}{2} - \dfrac{1}{2\sqrt{x}}$, $\dfrac{1}{2} - \dfrac{1}{2\sqrt{c}} = 0$, $c = 1$

7. $(f(5) - f(-3))/(5 - (-3)) = 1$; $f'(x) = 2x - 1$; $2c - 1 = 1$, $c = 1$

9. $f(0) = 1$, $f(3) = 2$, $f'(x) = \dfrac{1}{2\sqrt{x+1}}$, $\dfrac{1}{2\sqrt{c+1}} = \dfrac{2-1}{3-0} = \dfrac{1}{3}$, $\sqrt{c+1} = 3/2$, $c+1 = 9/4$, $c = 5/4$

11. $f(-5) = 0$, $f(3) = 4$, $f'(x) = -\dfrac{x}{\sqrt{25-x^2}}$, $-\dfrac{c}{\sqrt{25-c^2}} = \dfrac{4-0}{3-(-5)} = \dfrac{1}{2}$, $-2c = \sqrt{25-c^2}$,
$4c^2 = 25 - c^2$, $c^2 = 5$, $c = -\sqrt{5}$
(we reject $c = \sqrt{5}$ because it does not satisfy the equation $-2c = \sqrt{25-c^2}$)

13. (a) $f(-2) = f(1) = 0$
The interval is $[-2, 1]$

(b) $c = -1.29$

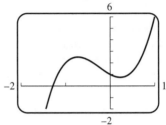

(c) $x_0 = -1$, $x_1 = -1.5$, $x_2 = -1.32$, $x_3 = -1.290$, $x_4 = -1.2885843$

15. (a) $f'(x) = \sec^2 x$, $\sec^2 c = 0$ has no solution   **(b)** $\tan x$ is not continuous on $[0, \pi]$

17. (a) Two x-intercepts of f determine two solutions a and b of $f(x) = 0$; by Rolle's Theorem there exists a point c between a and b such that $f'(c) = 0$, i.e. c is an x-intercept for f'.

(b) $f(x) = \sin x = 0$ at $x = n\pi$, and $f'(x) = \cos x = 0$ at $x = n\pi + \pi/2$, which lies between $n\pi$ and $(n+1)\pi$, $(n = 0, \pm 1, \pm 2, \ldots)$

19. Let $s(t)$ be the position function of the automobile for $0 \leq t \leq 5$, then by the Mean-Value Theorem there is at least one point c in $(0, 5)$ where
$s'(c) = v(c) = [s(5) - s(0)]/(5 - 0) = 4/5 = 0.8$ mi/min $= 48$ mi/h.

21. Let $f(t)$ and $g(t)$ denote the distances of the first and second runners to the starting point, and let $h(t) = f(t) - g(t)$. Since they start (at $t = 0$) at the same time and finish (at $t = t_1$) at the same time, $h(0) = h(t_1) = 0$, then by Rolle's Theorem there is a time t_2 for which $h'(t_2) = 0$, i.e. $f'(t_2) = g'(t_2)$; so they have the same velocity at time t_2.

23. (a) By the Constant Difference Theorem $f(x) - g(x) = k$ for some k; since $f(x_0) = g(x_0)$, $k = 0$, so $f(x) = g(x)$ for all x.

(b) Set $f(x) = \sin^2 x + \cos^2 x$, $g(x) = 1$; then $f'(x) = 2\sin x \cos x - 2\cos x \sin x = 0 = g'(x)$. Since $f(0) = 1 = g(0)$, $f(x) = g(x)$ for all x.

25. (a) If x, y belong to I and $x < y$ then for some c in I, $\dfrac{f(y) - f(x)}{y - x} = f'(c)$, so $|f(x) - f(y)| = |f'(c)||x - y| \leq M|x - y|$; if $x > y$ exchange x and y; if $x = y$ the inequality also holds.

(b) $f(x) = \sin x$, $f'(x) = \cos x$, $|f'(x)| \leq 1 = M$, so $|f(x) - f(y)| \leq |x - y|$ or $|\sin x - \sin y| \leq |x - y|$.

27. **(a)** Let $f(x) = \sqrt{x}$. By the Mean-Value Theorem there is a number c between x and y such that
$$\frac{\sqrt{y} - \sqrt{x}}{y - x} = \frac{1}{2\sqrt{c}} < \frac{1}{2\sqrt{x}} \text{ for } c \text{ in } (x, y), \text{ thus } \sqrt{y} - \sqrt{x} < \frac{y - x}{2\sqrt{x}}$$

     (b) multiply through and rearrange to get $\sqrt{xy} < \frac{1}{2}(x + y)$.

29. **(a)** If $f(x) = x^3 + 4x - 1$ then $f'(x) = 3x^2 + 4$ is never zero, so by Exercise 28 f has at most one real root; since f is a cubic polynomial it has at least one real root, so it has exactly one real root.

     (b) Let $f(x) = ax^3 + bx^2 + cx + d$. If $f(x) = 0$ has at least two distinct real solutions r_1 and r_2, then $f(r_1) = f(r_2) = 0$ and by Rolle's Theorem there is at least one number between r_1 and r_2 where $f'(x) = 0$. But $f'(x) = 3ax^2 + 2bx + c = 0$ for
$x = (-2b \pm \sqrt{4b^2 - 12ac})/(6a) = (-b \pm \sqrt{b^2 - 3ac})/(3a)$, which are not real if $b^2 - 3ac < 0$ so $f(x) = 0$ must have fewer than two distinct real solutions.

31. First we show $x > \sin x$ for $x > 0$. If $x > 1$ this is obvious, since $\sin x \leq 1$ for all x.
Let $F(x) = x - \sin x$ and $\pi/2 > x > 0$. By Exercise 25(a) with $y = 0$, there is a c with $0 < c < \pi/2$ such that $F(x)/x = F'(c) = 1 - \cos c > 0$ since $0 < x < \pi/2$. Thus $F(x) > 0$, i.e. $x > \sin x$.   (*)
Let $H(x) = 6\cos x + 3x^2$. Then $H(0) = 6$ and for $x > 0, H'(x) = -6\sin x + 6x > 0$ by (*). Thus $H(x) \geq 6$ for all $x > 0$.
Let $x > 0$ and set $G(x) = 6\sin x + x^3$. Observe that $G'(x) = 6\cos x + 3x^2 = 6$ when $x = 0$, and $\frac{d}{dx}G'(x) = 6(x - \sin x) > 0$ by the earlier result. Thus $G'(x)$ is strictly increasing for $x > 0$ and $G'(0) = 6$, hence $G'(x) > 6$ for all $x > 0$. Apply Exercise 25(a) to $G(x)$ with $x > 0, y = 0$. Then $G(x)/x = (G(x) - G(0))/(x - 0) = G'(c) > 6$ by the earlier result, so $G(x) > 6x, 6\sin x + x^3 > 6x, \sin x > x - x^3/6$ which proves the result.

33. **(a)** $\frac{d}{dx}[f^2(x) + g^2(x)] = 2f(x)f'(x) + 2g(x)g'(x) = 2f(x)g(x) + 2g(x)[-f(x)] = 0$,
so $f^2(x) + g^2(x)$ is constant.

     (b) $f(x) = \sin x$ and $g(x) = \cos x$

35.

37. **(a)** similar to the proof of Part (a) with $f'(c) < 0$

     (b) similar to the proof of Part (a) with $f'(c) = 0$

39. If f is differentiable at $x = 1$, then f is continuous there;
$$\lim_{x \to 1^+} f(x) = \lim_{x \to 1^-} f(x) = f(1) = 3, a + b = 3; \lim_{x \to 1^+} f'(x) = a \text{ and}$$
$$\lim_{x \to 1^-} f'(x) = 6 \text{ so } a = 6 \text{ and } b = 3 - 6 = -3.$$

41. From Section 3.2 a function has a vertical tangent line at a point of its graph if the slopes of secant lines through the point approach $+\infty$ or $-\infty$. Suppose f is continuous at $x = x_0$ and $\lim_{x \to x_0^+} f(x) = +\infty$. Then a secant line through $(x_1, f(x_1))$ and $(x_0, f(x_0))$, assuming $x_1 > x_0$, will have slope $\dfrac{f(x_1) - f(x_0)}{x_1 - x_0}$. By the Mean Value Theorem, this quotient is equal to $f'(c)$ for some

c between x_0 and x_1. But as x_1 approaches x_0, c must also approach x_0, and it is given that $\lim\limits_{c \to x_0^+} f'(c) = +\infty$, so the slope of the secant line approaches $+\infty$. The argument can be altered appropriately for $x_1 < x_0$, and/or for $f'(c)$ approaching $-\infty$.

EXERCISE SET 4.8

1. **(a)** positive, negative, slowing down   **(b)** positive, positive, speeding up
   (c) negative, positive, slowing down

3. **(a)** left because $v = ds/dt < 0$ at t_0
   (b) negative because $a = d^2s/dt^2$ and the curve is concave down at $t_0 (d^2s/dt^2 < 0)$
   (c) speeding up because v and a have the same sign
   (d) $v < 0$ and $a > 0$ at t_1 so the particle is slowing down because v and a have opposite signs.

5. s (m)

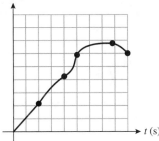

7.

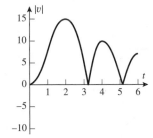

9. **(a)** At 60 mi/h the tangent line seems to pass through the points $(0, 20)$ and $(16, 100)$. Thus the acceleration would be $\dfrac{v_1 - v_0}{t_1 - t_0}\dfrac{88}{60} = \dfrac{100 - 20}{16 - 0}\dfrac{88}{60} \approx 7.3 \text{ ft/s}^2$.
   (b) The maximum acceleration occurs at maximum slope, so when $t = 0$.

11. **(a)**

t	1	2	3	4	5
s	0.71	1.00	0.71	0.00	-0.71
v	0.56	0.00	-0.56	-0.79	-0.56
a	-0.44	-0.62	-0.44	0.00	0.44

   (b) to the right at $t = 1$, stopped at $t = 2$, otherwise to the left
   (c) speeding up at $t = 3$; slowing down at $t = 1, 5$; neither at $t = 2, 4$

13. **(a)** $v(t) = 3t^2 - 6t$, $a(t) = 6t - 6$

(b) $s(1) = -2$ ft, $v(1) = -3$ ft/s, speed $= 3$ ft/s, $a(1) = 0$ ft/s^2

(c) $v = 0$ at $t = 0, 2$

(d) for $t \geq 0$, $v(t)$ changes sign at $t = 2$, and $a(t)$ changes sign at $t = 1$; so the particle is speeding up for $0 < t < 1$ and $2 < t$ and is slowing down for $1 < t < 2$

(e) total distance $= |s(2) - s(0)| + |s(5) - s(2)| = |-4 - 0| + |50 - (-4)| = 58$ ft

15. **(a)** $s(t) = 9 - 9\cos(\pi t/3)$, $v(t) = 3\pi \sin(\pi t/3)$, $a(t) = \pi^2 \cos(\pi t/3)$

(b) $s(1) = 9/2$ ft, $v(1) = 3\pi\sqrt{3}/2$ ft/s, speed $= 3\pi\sqrt{3}/2$ ft/s, $a(1) = \pi^2/2$ ft/s^2

(c) $v = 0$ at $t = 0, 3$

(d) for $0 < t < 5$, $v(t)$ changes sign at $t = 3$ and $a(t)$ changes sign at $t = 3/2, 9/2$; so the particle is speeding up for $0 < t < 3/2$ and $3 < t < 9/2$ and slowing down for $3/2 < t < 3$ and $9/2 < t < 5$

(e) total distance $= |s(3) - s(0)| + |s(5) - s(3)| = |18 - 0| + |9/2 - 18| = 18 + 27/2 = 63/2$ ft

17. $v(t) = \dfrac{5 - t^2}{(t^2 + 5)^2}$, $a(t) = \dfrac{2t(t^2 - 15)}{(t^2 + 5)^3}$

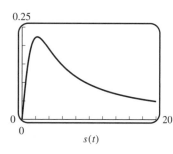

$s(t)$

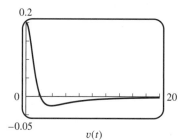

$v(t)$

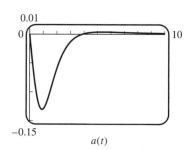

$a(t)$

(a) $v = 0$ at $t = \sqrt{5}$ **(b)** $s = \sqrt{5}/10$ at $t = \sqrt{5}$

(c) a changes sign at $t = \sqrt{15}$, so the particle is speeding up for $\sqrt{5} < t < \sqrt{15}$ and slowing down for $0 < t < \sqrt{5}$ and $\sqrt{15} < t$

19. $s = -4t + 3$
$v = -4$
$a = 0$

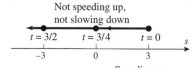

21. $s = t^3 - 9t^2 + 24t$
$v = 3(t - 2)(t - 4)$
$a = 6(t - 3)$

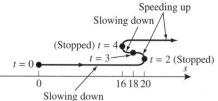

23. $s = \begin{cases} \cos t, & 0 \leq t \leq 2\pi \\ 1, & t > 2\pi \end{cases}$

$v = \begin{cases} -\sin t, & 0 \leq t \leq 2\pi \\ 0, & t > 2\pi \end{cases}$

$a = \begin{cases} -\cos t, & 0 \leq t < 2\pi \\ 0, & t > 2\pi \end{cases}$

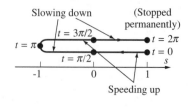

25. **(a)** $v = 10t - 22$, speed $= |v| = |10t - 22|$. $d|v|/dt$ does not exist at $t = 2.2$ which is the only critical point. If $t = 1, 2.2, 3$ then $|v| = 12, 0, 8$. The maximum speed is 12 ft/s.

(b) the distance from the origin is $|s| = |5t^2 - 22t| = |t(5t - 22)|$, but $t(5t - 22) < 0$ for $1 \le t \le 3$ so $|s| = -(5t^2 - 22t) = 22t - 5t^2$, $d|s|/dt = 22 - 10t$, thus the only critical point is $t = 2.2$. $d^2|s|/dt^2 < 0$ so the particle is farthest from the origin when $t = 2.2$. Its position is $s = 5(2.2)^2 - 22(2.2) = -24.2$.

27. $s = \sin 2t$; $v = 2\cos 2t$; $a = -4\sin 2t$

(a) $a = 0$ when $t = \pi/2$;
$s = 0, v = -2$ when $a = 0$

(b) $v = 0$ when $t = \pi/4$
$s(\pi/4) = 1; a(\pi/2) = -4$

29. (a)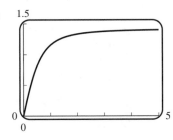

(b) $v = \dfrac{2t}{\sqrt{2t^2 + 1}}$, $\displaystyle\lim_{t \to +\infty} v = \dfrac{2}{\sqrt{2}} = \sqrt{2}$

31. (a) $s_1 = s_2$ if they collide, so $\frac{1}{2}t^2 - t + 3 = -\frac{1}{4}t^2 + t + 1$, $\frac{3}{4}t^2 - 2t + 2 = 0$ which has no real solution.

(b) Find the minimum value of $D = |s_1 - s_2| = |\frac{3}{4}t^2 - 2t + 2|$. From Part (a), $\frac{3}{4}t^2 - 2t + 2$

is never zero, and for $t = 0$ it is positive, hence it is always positive, so $D = \frac{3}{4}t^2 - 2t + 2$.

$\dfrac{dD}{dt} = \frac{3}{2}t - 2 = 0$ when $t = \frac{4}{3}$. $\dfrac{d^2D}{dt^2} > 0$ so D is minimum when $t = \frac{4}{3}$, $D = \frac{2}{3}$.

(c) $v_1 = t - 1$, $v_2 = -\dfrac{1}{2}t + 1$. $v_1 < 0$ if $0 \le t < 1$, $v_1 > 0$ if $t > 1$; $v_2 < 0$ if $t > 2$, $v_2 > 0$ if $0 \le t < 2$. They are moving in opposite directions during the intervals $0 \le t < 1$ and $t > 2$.

33. $r(t) = \sqrt{v^2(t)}$, $r'(t) = 2v(t)v'(t)/[2\sqrt{v(t)}] = v(t)a(t)/|v(t)|$ so $r'(t) > 0$ (speed is increasing) if v and a have the same sign, and $r'(t) < 0$ (speed is decreasing) if v and a have opposite signs.
If $v(t) > 0$ then $r(t) = v(t)$ and $r'(t) = a(t)$, so if $a(t) > 0$ then the particle is speeding up and a and v have the same sign; if $a(t) < 0$, then the particle is slowing down, and a and v have opposite signs.
If $v(t) < 0$ then $r(t) = -v(t)$, $r'(t) = -a(t)$, and if $a(t) > 0$ then the particle is speeding up and a and v have opposite signs; if $a(t) < 0$ then the particle is slowing down and a and v have the same sign.

REVIEW EXERCISES

3. $f'(x) = 2x - 5$ **(a)** $[5/2, +\infty)$ **(b)** $(-\infty, 5/2]$
$f''(x) = 2$ **(c)** $(-\infty, +\infty)$ **(d)** none
     (e) none

5. $f'(x) = \dfrac{4x}{(x^2 + 2)^2}$ $f''(x) = -4\dfrac{3x^2 - 2}{(x^2 + 2)^3}$

(a) $[0, +\infty)$ **(b)** $(-\infty, 0]$ **(c)** $(-\sqrt{2/3}, \sqrt{2/3})$
(d) $(-\infty, -\sqrt{2/3})$, $(\sqrt{2/3}, +\infty)$ **(e)** $-\sqrt{2/3}, \sqrt{2/3}$

7. $f'(x) = \dfrac{4(x+1)}{3x^{2/3}}$

$f''(x) = \dfrac{4(x-2)}{9x^{5/3}}$

(a) $[-1, +\infty)$ **(b)** $(-\infty, -1]$

(c) $(-\infty, 0), (2, +\infty)$ **(d)** $(0, 2)$

(e) $0, 2$

9. $f'(x) = -\sin x$

$f''(x) = -\cos x$

(a) $[\pi, 2\pi]$ **(b)** $[0, \pi]$

(c) $(\pi/2, 3\pi/2)$ **(d)** $(0, \pi/2), (3\pi/2, 2\pi)$

(e) $\pi/2, 3\pi/2$

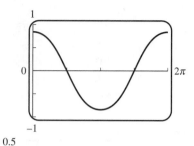

11. $f'(x) = \cos 2x$

$f''(x) = -2\sin 2x$

(a) $[0, \pi/4], [3\pi/4, \pi]$ **(b)** $[\pi/4, 3\pi/4]$

(c) $(\pi/2, \pi)$ **(d)** $(0, \pi/2)$

(e) $\pi/2$

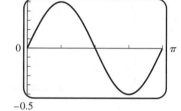

13. **(a)**

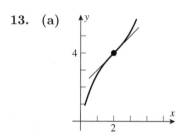

(b)

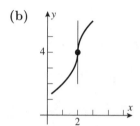

(c)

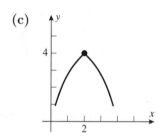

15. $f'(x) = 2ax + b$; $f'(x) > 0$ or $f'(x) < 0$ on $[0, +\infty)$ if $f'(x) = 0$ has no positive solution, so the polynomial is always increasing or always decreasing on $[0, +\infty)$ provided $-b/2a \le 0$.

21. **(a)** $f'(x) = (2 - x^2)/(x^2 + 2)^2$, $f'(x) = 0$ when $x = \pm\sqrt{2}$ (stationary points).

(b) $f'(x) = 8x/(x^2 + 1)^2$, $f'(x) = 0$ when $x = 0$ (stationary point).

23. **(a)** $f'(x) = \dfrac{7(x - 7)(x - 1)}{3x^{2/3}}$; critical numbers at $x = 0, 1, 7$;

neither at $x = 0$, relative maximum at $x = 1$, relative minimum at $x = 7$ (First Derivative Test)

(b) $f'(x) = 2\cos x(1 + 2\sin x)$; critical numbers at $x = \pi/2, 3\pi/2, 7\pi/6, 11\pi/6$; relative maximum at $x = \pi/2, 3\pi/2$, relative minimum at $x = 7\pi/6, 11\pi/6$

(c) $f'(x) = 3 - \dfrac{3\sqrt{x - 1}}{2}$; critical numbers at $x = 5$; relative maximum at $x = 5$

25. $\lim\limits_{x \to -\infty} f(x) = +\infty$, $\lim\limits_{x \to +\infty} f(x) = +\infty$

$f'(x) = x(4x^2 - 9x + 6)$, $f''(x) = 6(2x - 1)(x - 1)$

relative minimum at $x = 0$,

points of inflection when $x = 1/2, 1$,

no asymptotes

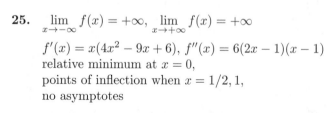

27. $\lim\limits_{x\to\pm\infty} f(x)$ doesn't exist

$f'(x) = 2x\sec^2(x^2+1)$,

$f''(x) = 2\sec^2(x^2+1)\left[1 + 4x^2\tan(x^2+1)\right]$

critical number at $x = 0$; relative minimum at $x = 0$

point of inflection when $1 + 4x^2\tan(x^2+1) = 0$

vertical asymptotes at $x = \pm\sqrt{\pi(n+\frac{1}{2})-1}$, $n = 0,1,2,\ldots$

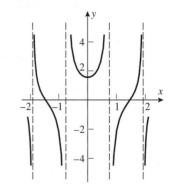

29. $f'(x) = 2\dfrac{x(x+5)}{(x^2+2x+5)^2}$, $f''(x) = -2\dfrac{2x^3+15x^2-25}{(x^2+2x+5)^3}$

critical numbers at $x = -5, 0$;

relative maximum at $x = -5$,

relative minimum at $x = 0$

points of inflection at $x = -7.26, -1.44, 1.20$

horizontal asymptote $y = 1$ as $x \to \pm\infty$

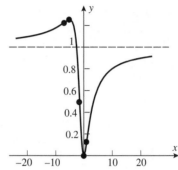

31. $\lim\limits_{x\to-\infty} f(x) = +\infty$, $\lim\limits_{x\to+\infty} f(x) = -\infty$

$f'(x) = \begin{cases} x \\ -2x \end{cases}$ if $\begin{cases} x \le 0 \\ x > 0 \end{cases}$

critical number at $x = 0$, no extrema

inflection point at $x = 0$ (f changes concavity)

no asymptotes

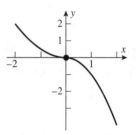

33. $f'(x) = 3x^2 + 5$; no relative extrema because there are no critical numbers.

35. $f'(x) = \frac{4}{5}x^{-1/5}$; critical number $x = 0$; relative minimum of 0 at $x = 0$ (first derivative test)

37. $f'(x) = 2x/(x^2+1)^2$; critical number $x = 0$; relative minimum of 0 at $x = 0$

39. (a)

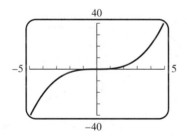

(b) $f'(x) = x^2 - \dfrac{1}{400}$, $f''(x) = 2x$

critical points at $x = \pm\dfrac{1}{20}$;

relative maximum at $x = -\dfrac{1}{20}$,

relative minimum at $x = \dfrac{1}{20}$

(c) The finer details can be seen when graphing over a much smaller x-window.

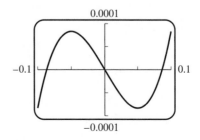

41. **(a)**

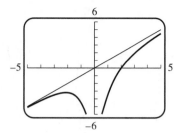

(b) Divide $y = x^2 + 1$ into $y = x^3 - 8$ to get the asymptote $ax + b = x$

43. $f(x) = \dfrac{(2x-1)(x^2+x-7)}{(2x-1)(3x^2+x-1)} = \dfrac{x^2+x-7}{3x^2+x-1}, \qquad x \neq 1/2$

horizontal asymptote: $y = 1/3$,
vertical asymptotes: $x = (-1 \pm \sqrt{13})/6$

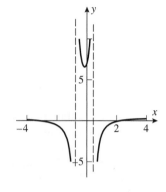

47. **(a)** If f has an absolute extremum at a point of (a, b) then it must, by Theorem 4.4.3, be at a critical point of f; since f is differentiable on (a, b) the critical point is a stationary point.

(b) It could occur at a critical point which is not a stationary point: for example, $f(x) = |x|$ on $[-1, 1]$ has an absolute minimum at $x = 0$ but is not differentiable there.

49. **(a)** $f'(x) = 2x - 3$; critical point $x = 3/2$. Minimum value $f(3/2) = -13/4$, no maximum.

(b) No maximum or minimum because $\lim\limits_{x \to +\infty} f(x) = +\infty$ and $\lim\limits_{x \to -\infty} f(x) = -\infty$.

(c) By observation f has an absolute maximum at $x = 0$ and at $x = 2$ and $f(x) \leq 0$ for all x. Since the only two critical points occur at $x = 0, 2$ there can be no relative or absolute minimum.

51. **(a)** $(x^2 - 1)^2$ can never be less than zero because it is the square of $x^2 - 1$; the minimum value is 0 for $x = \pm 1$, no maximum because $\lim\limits_{x \to +\infty} f(x) = +\infty$.

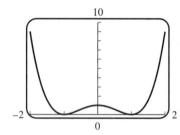

(b) $f'(x) = (1 - x^2)/(x^2 + 1)^2$; critical point $x = 1$. Maximum value $f(1) = 1/2$, minimum value 0 because $f(x)$ is never less than zero on $[0, +\infty)$ and $f(0) = 0$.

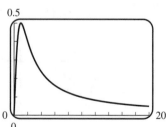

(c) $f'(x) = 2\sec x \tan x - \sec^2 x = (2\sin x - 1)/\cos^2 x$, $f'(x) = 0$ for x in $(0, \pi/4)$ when $x = \pi/6$; $f(0) = 2$, $f(\pi/6) = \sqrt{3}$, $f(\pi/4) = 2\sqrt{2} - 1$ so the maximum value is 2 at $x = 0$ and the minimum value is $\sqrt{3}$ at $x = \pi/6$.

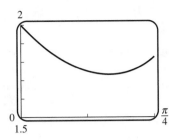

53. (a)

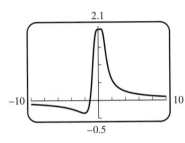

(b) minimum: $(-2.111985, -0.355116)$
maximum: $(0.372591, 2.012931)$

55. If one corner of the rectangle is at (x, y) with $x > 0$, $y > 0$, then $A = 4xy$, $y = 3\sqrt{1 - (x/4)^2}$, $A = 12x\sqrt{1 - (x/4)^2} = 3x\sqrt{16 - x^2}$, $\dfrac{dA}{dx} = 6\dfrac{8 - x^2}{\sqrt{16 - x^2}}$, critical point at $x = 2\sqrt{2}$. Since $A = 0$ when $x = 0, 4$ and $A > 0$ otherwise, there is an absolute maximum $A = 24$ at $x = 2\sqrt{2}$.

57. $V = x(12 - 2x)^2$ for $0 \le x \le 6$; $dV/dx = 12(x - 2)(x - 6)$, $dV/dx = 0$ when $x = 2$ for $0 < x < 6$. If $x = 0, 2, 6$ then $V = 0, 128, 0$ so the volume is largest when $x = 2$ in.

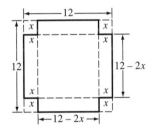

59. $x = -2.11491, 0.25410, 1.86081$

61. At the point of intersection, $x^3 = 0.5x - 1$, $x^3 - 0.5x + 1 = 0$. Let $f(x) = x^3 - 0.5x + 1$. By graphing $y = x^3$ and $y = 0.5x - 1$ it is evident that there is only one point of intersection and it occurs in the interval $[-2, -1]$; note that $f(-2) < 0$ and $f(-1) > 0$. $f'(x) = 3x^2 - 0.5$ so

$$x_{n+1} = x_n - \frac{x_n^3 - 0.5x + 1}{3x_n^2 - 0.5};$$

$x_1 = -1$, $x_2 = -1.2$,
$x_3 = -1.166492147, \ldots$,
$x_5 = x_6 = -1.165373043$

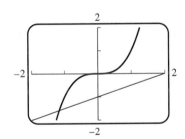

63. Solve $\phi - 0.0934 \sin \phi = 2\pi(1)/1.88$ to get $\phi = 3.325078$ so $r = 228 \times 10^6(1 - 0.0934 \cos \phi) = 248.938 \times 10^6$ km.

65. (a) yes; $f'(0) = 0$

(b) no, f is not differentiable on $(-1, 1)$

(c) yes, $f'(\sqrt{\pi/2}) = 0$

67. $f(x) = x^6 - 2x^2 + x$ satisfies $f(0) = f(1) = 0$, so by Rolle's Theorem $f'(c) = 0$ for some c in $(0, 1)$.

69. **(a)** If $a = k$, a constant, then $v = kt + b$ where b is constant; so the velocity changes sign at $t = -b/k$.

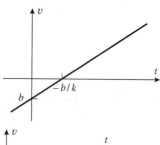

(b) Consider the equation $s = 5 - t^3/6$, $v = -t^2/2$, $a = -t$. Then for $t > 0$, a is decreasing and $av > 0$, so the particle is speeding up.

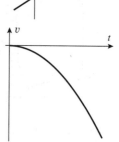

71. **(a)** $v = -2\dfrac{t(t^4 + 2t^2 - 1)}{(t^4 + 1)^2}$, $a = 2\dfrac{3t^8 + 10t^6 - 12t^4 - 6t^2 + 1}{(t^4 + 1)^3}$

(b)

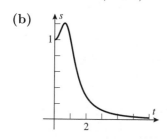 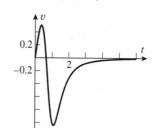 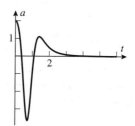

(c) It is farthest from the origin at approximately $t = 0.64$ (when $v = 0$) and $s = 1.2$

(d) Find t so that the velocity $v = ds/dt > 0$. The particle is moving in the positive direction for $0 \le t \le 0.64$ s.

(e) It is speeding up when $a, v > 0$ or $a, v < 0$, so for $0 \le t < 0.36$ and $0.64 < t < 1.1$, otherwise it is slowing down.

(f) Find the maximum value of $|v|$ to obtain: maximum speed $= 1.05$ m/s when $t = 1.10$ s.

CHAPTER 5

Integration

EXERCISE SET 5.1

1. Endpoints $0, \dfrac{1}{n}, \dfrac{2}{n}, \ldots, \dfrac{n-1}{n}, 1$; using right endpoints,

$$A_n = \left[\sqrt{\frac{1}{n}} + \sqrt{\frac{2}{n}} + \cdots + \sqrt{\frac{n-1}{n}} + 1\right]\frac{1}{n}$$

n	2	5	10	50	100
A_n	0.853553	0.749739	0.710509	0.676095	0.671463

3. Endpoints $0, \dfrac{\pi}{n}, \dfrac{2\pi}{n}, \ldots, \dfrac{(n-1)\pi}{n}, \pi$; using right endpoints,

$$A_n = [\sin(\pi/n) + \sin(2\pi/n) + \cdots + \sin(\pi(n-1)/n) + \sin\pi]\frac{\pi}{n}$$

n	2	5	10	50	100
A_n	1.57080	1.93376	1.98352	1.99935	1.99984

5. Endpoints $1, \dfrac{n+1}{n}, \dfrac{n+2}{n}, \ldots, \dfrac{2n-1}{n}, 2$; using right endpoints,

$$A_n = \left[\frac{n}{n+1} + \frac{n}{n+2} + \cdots + \frac{n}{2n-1} + \frac{1}{2}\right]\frac{1}{n}$$

n	2	5	10	50	100
A_n	0.583333	0.645635	0.668771	0.688172	0.690653

7. Endpoints $0, \dfrac{1}{n}, \dfrac{2}{n}, \ldots, \dfrac{n-1}{n}, 1$; using right endpoints,

$$A_n = \left[\sqrt{1 - \left(\frac{1}{n}\right)^2} + \sqrt{1 - \left(\frac{2}{n}\right)^2} + \cdots + \sqrt{1 - \left(\frac{n-1}{n}\right)^2} + 0\right]\frac{1}{n}$$

n	2	5	10	50	100
A_n	0.433013	0.659262	0.726130	0.774567	0.780106

9. $3(x-1)$ **11.** $x(x+2)$ **13.** $(x+3)(x-1)$

15. The area in Exercise 13 is always 3 less than the area in Exercise 11. The regions are identical except that the area in Exercise 11 has the extra trapezoid with vertices at $(0,0), (1,0), (0,2), (1,4)$ (with area 3).

17. $A(6)$ represents the area between $x = 0$ and $x = 6$; $A(3)$ represents the area between $x = 0$ and $x = 3$; their difference $A(6) - A(3)$ represents the area between $x = 3$ and $x = 6$, and $A(6) - A(3) = \dfrac{1}{3}(6^3 - 3^3) = 63$.

19. $f(x) = A'(x) = 2x; 0 = A(a) = a^2 - 4$, so take $a = 2$ (or $a = -2$), $A(x) = x^2 - 4$, $A'(x) = 2x = f(x)$, so $a = \pm 2, f(x) = 2x$

21. B is also the area between the graph of $f(x) = \sqrt{x}$ and the interval $[0, 1]$ on the y−axis, so $A + B$ is the area of the square.

EXERCISE SET 5.2

1. (a) $\displaystyle\int \frac{x}{\sqrt{1+x^2}}dx = \sqrt{1+x^2} + C$ (b) $\displaystyle\int x^2 \cos(1+x^3)dx = \frac{1}{3}\sin(1+x^3) + C$

5. $\displaystyle\frac{d}{dx}\left[\sqrt{x^3+5}\right] = \frac{3x^2}{2\sqrt{x^3+5}}$ so $\displaystyle\int \frac{3x^2}{2\sqrt{x^3+5}}dx = \sqrt{x^3+5} + C$

7. $\displaystyle\frac{d}{dx}\left[\sin\left(2\sqrt{x}\right)\right] = \frac{\cos\left(2\sqrt{x}\right)}{\sqrt{x}}$ so $\displaystyle\int \frac{\cos\left(2\sqrt{x}\right)}{\sqrt{x}}dx = \sin\left(2\sqrt{x}\right) + C$

9. (a) $x^9/9 + C$ (b) $\displaystyle\frac{7}{12}x^{12/7} + C$ (c) $\displaystyle\frac{2}{9}x^{9/2} + C$

11. $\displaystyle\int \left[5x + \frac{2}{3x^5}\right]dx = \int 5x\,dx + \frac{2}{3}\int \frac{1}{x^5}dx = \frac{5}{2}x^2 + \frac{2}{3}\left(\frac{-1}{4}\right)\frac{1}{x^4}C = \frac{5}{2}x^2 - \frac{1}{6x^4} + C$

13. $\displaystyle\int \left[x^{-3} - 3x^{1/4} + 8x^2\right]dx = \int x^{-3}dx - 3\int x^{1/4}dx + 8\int x^2\,dx = -\frac{1}{2}x^{-2} - \frac{12}{5}x^{5/4} + \frac{8}{3}x^3 + C$

15. $\displaystyle\int (x + x^4)dx = x^2/2 + x^5/5 + C$

17. $\displaystyle\int x^{1/3}(4 - 4x + x^2)dx = \int (4x^{1/3} - 4x^{4/3} + x^{7/3})dx = 3x^{4/3} - \frac{12}{7}x^{7/3} + \frac{3}{10}x^{10/3} + C$

19. $\displaystyle\int (x + 2x^{-2} - x^{-4})dx = x^2/2 - 2/x + 1/(3x^3) + C$

21. $\displaystyle\int [3\sin x - 2\sec^2 x]\,dx = -3\cos x - 2\tan x + C$

23. $\displaystyle\int (\sec^2 x + \sec x \tan x)dx = \tan x + \sec x + C$

25. $\displaystyle\int \frac{\sec\theta}{\cos\theta}d\theta = \int \sec^2\theta\,d\theta = \tan\theta + C$ 27. $\displaystyle\int \sec x \tan x\,dx = \sec x + C$

29. $\displaystyle\int (1 + \sin\theta)d\theta = \theta - \cos\theta + C$

31. $\displaystyle\int \frac{1 - \sin x}{1 - \sin^2 x}dx = \int \frac{1 - \sin x}{\cos^2 x}dx = \int (\sec^2 x - \sec x \tan x)\,dx = \tan x - \sec x + C$

33.
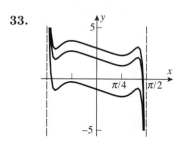

35. $f'(x) = m = -\sin x$ so $f(x) = \int(-\sin x)dx = \cos x + C$; $f(0) = 2 = 1 + C$

so $C = 1$, $f(x) = \cos\ x\ +\ 1$

37. (a) $y(x) = \int x^{1/3}dx = \dfrac{3}{4}x^{4/3} + C$, $y(1) = \dfrac{3}{4} + C = 2$, $C = \dfrac{5}{4}$; $y(x) = \dfrac{3}{4}x^{4/3} + \dfrac{5}{4}$

(b) $y(t) = \int(\sin t + 1)\,dt = -\cos t + t + C$, $y\left(\dfrac{\pi}{3}\right) = -\dfrac{1}{2} + \dfrac{\pi}{3} + C = 1/2$, $C = 1 - \dfrac{\pi}{3}$;

$y(t) = -\cos t + t + 1 - \dfrac{\pi}{3}$

(c) $y(x) = \int(x^{1/2} + x^{-1/2})dx = \dfrac{2}{3}x^{3/2} + 2x^{1/2} + C$, $y(1) = 0 = \dfrac{8}{3} + C$, $C = -\dfrac{8}{3}$,

$y(x) = \dfrac{2}{3}x^{3/2} + 2x^{1/2} - \dfrac{8}{3}$

39. $f'(x) = \dfrac{2}{3}x^{3/2} + C_1$; $f(x) = \dfrac{4}{15}x^{5/2} + C_1x + C_2$

41. $dy/dx = 2x + 1$, $y = \int(2x + 1)dx = x^2 + x + C$; $y = 0$ when $x = -3$

so $(-3)^2 + (-3) + C = 0$, $C = -6$ thus $y = x^2 + x - 6$

43. $dy/dx = \int 6x\,dx = 3x^2 + C_1$. The slope of the tangent line is -3 so $dy/dx = -3$ when $x = 1$.

Thus $3(1)^2 + C_1 = -3$, $C_1 = -6$ so $dy/dx = 3x^2 - 6$, $y = \int(3x^2 - 6)dx = x^3 - 6x + C_2$. If $x = 1$,

then $y = 5 - 3(1) = 2$ so $(1)^2 - 6(1) + C_2 = 2$, $C_2 = 7$ thus $y = x^3 - 6x + 7$.

45. (a)

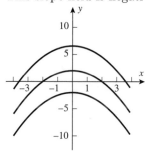

(b)

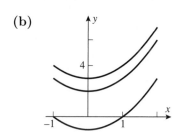

(c) $f(x) = x^2/2 - 1$

47. This slope field is negative when x is positive, and vice versa, and so corresponds to (b).

49. This slope field is zero along the lines $x = \pm 2$, and thus corresponds to (c).

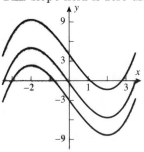

51. **(a)** $F'(x) = G'(x) = 3x + 4$

(b) $F(0) = 16/6 = 8/3$, $G(0) = 0$, so $F(0) - G(0) = 8/3$

(c) $F(x) = (9x^2 + 24x + 16)/6 = 3x^2/2 + 4x + 8/3 = G(x) + 8/3$

53. **(a)** For $x \neq 0$, $F'(x) = G'(x) = 1$. But if I is an interval containing 0 then neither F nor G has a derivative at 0, so neither F nor G is an antiderivative on I.

(b) Suppose $G(x) = F(x) + C$ for some C. Then $F(1) = 4$ and $G(1) = 4 + C$, so $C = 0$, but $F(-1) = -2$ and $G(-1) = -1$, a contradiction.

(c) No, because neither F nor G is an antiderivative on $(-\infty, +\infty)$.

55. $\displaystyle\int (\sec^2 x - 1)dx = \tan x - x + C$
    57. $\displaystyle\frac{1}{2}\int (1 - \cos x)dx = \frac{1}{2}(x - \sin x) + C$

59. $v = \dfrac{1087}{2\sqrt{273}}\displaystyle\int T^{-1/2}\,dT = \dfrac{1087}{\sqrt{273}}T^{1/2} + C$, $v(273) = 1087 = 1087 + C$ so $C = 0$, $v = \dfrac{1087}{\sqrt{273}}T^{1/2}$ ft/s

EXERCISE SET 5.3

1. **(a)** $\displaystyle\int u^{23}du = u^{24}/24 + C = (x^2 + 1)^{24}/24 + C$

(b) $-\displaystyle\int u^3 du = -u^4/4 + C = -(\cos^4 x)/4 + C$

(c) $2\displaystyle\int \sin u\,du = -2\cos u + C = -2\cos\sqrt{x} + C$

(d) $\dfrac{3}{8}\displaystyle\int u^{-1/2}du = \dfrac{3}{4}u^{1/2} + C = \dfrac{3}{4}\sqrt{4x^2 + 5} + C$

3. **(a)** $-\displaystyle\int u\,du = -\dfrac{1}{2}u^2 + C = -\dfrac{1}{2}\cot^2 x + C$

(b) $\displaystyle\int u^9 du = \dfrac{1}{10}u^{10} + C = \dfrac{1}{10}(1 + \sin t)^{10} + C$

(c) $\dfrac{1}{2}\displaystyle\int \cos u\,du = \dfrac{1}{2}\sin u + C = \dfrac{1}{2}\sin 2x + C$

(d) $\dfrac{1}{2}\displaystyle\int \sec^2 u\,du = \dfrac{1}{2}\tan u + C = \dfrac{1}{2}\tan x^2 + C$

7. $u = 4x - 3$, $\dfrac{1}{4}\displaystyle\int u^9\,du = \dfrac{1}{40}u^{10} + C = \dfrac{1}{40}(4x - 3)^{10} + C$

9. $u = 7x$, $\quad \dfrac{1}{7}\displaystyle\int \sin u\,du = -\dfrac{1}{7}\cos u + C = -\dfrac{1}{7}\cos 7x + C$

11. $u = 4x$, $du = 4dx$; $\dfrac{1}{4}\displaystyle\int \sec u \tan u\,du = \dfrac{1}{4}\sec u + C = \dfrac{1}{4}\sec 4x + C$

13. $u = 7t^2 + 12$, $du = 14t\,dt$; $\dfrac{1}{14}\displaystyle\int u^{1/2}du = \dfrac{1}{21}u^{3/2} + C = \dfrac{1}{21}(7t^2 + 12)^{3/2} + C$

15. $u = 1 - 2x$, $du = -2dx$, $-3\displaystyle\int \dfrac{1}{u^3}\,du = (-3)\left(-\dfrac{1}{2}\right)\dfrac{1}{u^2} + C = \dfrac{3}{2}\dfrac{1}{(1-2x)^2} + C$

17. $u = 5x^4 + 2$, $du = 20x^3\,dx$, $\dfrac{1}{20}\displaystyle\int \dfrac{du}{u^3} = -\dfrac{1}{40}\dfrac{1}{u^2} + C = -\dfrac{1}{40(5x^4 + 2)^2} + C$

19. $u = 5/x$, $du = -(5/x^2)dx$; $-\dfrac{1}{5}\displaystyle\int \sin u\,du = \dfrac{1}{5}\cos u + C = \dfrac{1}{5}\cos(5/x) + C$

21. $u = \cos 3t$, $du = -3\sin 3t\,dt$, $-\dfrac{1}{3}\displaystyle\int u^4\,du = -\dfrac{1}{15}u^5 + C = -\dfrac{1}{15}\cos^5 3t + C$

23. $u = x^2$, $du = 2x\,dx$; $\dfrac{1}{2}\displaystyle\int \sec^2 u\,du = \dfrac{1}{2}\tan u + C = \dfrac{1}{2}\tan\left(x^2\right) + C$

25. $u = 2 - \sin 4\theta$, $du = -4\cos 4\theta\,d\theta$; $-\dfrac{1}{4}\displaystyle\int u^{1/2}du = -\dfrac{1}{6}u^{3/2} + C = -\dfrac{1}{6}(2 - \sin 4\theta)^{3/2} + C$

27. $u = \sec 2x$, $du = 2\sec 2x \tan 2x\,dx$; $\dfrac{1}{2}\displaystyle\int u^2du = \dfrac{1}{6}u^3 + C = \dfrac{1}{6}\sec^3 2x + C$

29. $u = 2y + 1$, $du = 2dy$;
$\displaystyle\int \dfrac{1}{4}(u-1)\dfrac{1}{\sqrt{u}}\,du = \dfrac{1}{6}u^{3/2} - \dfrac{1}{2}\sqrt{u} + C = \dfrac{1}{6}(2y+1)^{3/2} - \dfrac{1}{2}\sqrt{2y+1} + C$

31. $\displaystyle\int \sin^2 2\theta \sin 2\theta\,d\theta = \int (1 - \cos^2 2\theta)\sin 2\theta\,d\theta$; $u = \cos 2\theta$, $du = -2\sin 2\theta\,d\theta$,

$-\dfrac{1}{2}\displaystyle\int (1 - u^2)du = -\dfrac{1}{2}u + \dfrac{1}{6}u^3 + C = -\dfrac{1}{2}\cos 2\theta + \dfrac{1}{6}\cos^3 2\theta + C$

33. $u = a + bx$, $du = b\,dx$,
$\displaystyle\int (a + bx)^n\,dx = \dfrac{1}{b}\int u^n du = \dfrac{(a + bx)^{n+1}}{b(n+1)} + C$

35. $u = \sin(a + bx)$, $du = b\cos(a + bx)dx$
$\dfrac{1}{b}\displaystyle\int u^n du = \dfrac{1}{b(n+1)}u^{n+1} + C = \dfrac{1}{b(n+1)}\sin^{n+1}(a + bx) + C$

37. (a) with $u = \sin x$, $du = \cos x\,dx$; $\displaystyle\int u\,du = \dfrac{1}{2}u^2 + C_1 = \dfrac{1}{2}\sin^2 x + C_1$;

with $u = \cos x$, $du = -\sin x\,dx$; $-\displaystyle\int u\,du = -\dfrac{1}{2}u^2 + C_2 = -\dfrac{1}{2}\cos^2 x + C_2$

(b) because they differ by a constant:

$$\left(\dfrac{1}{2}\sin^2 x + C_1\right) - \left(-\dfrac{1}{2}\cos^2 x + C_2\right) = \dfrac{1}{2}(\sin^2 x + \cos^2 x) + C_1 - C_2 = 1/2 + C_1 - C_2$$

39. $y = \displaystyle\int \sqrt{5x+1}\,dx = \frac{2}{15}(5x+1)^{3/2} + C;\ -2 = y(3) = \frac{2}{15}64 + C,$

so $C = -2 - \dfrac{2}{15}64 = -\dfrac{158}{15}$, and $y = \dfrac{2}{15}(5x+1)^{3/2} - \dfrac{158}{15}$

41. **(a)** $u = x^2 + 1, du = 2x\,dx;\ \dfrac{1}{2}\displaystyle\int \dfrac{1}{\sqrt{u}}\,du = \sqrt{u} + C = \sqrt{x^2+1} + C$

(b)

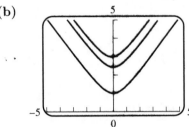

43. $f'(x) = m = \sqrt{3x+1},\ f(x) = \displaystyle\int (3x+1)^{1/2}dx = \frac{2}{9}(3x+1)^{3/2} + C$

$f(0) = 1 = \dfrac{2}{9} + C,\ C = \dfrac{7}{9}$, so $f(x) = \dfrac{2}{9}(3x+1)^{3/2} + \dfrac{7}{9}$

EXERCISE SET 5.4

1. **(a)** $1 + 8 + 27 = 36$ **(b)** $5 + 8 + 11 + 14 + 17 = 55$
 (c) $20 + 12 + 6 + 2 + 0 + 0 = 40$ **(d)** $1 + 1 + 1 + 1 + 1 + 1 = 6$
 (e) $1 - 2 + 4 - 8 + 16 = 11$ **(f)** $0 + 0 + 0 + 0 + 0 + 0 = 0$

3. $\displaystyle\sum_{k=1}^{10} k$ **5.** $\displaystyle\sum_{k=1}^{10} 2k$ **7.** $\displaystyle\sum_{k=1}^{6} (-1)^{k+1}(2k-1)$

9. **(a)** $\displaystyle\sum_{k=1}^{50} 2k$ **(b)** $\displaystyle\sum_{k=1}^{50} (2k-1)$

11. $\dfrac{1}{2}(100)(100+1) = 5050$ **13.** $\dfrac{1}{6}(20)(21)(41) = 2870$

15. $\displaystyle\sum_{k=1}^{30} k(k^2-4) = \sum_{k=1}^{30}(k^3-4k) = \sum_{k=1}^{30}k^3 - 4\sum_{k=1}^{30}k = \frac{1}{4}(30)^2(31)^2 - 4\cdot\frac{1}{2}(30)(31) = 214{,}365$

17. $\displaystyle\sum_{k=1}^{n} \frac{3k}{n} = \frac{3}{n}\sum_{k=1}^{n}k = \frac{3}{n}\cdot\frac{1}{2}n(n+1) = \frac{3}{2}(n+1)$

19. $\displaystyle\sum_{k=1}^{n-1} \frac{k^3}{n^2} = \frac{1}{n^2}\sum_{k=1}^{n-1}k^3 = \frac{1}{n^2}\cdot\frac{1}{4}(n-1)^2 n^2 = \frac{1}{4}(n-1)^2$

23. $\dfrac{1+2+3+\cdots+n}{n^2} = \displaystyle\sum_{k=1}^{n}\frac{k}{n^2} = \frac{1}{n^2}\sum_{k=1}^{n}k = \frac{1}{n^2}\cdot\frac{1}{2}n(n+1) = \frac{n+1}{2n};\ \lim_{n\to+\infty}\frac{n+1}{2n} = \frac{1}{2}$

25. $\displaystyle\sum_{k=1}^{n} \frac{5k}{n^2} = \frac{5}{n^2}\sum_{k=1}^{n}k = \frac{5}{n^2}\cdot\frac{1}{2}n(n+1) = \frac{5(n+1)}{2n};\ \lim_{n\to+\infty}\frac{5(n+1)}{2n} = \frac{5}{2}$

27. **(a)** $\displaystyle\sum_{j=0}^{5} 2^j$　　　　　**(b)** $\displaystyle\sum_{j=1}^{6} 2^{j-1}$　　　　　**(c)** $\displaystyle\sum_{j=2}^{7} 2^{j-2}$

29. **(a)** $\left(2+\dfrac{3}{n}\right)^4 \dfrac{3}{n}, \left(2+\dfrac{6}{n}\right)^4 \dfrac{3}{n}, \left(2+\dfrac{9}{n}\right)^4 \dfrac{3}{n}, \ldots, \left(2+\dfrac{3(n-1)}{n}\right)^4 \dfrac{3}{n}, (2+3)^4 \dfrac{3}{n}$

When $[2,5]$ is subdivided into n equal intervals, the endpoints are $2, 2+\dfrac{3}{n}, 2+2\cdot\dfrac{3}{n}, 2+3\cdot$

$\dfrac{3}{n}, \ldots, 2+(n-1)\dfrac{3}{n}, 2+3 = 5$, and the right endpoint approximation to the area under the

curve $y = x^4$ is given by the summands above.

(b) $\displaystyle\sum_{k=0}^{n-1} \left(2 + k\cdot\dfrac{3}{n}\right)^4 \dfrac{3}{n}$ gives the left endpoint approximation.

31. Endpoints $2, 3, 4, 5, 6; \Delta x = 1$;

(a) Left endpoints: $\displaystyle\sum_{k=1}^{4} f(x_k^*)\Delta x = 7 + 10 + 13 + 16 = 46$

(b) Midpoints: $\displaystyle\sum_{k=1}^{4} f(x_k^*)\Delta x = 8.5 + 11.5 + 14.5 + 17.5 = 52$

(c) Right endpoints: $\displaystyle\sum_{k=1}^{4} f(x_k^*)\Delta x = 10 + 13 + 16 + 19 = 58$

33. Endpoints: $0, \pi/4, \pi/2, 3\pi/4, \pi; \Delta x = \pi/4$

(a) Left endpoints: $\displaystyle\sum_{k=1}^{4} f(x_k^*)\Delta x = \left(1 + \sqrt{2}/2 + 0 - \sqrt{2}/2\right)(\pi/4) = \pi/4$

(b) Midpoints: $\displaystyle\sum_{k=1}^{4} f(x_k^*)\Delta x = [\cos(\pi/8) + \cos(3\pi/8) + \cos(5\pi/8) + \cos(7\pi/8)](\pi/4)$

$$= [\cos(\pi/8) + \cos(3\pi/8) - \cos(3\pi/8) - \cos(\pi/8)](\pi/4) = 0$$

(c) Right endpoints: $\displaystyle\sum_{k=1}^{4} f(x_k^*)\Delta x = \left(\sqrt{2}/2 + 0 - \sqrt{2}/2 - 1\right)(\pi/4) = -\pi/4$

35. **(a)** $0.718771403, 0.705803382, 0.698172179$
(b) $0.668771403, 0.680803382, 0.688172179$
(c) $0.692835360, 0.693069098, 0.693134682$

37. **(a)** $4.884074734, 5.115572731, 5.248762738$
(b) $5.684074734, 5.515572731, 5.408762738$
(c) $5.34707029, 5.338362719, 5.334644416$

39. $\Delta x = \dfrac{3}{n}, x_k^* = 1 + \dfrac{3}{n}k; f(x_k^*)\Delta x = \dfrac{1}{2}x_k^*\Delta x = \dfrac{1}{2}\left(1 + \dfrac{3}{n}k\right)\dfrac{3}{n} = \dfrac{3}{2}\left[\dfrac{1}{n} + \dfrac{3}{n^2}k\right]$

$\displaystyle\sum_{k=1}^{n} f(x_k^*)\Delta x = \dfrac{3}{2}\left[\sum_{k=1}^{n}\dfrac{1}{n} + \sum_{k=1}^{n}\dfrac{3}{n^2}k\right] = \dfrac{3}{2}\left[1 + \dfrac{3}{n^2}\cdot\dfrac{1}{2}n(n+1)\right] = \dfrac{3}{2}\left[1 + \dfrac{3}{2}\dfrac{n+1}{n}\right]$

$A = \displaystyle\lim_{n\to+\infty} \dfrac{3}{2}\left[1 + \dfrac{3}{2}\left(1 + \dfrac{1}{n}\right)\right] = \dfrac{3}{2}\left(1 + \dfrac{3}{2}\right) = \dfrac{15}{4}$

41. $\Delta x = \dfrac{3}{n}$, $x_k^* = 0 + k\dfrac{3}{n}$; $f(x_k^*)\Delta x = \left(9 - 9\dfrac{k^2}{n^2}\right)\dfrac{3}{n}$

$$\sum_{k=1}^{n} f(x_k^*)\Delta x = \sum_{k=1}^{n}\left(9 - 9\dfrac{k^2}{n^2}\right)\dfrac{3}{n} = \dfrac{27}{n}\sum_{k=1}^{n}\left(1 - \dfrac{k^2}{n^2}\right) = 27 - \dfrac{27}{n^3}\sum_{k=1}^{n}k^2$$

$$A = \lim_{n\to+\infty}\left[27 - \dfrac{27}{n^3}\sum_{k=1}^{n}k^2\right] = 27 - 27\left(\dfrac{1}{3}\right) = 18$$

43. $\Delta x = \dfrac{4}{n}$, $x_k^* = 2 + k\dfrac{4}{n}$

$$f(x_k^*)\Delta x = (x_k^*)^3\Delta x = \left[2 + \dfrac{4}{n}k\right]^3\dfrac{4}{n} = \dfrac{32}{n}\left[1 + \dfrac{2}{n}k\right]^3 = \dfrac{32}{n}\left[1 + \dfrac{6}{n}k + \dfrac{12}{n^2}k^2 + \dfrac{8}{n^3}k^3\right]$$

$$\sum_{k=1}^{n} f(x_k^*)\Delta x = \dfrac{32}{n}\left[\sum_{k=1}^{n}1 + \dfrac{6}{n}\sum_{k=1}^{n}k + \dfrac{12}{n^2}\sum_{k=1}^{n}k^2 + \dfrac{8}{n^3}\sum_{k=1}^{n}k^3\right]$$

$$= \dfrac{32}{n}\left[n + \dfrac{6}{n}\cdot\dfrac{1}{2}n(n+1) + \dfrac{12}{n^2}\cdot\dfrac{1}{6}n(n+1)(2n+1) + \dfrac{8}{n^3}\cdot\dfrac{1}{4}n^2(n+1)^2\right]$$

$$= 32\left[1 + 3\dfrac{n+1}{n} + 2\dfrac{(n+1)(2n+1)}{n^2} + 2\dfrac{(n+1)^2}{n^2}\right]$$

$$A = \lim_{n\to+\infty} 32\left[1 + 3\left(1 + \dfrac{1}{n}\right) + 2\left(1 + \dfrac{1}{n}\right)\left(2 + \dfrac{1}{n}\right) + 2\left(1 + \dfrac{1}{n}\right)^2\right]$$

$$= 32[1 + 3(1) + 2(1)(2) + 2(1)^2] = 320$$

45. $\Delta x = \dfrac{3}{n}$, $x_k^* = 1 + (k-1)\dfrac{3}{n}$

$$f(x_k^*)\Delta x = \dfrac{1}{2}x_k^*\Delta x = \dfrac{1}{2}\left[1 + (k-1)\dfrac{3}{n}\right]\dfrac{3}{n} = \dfrac{1}{2}\left[\dfrac{3}{n} + (k-1)\dfrac{9}{n^2}\right]$$

$$\sum_{k=1}^{n} f(x_k^*)\Delta x = \dfrac{1}{2}\left[\sum_{k=1}^{n}\dfrac{3}{n} + \dfrac{9}{n^2}\sum_{k=1}^{n}(k-1)\right] = \dfrac{1}{2}\left[3 + \dfrac{9}{n^2}\cdot\dfrac{1}{2}(n-1)n\right] = \dfrac{3}{2} + \dfrac{9}{4}\dfrac{n-1}{n}$$

$$A = \lim_{n\to+\infty}\left[\dfrac{3}{2} + \dfrac{9}{4}\left(1 - \dfrac{1}{n}\right)\right] = \dfrac{3}{2} + \dfrac{9}{4} = \dfrac{15}{4}$$

47. $\Delta x = \dfrac{3}{n}$, $x_k^* = 0 + (k-1)\dfrac{3}{n}$; $f(x_k^*)\Delta x = \left(9 - 9\dfrac{(k-1)^2}{n^2}\right)\dfrac{3}{n}$

$$\sum_{k=1}^{n} f(x_k^*)\Delta x = \sum_{k=1}^{n}\left[9 - 9\dfrac{(k-1)^2}{n^2}\right]\dfrac{3}{n} = \dfrac{27}{n}\sum_{k=1}^{n}\left(1 - \dfrac{(k-1)^2}{n^2}\right) = 27 - \dfrac{27}{n^3}\sum_{k=1}^{n}k^2 + \dfrac{54}{n^3}\sum_{k=1}^{n}k - \dfrac{27}{n^2}$$

$$A = \lim_{n\to+\infty} = 27 - 27\left(\dfrac{1}{3}\right) + 0 + 0 = 18$$

49. Endpoints $0, \dfrac{4}{n}, \dfrac{8}{n}, \ldots, \dfrac{4(n-1)}{n}, \dfrac{4n}{n} = 4$, and midpoints $\dfrac{2}{n}, \dfrac{6}{n}, \dfrac{10}{n}, \ldots, \dfrac{4n-6}{n}, \dfrac{4n-2}{n}$. Approximate the area with the sum $\displaystyle\sum_{k=1}^{n} 2\left(\dfrac{4k-2}{n}\right)\dfrac{4}{n} = \dfrac{16}{n^2}\left[2\dfrac{n(n+1)}{2} - n\right] \to 16$ as $n \to +\infty$.

51. $\Delta x = \dfrac{1}{n}$, $x_k^* = \dfrac{2k-1}{2n}$

$$f(x_k^*)\Delta x = \dfrac{(2k-1)^2}{(2n)^2}\dfrac{1}{n} = \dfrac{k^2}{n^3} - \dfrac{k}{n^3} + \dfrac{1}{4n^3}$$

$$\sum_{k=1}^{n} f(x_k^*)\Delta x = \frac{1}{n^3}\sum_{k=1}^{n}k^2 - \frac{1}{n^3}\sum_{k=1}^{n}k + \frac{1}{4n^3}\sum_{k=1}^{n}1$$

Using Theorem 5.4.4,

$$A = \lim_{n\to+\infty}\sum_{k=1}^{n} f(x_k^*)\Delta x = \frac{1}{3} + 0 + 0 = \frac{1}{3}$$

53. $\Delta x = \dfrac{2}{n},\ x_k^* = -1 + \dfrac{2k}{n}$

$$f(x_k^*)\Delta x = \left(-1 + \frac{2k}{n}\right)\frac{2}{n} = -\frac{2}{n} + 4\frac{k}{n^2}$$

$$\sum_{k=1}^{n} f(x_k^*)\Delta x = -2 + \frac{4}{n^2}\sum_{k=1}^{n}k = -2 + \frac{4}{n^2}\frac{n(n+1)}{2} = -2 + 2 + \frac{2}{n}$$

$$A = \lim_{n\to+\infty}\sum_{k=1}^{n} f(x_k^*)\Delta x = 0$$

The area below the x-axis cancels the area above the x-axis.

55. $\Delta x = \dfrac{2}{n},\ x_k^* = \dfrac{2k}{n}$

$$f(x_k^*) = \left[\left(\frac{2k}{n}\right)^2 - 1\right]\frac{2}{n} = \frac{8k^2}{n^3} - \frac{2}{n}$$

$$\sum_{k=1}^{n} f(x_k^*)\Delta x = \frac{8}{n^3}\sum_{k=1}^{n}k^2 - \frac{2}{n}\sum_{k=1}^{n}1 = \frac{8}{n^3}\frac{n(n+1)(2n+1)}{6} - 2$$

$$A = \lim_{n\to+\infty}\sum_{k=1}^{n} f(x_k^*)\Delta x = \frac{16}{6} - 2 = \frac{2}{3}$$

57. $\Delta x = \dfrac{b-a}{n},\ x_k^* = a + \dfrac{b-a}{n}(k-1)$

$$f(x_k^*)\Delta x = mx_k^*\Delta x = m\left[a + \frac{b-a}{n}(k-1)\right]\frac{b-a}{n} = m(b-a)\left[\frac{a}{n} + \frac{b-a}{n^2}(k-1)\right]$$

$$\sum_{k=1}^{n} f(x_k^*)\Delta x = m(b-a)\left[a + \frac{b-a}{2}\cdot\frac{n-1}{n}\right]$$

$$A = \lim_{n\to+\infty} m(b-a)\left[a + \frac{b-a}{2}\left(1 - \frac{1}{n}\right)\right] = m(b-a)\frac{b+a}{2} = \frac{1}{2}m(b^2 - a^2)$$

59. **(a)** With x_k^* as the right endpoint, $\Delta x = \dfrac{b}{n},\ x_k^* = \dfrac{b}{n}k$

$$f(x_k^*)\Delta x = (x_k^*)^3\Delta x = \frac{b^4}{n^4}k^3,\quad \sum_{k=1}^{n} f(x_k^*)\Delta x = \frac{b^4}{n^4}\sum_{k=1}^{n}k^3 = \frac{b^4}{4}\frac{(n+1)^2}{n^2}$$

$$A = \lim_{n\to+\infty} \frac{b^4}{4}\left(1 + \frac{1}{n}\right)^2 = b^4/4$$

(b) $\Delta x = \dfrac{b-a}{n}, \; x_k^* = a + \dfrac{b-a}{n}k$

$$f(x_k^*)\Delta x = (x_k^*)^3 \Delta x = \left[a + \frac{b-a}{n}k\right]^3 \frac{b-a}{n}$$

$$= \frac{b-a}{n}\left[a^3 + \frac{3a^2(b-a)}{n}k + \frac{3a(b-a)^2}{n^2}k^2 + \frac{(b-a)^3}{n^3}k^3\right]$$

$$\sum_{k=1}^{n} f(x_k^*)\Delta x = (b-a)\left[a^3 + \frac{3}{2}a^2(b-a)\frac{n+1}{n} + \frac{1}{2}a(b-a)^2\frac{(n+1)(2n+1)}{n^2}\right.$$

$$\left. + \frac{1}{4}(b-a)^3\frac{(n+1)^2}{n^2}\right]$$

$$A = \lim_{n\to+\infty}\sum_{k=1}^{n} f(x_k^*)\Delta x$$

$$= (b-a)\left[a^3 + \frac{3}{2}a^2(b-a) + a(b-a)^2 + \frac{1}{4}(b-a)^3\right] = \frac{1}{4}(b^4 - a^4)$$

61. If $n = 2m$ then $2m + 2(m-1) + \cdots + 2\cdot 2 + 2 = 2\sum_{k=1}^{m} k = 2\cdot\frac{m(m+1)}{2} = m(m+1) = \frac{n^2+2n}{4}$;

if $n = 2m+1$ then $(2m+1) + (2m-1) + \cdots + 5 + 3 + 1 = \sum_{k=1}^{m+1}(2k-1)$

$$= 2\sum_{k=1}^{m+1} k - \sum_{k=1}^{m+1} 1 = 2\cdot\frac{(m+1)(m+2)}{2} - (m+1) = (m+1)^2 = \frac{n^2+2n+1}{4}$$

63. $(3^5 - 3^4) + (3^6 - 3^5) + \cdots + (3^{17} - 3^{16}) = 3^{17} - 3^4$

65. $\left(\dfrac{1}{2^2} - \dfrac{1}{1^2}\right) + \left(\dfrac{1}{3^2} - \dfrac{1}{2^2}\right) + \cdots + \left(\dfrac{1}{20^2} - \dfrac{1}{19^2}\right) = \dfrac{1}{20^2} - 1 = -\dfrac{399}{400}$

67. **(a)** $\displaystyle\sum_{k=1}^{n}\frac{1}{(2k-1)(2k+1)} = \frac{1}{2}\sum_{k=1}^{n}\left(\frac{1}{2k-1} - \frac{1}{2k+1}\right)$

$$= \frac{1}{2}\left[\left(1 - \frac{1}{3}\right) + \left(\frac{1}{3} - \frac{1}{5}\right) + \left(\frac{1}{5} - \frac{1}{7}\right) + \cdots + \left(\frac{1}{2n-1} - \frac{1}{2n+1}\right)\right]$$

$$= \frac{1}{2}\left[1 - \frac{1}{2n+1}\right] = \frac{n}{2n+1}$$

(b) $\displaystyle\lim_{n\to+\infty}\frac{n}{2n+1} = \frac{1}{2}$

69. $\displaystyle\sum_{i=1}^{n}(x_i - \bar{x}) = \sum_{i=1}^{n} x_i - \sum_{i=1}^{n}\bar{x} = \sum_{i=1}^{n} x_i - n\bar{x}$ but $\bar{x} = \frac{1}{n}\sum_{i=1}^{n} x_i$ thus

$$\sum_{i=1}^{n} x_i = n\bar{x} \;\; \text{so} \; \sum_{i=1}^{n}(x_i - \bar{x}) = n\bar{x} - n\bar{x} = 0$$

71. **(a)** $\displaystyle\sum_{k=0}^{19} 3^{k+1} = \sum_{k=0}^{19} 3(3^k) = \frac{3(1 - 3^{20})}{1 - 3} = \frac{3}{2}(3^{20} - 1)$

(b) $\displaystyle\sum_{k=0}^{25} 2^{k+5} = \sum_{k=0}^{25} 2^5 2^k = \frac{2^5(1-2^{26})}{1-2} = 2^{31} - 2^5$

(c) $\displaystyle\sum_{k=0}^{100}(-1)\left(\frac{-1}{2}\right)^k = \frac{(-1)(1-(-1/2)^{101})}{1-(-1/2)} = -\frac{2}{3}(1+1/2^{101})$

73. both are valid

75. $\displaystyle\sum_{k=1}^{n}(a_k - b_k) = (a_1 - b_1) + (a_2 - b_2) + \cdots + (a_n - b_n)$

$\qquad\qquad = (a_1 + a_2 + \cdots + a_n) - (b_1 + b_2 + \cdots + b_n) = \displaystyle\sum_{k=1}^{n} a_k - \sum_{k=1}^{n} b_k$

EXERCISE SET 5.5

1. (a) $(4/3)(1) + (5/2)(1) + (4)(2) = 71/6$ (b) 2

3. (a) $(-9/4)(1) + (3)(2) + (63/16)(1) + (-5)(3) = -117/16$
(b) 3

5. $\displaystyle\int_{-1}^{2} x^2\, dx$
7. $\displaystyle\int_{-3}^{3} 4x(1-3x)\,dx$

9. (a) $\displaystyle\lim_{\max \Delta x_k \to 0} \sum_{k=1}^{n} 2x_k^* \Delta x_k;\ a = 1,\ b = 2$
(b) $\displaystyle\lim_{\max \Delta x_k \to 0} \sum_{k=1}^{n} \frac{x_k^*}{x_k^* + 1}\Delta x_k;\ a = 0,\ b = 1$

11. (a) $A = \dfrac{1}{2}(3)(3) = 9/2$
(b) $-A = -\dfrac{1}{2}(1)(1+2) = -3/2$

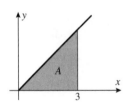 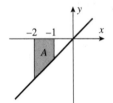

(c) $-A_1 + A_2 = -\dfrac{1}{2} + 8 = 15/2$
(d) $-A_1 + A_2 = 0$

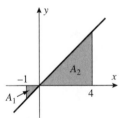 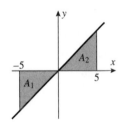

13. (a) $A = 2(5) = 10$
(b) $0;\ A_1 = A_2$ by symmetry

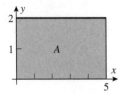

(c) $A_1 + A_2 = \dfrac{1}{2}(5)(5/2) + \dfrac{1}{2}(1)(1/2)$

$\qquad\qquad = 13/2$

(d) $\dfrac{1}{2}[\pi(1)^2] = \pi/2$

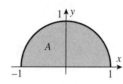

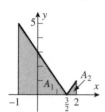

15. (a) $\displaystyle\int_{-2}^{0} f(x)\,dx = \int_{-2}^{0} (x+2)\,dx$

Triangle of height 2 and width 2, above x-axis, so answer is 2.

(b) $\displaystyle\int_{-2}^{2} f(x)\,dx = \int_{-2}^{0} (x+2)\,dx + \int_{2}^{0} (2-x)\,dx$

Two triangles of height 2 and base 2; answer is 4.

(c) $\displaystyle\int_{0}^{6} |x-2|\,dx = \int_{0}^{2} (2-x)\,dx + \int_{2}^{6} (x-2)\,dx$

Triangle of height 2 and base 2 together with a triangle with height 4 and base 4, so $2+8=10$.

(d) $\displaystyle\int_{-4}^{6} f(x)\,dx = \int_{-4}^{-2} (x+2)\,dx + \int_{-2}^{0} (x+2)\,dx + \int_{0}^{2} (2-x)\,dx + \int_{2}^{6} (x-2)\,dx$

Triangle of height 2 and base 2, below axis, plus a triangle of height 2, base 2 above axis, another of height 2 and base 2 above axis, and a triangle of height 4 and base 4, above axis. Thus $\int f(x) = -2+2+2+8 = 10$.

17. (a) 0.8   **(b)** -2.6 **(c)** -1.8 **(d)** -0.3

19. $\displaystyle\int_{-1}^{2} f(x)dx + 2\int_{-1}^{2} g(x)dx = 5 + 2(-3) = -1$

21. $\displaystyle\int_{1}^{5} f(x)dx = \int_{0}^{5} f(x)dx - \int_{0}^{1} f(x)dx = 1 - (-2) = 3$

23. $4\displaystyle\int_{-1}^{3} dx - 5\int_{-1}^{3} x\,dx = 4\cdot 4 - 5(-1/2 + (3\cdot 3)/2) = -4$

25. $\displaystyle\int_{0}^{1} x\,dx + 2\int_{0}^{1} \sqrt{1-x^2}\,dx = 1/2 + 2(\pi/4) = (1+\pi)/2$

27. (a) $\sqrt{x} > 0,\ 1-x < 0$ on $[2,3]$ so the integral is negative

    (b) $x^2 > 0,\ 3 - \cos x > 0$ for all x so the integral is positive

29. $\displaystyle\int_{0}^{10} \sqrt{25 - (x-5)^2}\,dx = \pi(5)^2/2 = 25\pi/2$ **31.** $\displaystyle\int_{0}^{1} (3x+1)\,dx = 5/2$

33. (a) The graph of the integrand is the horizontal line $y = C$. At first, assume that $C > 0$. Then the region is a rectangle of height C whose base extends from $x = a$ to $x = b$. Thus

$$\int_{a}^{b} C\,dx = (\text{area of rectangle}) = C(b-a).$$

If $C \le 0$ then the rectangle lies below the axis and its integral is the negative area, i.e. $-|C|(b-a) = C(b-a)$.

(b) Since $f(x) = C$, the Riemann sum becomes

$$\lim_{\max \Delta x_k \to 0} \sum_{k=1}^{n} f(x_k^*) \Delta x_k = \lim_{\max \Delta x_k \to 0} \sum_{k=1}^{n} C = \lim_{\max \Delta x_k \to 0} C(b-a) = C(b-a).$$

By Definition 5.5.1, $\int_a^b f(x)\,dx = C(b-a)$.

35. For any partition of $[0,1]$ we have

$$\sum_{k=1}^{n} f(x_k^*) \Delta x_k = \sum_{k=2}^{n} \Delta x_k = 1 - \Delta x_1 \text{ or we have}$$

$$\sum_{k=1}^{n} f(x_k^*) \Delta x_k = \sum_{k=1}^{n} \Delta x_k = 1.$$

This is because $f(x) = 1$ for all x except possibly x_1^*, which lies in the interval $[0, x_1]$ and could be 0. In any event, since in the limit the maximum size of the Δx_k goes to zero, the two possibilities are 1 in the limit, and thus $\int_0^1 f(x)\,dx = 1$.

37. On $[0, \frac{\pi}{4}]$ the minimum value of the integrand is 0 and the maximum is $\sin\left(\frac{\pi}{4}\right) = \frac{\sqrt{2}}{2}$. On $\left[\frac{\pi}{4}, \frac{5\pi}{6}\right]$ the minimum value is $\sin\left(\frac{5\pi}{6}\right) = \frac{1}{2}$ and the maximum is $\sin\left(\frac{\pi}{2}\right) = 1$. On $\left[\frac{5\pi}{6}, \pi\right]$ the minimum value is 0 and the maximum value is $\frac{1}{2}$. Thus the minimum value of the Riemann sums is $0 \cdot \frac{\pi}{4} + \left(\frac{1}{2}\right) \cdot \frac{7\pi}{12} + 0 \cdot \frac{\pi}{6} = \frac{7\pi}{24}$, and the maximum is $\frac{\sqrt{2}}{2} \cdot \frac{\pi}{4} + 1 \cdot \frac{7\pi}{12} + \frac{1}{2} \cdot \frac{\pi}{6} = \left(\frac{\sqrt{2}}{8} + \frac{2}{3}\right)\pi$.

39. $\Delta x_k = \frac{4k^2}{n^2} - \frac{4(k-1)^2}{n^2} = \frac{4}{n^2}(2k-1)$, $x_k^* = \frac{4k^2}{n^2}$,

$f(x_k^*) = \frac{2k}{n}$, $f(x_k^*)\Delta x_k = \frac{8k}{n^3}(2k-1) = \frac{8}{n^3}(2k^2 - k)$,

$$\sum_{k=1}^{n} f(x_k^*) \Delta x_k = \frac{8}{n^3} \sum_{k=1}^{n} (2k^2 - k) = \frac{8}{n^3}\left[\frac{1}{3}n(n+1)(2n+1) - \frac{1}{2}n(n+1)\right] = \frac{4}{3}\frac{(n+1)(4n-1)}{n^2},$$

$$\lim_{n \to +\infty} \sum_{k=1}^{n} f(x_k^*) \Delta x_k = \lim_{n \to +\infty} \frac{4}{3}\left(1 + \frac{1}{n}\right)\left(4 - \frac{1}{n}\right) = \frac{16}{3}.$$

41. With $f(x) = g(x)$ then $f(x) - g(x) = 0$ for $a < x \le b$. By Theorem 5.5.4(b)

$$\int_a^b f(x)\,dx = \int_a^b [(f(x) - g(x) + g(x)]dx = \int_a^b [f(x) - g(x)]dx + \int_a^b g(x)dx.$$

But the first term on the right hand side is zero (from Exercise 40), so

$$\int_a^b f(x)\,dx = \int_a^b g(x)\,dx$$

EXERCISE SET 5.6

1. **(a)** $\int_0^2 (2-x)dx = (2x - x^2/2)\Big]_0^2 = 4 - 4/2 = 2$

(b) $\int_{-1}^1 2dx = 2x\Big]_{-1}^1 = 2(1) - 2(-1) = 4$

(c) $\int_1^3 (x+1)dx = (x^2/2 + x)\Big]_1^3 = 9/2 + 3 - (1/2 + 1) = 6$

3. $\int_2^3 x^3 dx = x^4/4\Big]_2^3 = 81/4 - 16/4 = 65/4$ **5.** $\int_1^4 3\sqrt{x}\,dx = 2x^{3/2}\Big]_1^4 = 16 - 2 = 14$

7. $\int_{-2}^1 (x^2 - 6x + 12)\,dx = \left[\frac{1}{3}x^3 - 3x^2 + 12x\right]\Big]_{-2}^1 = \frac{1}{3} - 3 + 12 - \left(-\frac{8}{3} - 12 - 24\right) = 48$

9. $\int_1^4 \frac{1}{x^2}\,dx = -x^{-1}\Big]_1^4 = -\frac{1}{4} + 1 = \frac{3}{4}$ **11.** $\frac{4}{5}x^{5/2}\Big]_4^9 = 844/5$

13. $-\cos\theta\Big]_{-\pi/2}^{\pi/2} = 0$ **15.** $\sin x\Big]_{-\pi/4}^{\pi/4} = \sqrt{2}$

17. $\left(2\sqrt{t} - 2t^{3/2}\right)\Big]_1^4 = -12$ **19.** $\left(\frac{1}{2}x^2 - 2\cot x\right)\Big]_{\pi/6}^{\pi/2} = \pi^2/9 + 2\sqrt{3}$

21. (a) $\int_{-1}^1 |2x - 1|\,dx = \int_{-1}^{1/2} (1 - 2x)\,dx + \int_{1/2}^1 (2x - 1)\,dx = (x - x^2)\Big]_{-1}^{1/2} + (x^2 - x)\Big]_{1/2}^1 = \frac{5}{2}$

     (b) $\int_0^{\pi/2} \cos x\,dx + \int_{\pi/2}^{3\pi/4} (-\cos x)dx = \sin x\Big]_0^{\pi/2} - \sin x\Big]_{\pi/2}^{3\pi/4} = 2 - \sqrt{2}/2$

23. (a) $17/6$ **(b)** $F(x) = \begin{cases} \dfrac{1}{2}x^2, & x \le 1 \\ \dfrac{1}{3}x^3 + \dfrac{1}{6}, & x > 1 \end{cases}$

25. $0.665867079;\ \int_1^3 \frac{1}{x^2}\,dx = -\frac{1}{x}\Big]_1^3 = 2/3$

27. $3.106017890;\ \int_{-1}^1 \sec^2 x\,dx = \tan x\Big|_{-1}^1 = 2\tan 1 \approx 3.114815450$

29. $A = \int_0^3 (x^2 + 1)dx = \left(\frac{1}{3}x^3 + x\right)\Big]_0^3 = 12$

31. $A = \int_0^{2\pi/3} 3\sin x\,dx = -3\cos x\Big]_0^{2\pi/3} = 9/2$ **32.** $A = -\int_{-2}^{-1} x^3 dx = -\frac{1}{4}x^4\Big]_{-2}^{-1} = 15/4$

33. $\text{Area} = -\int_0^1 (x^2 - x)\,dx + \int_1^2 (x^2 - x)\,dx = 5/6 + 1/6 = 1$

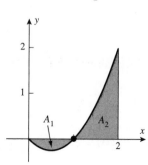

35. $\text{Area} = -\int_0^{5/4} (2\sqrt{x+1} - 3)\, dx + \int_{5/4}^3 (2\sqrt{x+1} - 3)\, dx$

$= 7/12 + 11/12 = 3/2$

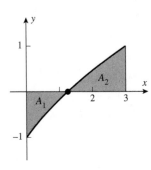

37. **(a)** $A = \int_0^{0.8} \cos x\, dx = -\sin x \Big]_0^{0.8} = \sin 0.8$

(b) The calculator was in degree mode instead of radian mode; the correct answer is 0.717.

39. **(a)** the area between the curve and the x-axis breaks into equal parts, one above and one below the x-axis, so the integral is zero

(b) $\int_{-1}^1 x^3\, dx = \frac{1}{4}x^4 \Big]_{-1}^1 = \frac{1}{4}(1^4 - (-1)^4) = 0;$

$\int_{-\pi/2}^{\pi/2} \sin x\, dx = -\cos x \Big]_{-\pi/2}^{\pi/2} = -\cos(\pi/2) + \cos(-\pi/2) = 0 + 0 = 0$

(c) The area on the left side of the y-axis is equal to the area on the right side, so

$\int_{-a}^a f(x)\, dx = 2\int_0^a f(x)\, dx$

(d) $\int_{-1}^1 x^2\, dx = \frac{1}{3}x^3 \Big]_{-1}^1 = \frac{1}{3}(1^3 - (-1)^3) = \frac{2}{3} = 2\int_0^1 x^2\, dx;$

$\int_{-\pi/2}^{\pi/2} \cos x\, dx = \sin x \Big]_{-\pi/2}^{\pi/2} = \sin(\pi/2) - \sin(-\pi/2) = 1 + 1 = 2 = 2\int_0^{\pi/2} \cos x\, dx$

41. **(a)** $F'(x) = 3x^2 - 3$

(b) $\int_1^x (3t^2 - 3)\, dt = (t^3 - 3t) \Big]_1^x = x^3 - 3x + 2$, and $\frac{d}{dx}(x^3 - 3x + 2) = 3x^2 - 3$

43. **(a)** $\sin x^2$ **(b)** $\sqrt{1 - \cos(x)}$

45. $-\dfrac{x}{\cos x}$

47. $F'(x) = \sqrt{x^2 + 9}$, $\quad F''(x) = \dfrac{x}{\sqrt{x^2 + 9}}$

(a) 0   **(b)** 5   **(c)** $\dfrac{4}{5}$

49. **(a)** $F'(x) = \dfrac{x - 3}{x^2 + 7} = 0$ when $x = 3$, which is a relative minimum, and hence the absolute minimum, by the first derivative test.

(b) increasing on $[3, +\infty)$, decreasing on $(-\infty, 3]$

(c) $F''(x) = \dfrac{7 + 6x - x^2}{(x^2 + 7)^2} = \dfrac{(7 - x)(1 + x)}{(x^2 + 7)^2}$; concave up on $(-1, 7)$, concave down on $(-\infty, -1)$ and on $(7, +\infty)$

51. (a) $(0, +\infty)$ because f is continuous there and 1 is in $(0, +\infty)$

(b) at $x = 1$ because $F(1) = 0$

53. (a) $\displaystyle\int_0^3 \sqrt{x}\,dx = \frac{2}{3}x^{3/2}\Big]_0^3 = 2\sqrt{3} = f(x^*)(3-0)$, so $f(x^*) = \dfrac{2}{\sqrt{3}}, x^* = \dfrac{4}{3}$

(b) $\displaystyle\int_{-12}^0 (x^2 + x)\,dx = \frac{1}{3}x^3 + \frac{1}{2}x^2\Big]_{-12}^0 = 504$, so $f(x^*)(0 - (-12)) = 504, x^2 + x = 42, x^* = 6$

55. $\sqrt{2} \le \sqrt{x^3 + 2} \le \sqrt{29}$, so $3\sqrt{2} \le \displaystyle\int_0^3 \sqrt{x^3 + 2}\,dx \le 3\sqrt{29}$

57. (a) $\left[cF(x)\right]_a^b = cF(b) - cF(a) = c[F(b) - F(a)] = c\left[F(x)\right]_a^b$

(b) $\left[F(x) + G(x)\right]_a^b = [F(b) + G(b)] - [F(a) + G(a)]$

$\qquad\qquad = [F(b) - F(a)] + [G(b) - G(a)] = F(x)\Big]_a^b + G(x)\Big]_a^b$

(c) $\left[F(x) - G(x)\right]_a^b = [F(b) - G(b)] - [F(a) - G(a)]$

$\qquad\qquad = [F(b) - F(a)] - [G(b) - G(a)] = F(x)\Big]_a^b - G(x)\Big]_a^b$

59. (a) the increase in height in inches, during the first ten years

(b) the change in the radius in centimeters, during the time interval $t = 1$ to $t = 2$ seconds

(c) the change in the speed of sound in ft/s, during an increase in temperature from $t = 32°F$ to $t = 100°F$

(d) the displacement of the particle in cm, during the time interval $t = t_1$ to $t = t_2$ seconds

61. (a) amount of water = (rate of flow)(time) = $4t$ gal, total amount = $4(30) = 120$ gal

(b) amount of water = $\displaystyle\int_0^{60} (4 + t/10)\,dt = 420$ gal

(c) amount of water = $\displaystyle\int_0^{120} (10 + \sqrt{t})\,dt = 1200 + 160\sqrt{30} \approx 2076.36$ gal

63. $\displaystyle\sum_{k=1}^n \frac{\pi}{4n}\sec^2\left(\frac{\pi k}{4n}\right) = \sum_{k=1}^n f(x_k^*)\Delta x$ where $f(x) = \sec^2 x, x_k^* = \dfrac{\pi k}{4n}$ and $\Delta x = \dfrac{\pi}{4n}$ for $0 \le x \le \dfrac{\pi}{4}$.

Thus $\displaystyle\lim_{n\to+\infty}\sum_{k=1}^n \frac{\pi}{4n}\sec^2\left(\frac{\pi k}{4n}\right) = \lim_{n\to+\infty}\sum_{k=1}^n f(x_k^*)\Delta x = \int_0^{\pi/4}\sec^2 x\,dx = \tan x\Big]_0^{\pi/4} = 1$

65. Let f be continuous on a closed interval $[a, b]$ and let F be an antiderivative of f on $[a, b]$. By Theorem 4.7.2, $\dfrac{F(b) - F(a)}{b - a} = F'(x^*)$ for some x^* in (a, b). By Theorem 5.6.1,

$\displaystyle\int_a^b f(x)\,dx = F(b) - F(a)$, i.e. $\displaystyle\int_a^b f(x)\,dx = F'(x^*)(b - a) = f(x^*)(b - a)$.

EXERCISE SET 5.7

1. (a) displ = $s(3) - s(0)$

$\qquad = \displaystyle\int_0^3 v(t)\,dt = \int_0^2 (1 - t)\,dt + \int_2^3 (t - 3)\,dt = (t - t^2/2)\Big]_0^2 + (t^2/2 - 3t)\Big]_2^3 = -1/2;$

$\quad$ dist $= \displaystyle\int_0^3 |v(t)|\,dt = (t - t^2/2)\Big]_0^1 + (t^2/2 - t)\Big]_1^2 - (t^2/2 - 3t)\Big]_2^3 = 3/2$

(b) $\text{displ} = s(3) - s(0)$

$$= \int_0^3 v(t)dt = \int_0^1 t\,dt + \int_1^2 dt + \int_2^3 (5 - 2t)dt = t^2/2\Big]_0^1 + t\Big]_1^2 + (5t - t^2)\Big]_2^3 = 3/2;$$

$$\text{dist} = \int_0^1 t\,dt + \int_1^2 dt + \int_2^{5/2} (5 - 2t)dt + \int_{5/2}^3 (2t - 5)dt$$

$$= t^2/2\Big]_0^1 + t\Big]_1^2 + (5t - t^2)\Big]_2^{5/2} + (t^2 - 5t)\Big]_{5/2}^3 = 2$$

3. **(a)** $v(t) = 20 + \int_0^t a(u)du$; add areas of the small blocks to get

$$v(4) \approx 20 + 1.4 + 3.0 + 4.7 + 6.2 = 35.3 \text{ m/s}$$

(b) $v(6) = v(4) + \int_4^6 a(u)du \approx 35.3 + 7.5 + 8.6 = 51.4 \text{ m/s}$

5. **(a)** $s(t) = t^3 - t^2 + C; 1 = s(0) = C$, so $s(t) = t^3 - t^2 + 1$

(b) $v(t) = -\cos 3t + C_1; 3 = v(0) = -1 + C_1, C_1 = 4$, so $v(t) = -\cos 3t + 4$. Then

$$s(t) = -\frac{1}{3}\sin 3t + 4t + C_2; 3 = s(0) = C_2, \text{ so } s(t) = -\frac{1}{3}\sin 3t + 4t + 3$$

7. **(a)** $s(t) = \frac{3}{2}t^2 + t + C; 4 = s(2) = 6 + 2 + C, C = -4$ and $s(t) = \frac{3}{2}t^2 + t - 4$

(b) $v(t) = -t^{-2} + C_1, 0 = v(1) = -1 + C_1, C_1 = 1$ and
$v(t) = -t^{-2} + 1$ so $s(t) = t^{-1} + t + C_2, 2 = s(1) = 2 + C_2,$
$C_2 = 0$ and $s(t) = t^{-1} + t$

9. **(a)** $\text{displacement} = s(\pi/2) - s(0) = \int_0^{\pi/2} \sin t\,dt = -\cos t\Big]_0^{\pi/2} = 1 \text{ m}$

$$\text{distance} = \int_0^{\pi/2} |\sin t|dt = 1 \text{ m}$$

(b) $\text{displacement} = s(2\pi) - s(\pi/2) = \int_{\pi/2}^{2\pi} \cos t\,dt = \sin t\Big]_{\pi/2}^{2\pi} = -1 \text{ m}$

$$\text{distance} = \int_{\pi/2}^{2\pi} |\cos t|dt = -\int_{\pi/2}^{3\pi/2} \cos t\,dt + \int_{3\pi/2}^{2\pi} \cos t\,dt = 3 \text{ m}$$

11. **(a)** $v(t) = t^3 - 3t^2 + 2t = t(t - 1)(t - 2)$

$$\text{displacement} = \int_0^3 (t^3 - 3t^2 + 2t)dt = 9/4 \text{ m}$$

$$\text{distance} = \int_0^3 |v(t)|dt = \int_0^1 v(t)dt + \int_1^2 -v(t)dt + \int_2^3 v(t)dt = 11/4 \text{ m}$$

(b) $\text{displacement} = \int_0^3 (\sqrt{t} - 2)dt = 2\sqrt{3} - 6 \text{ m}$

$$\text{distance} = \int_0^3 |v(t)|dt = -\int_0^3 v(t)dt = 6 - 2\sqrt{3} \text{ m}$$

13. $v = 3t - 1$

$$\text{displacement} = \int_0^2 (3t - 1)\,dt = 4 \text{ m}$$

$$\text{distance} = \int_0^2 |3t - 1|\,dt = \frac{13}{3} \text{ m}$$

15. $v = \int (1/\sqrt{3t+1}\, dt = \frac{2}{3}\sqrt{3t+1} + C; v(0) = 4/3$ so $C = 2/3$, $v = \frac{2}{3}\sqrt{3t+1} + 2/3$

$\qquad$ displacement $= \int_1^5 \frac{2}{3}\sqrt{3t+1}\, dt = \frac{296}{27}$ m

$\qquad$ distance $= \int_1^5 \frac{2}{3}\sqrt{3t+1}\, dt = \frac{296}{27}$ m

17. (a) $s = \int \sin\frac{1}{2}\pi t\, dt = -\frac{2}{\pi}\cos\frac{1}{2}\pi t + C$

$\qquad$ $s = 0$ when $t = 0$ which gives $C = \frac{2}{\pi}$ so $s = -\frac{2}{\pi}\cos\frac{1}{2}\pi t + \frac{2}{\pi}$.

$\qquad$ $a = \dfrac{dv}{dt} = \dfrac{\pi}{2}\cos\dfrac{1}{2}\pi t$. When $t = 1$: $s = 2/\pi$, $v = 1$, $|v| = 1$, $a = 0$.

$\quad$ **(b)** $v = -3\int t\, dt = -\frac{3}{2}t^2 + C_1$, $v = 0$ when $t = 0$ which gives $C_1 = 0$ so $v = -\frac{3}{2}t^2$

$\qquad$ $s = -\frac{3}{2}\int t^2\, dt = -\frac{1}{2}t^3 + C_2$, $s = 1$ when $t = 0$ which gives $C_2 = 1$ so $s = -\frac{1}{2}t^3 + 1$.

$\qquad$ When $t = 1$: $s = 1/2$, $v = -3/2$, $|v| = 3/2$, $a = -3$.

19. By inspection $s = 4$ cannot happen during the first second of travel, and at the end of that second
the particle has traveled to $s = 5/2$ cm from the starting position. For $t > 1$ the displacement of
the particle during the time interval $[0, t]$ is given by

$$s(t) = \int_0^t v(\tau)\, d\tau = 5/2 + \int_1^t (6\sqrt{\tau} - 1)\, dt = 5/2 + \left(4\tau^{3/2} - \tau\right)\Big]_1^t = -3 + 4t^{3/2} - t,$$

and the displacement equals 4 cm if $4t^{3/2} - t = 7$, $t \approx 1.676$ s.

21. $s(t) = \int (20t^2 - 110t + 120)\, dt = \frac{20}{3}t^3 - 55t^2 + 120t + C$. But $s = 0$ when $t = 0$, so $C = 0$ and

$\quad s = \dfrac{20}{3}t^3 - 55t^2 + 120t$. Moreover, $a(t) = \dfrac{d}{dt}v(t) = 40t - 110$.

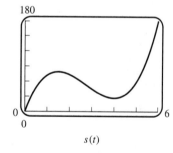

$s(t)$

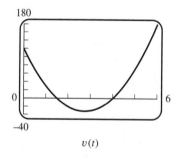

$v(t)$

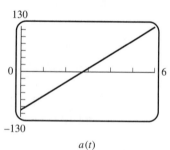

$a(t)$

23. (a) positive on $(0, 0.74)$ and $(2.97, 5)$, negative on $(0.74, 2.97)$

$\quad$ **(b)** For $0 < T < 5$ the displacement is

$\qquad$ disp $= T/2 - \sin(T) + T\cos(T)$

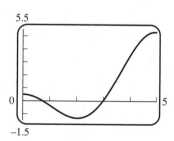

25. **(a)** $a(t) = \begin{cases} 0, & t < 4 \\ -10, & t > 4 \end{cases}$ **(b)** $v(t) = \begin{cases} 25, & t < 4 \\ 65 - 10t, & t > 4 \end{cases}$

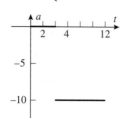

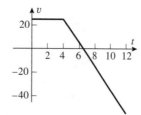

(c) $x(t) = \begin{cases} 25t, & t < 4 \\ 65t - 5t^2 - 80, & t > 4 \end{cases}$, so $x(8) = 120$, $x(12) = -20$

(d) $x(6.5) = 131.25$

27. **(a)** $a = -1.5$ mi/h/s $= -33/15$ ft/s^2 **(b)** $a = 30$ km/h/min $= 1/7200$ km/s^2

29. $a = a_0$ ft/s^2, $v = a_0 t + v_0 = a_0 t + 132$ ft/s, $s = a_0 t^2/2 + 132t + s_0 = a_0 t^2/2 + 132t$ ft; $s = 200$ ft when $v = 88$ ft/s. Solve $88 = a_0 t + 132$ and $200 = a_0 t^2/2 + 132t$ to get $a_0 = -\dfrac{121}{5}$ when $t = \dfrac{20}{11}$, so $s = -12.1t^2 + 132t$, $v = -\dfrac{121}{5}t + 132$.

(a) $a_0 = -\dfrac{121}{5}$ ft/s^2 **(b)** $v = 55$ mi/h $= \dfrac{242}{3}$ ft/s when $t = \dfrac{70}{33}$ s

(c) $v = 0$ when $t = \dfrac{60}{11}$ s

31. Suppose $s = s_0 = 0$, $v = v_0 = 0$ at $t = t_0 = 0$; $s = s_1 = 120$, $v = v_1$ at $t = t_1$; and $s = s_2$, $v = v_2 = 12$ at $t = t_2$. From Exercise 28(a),

$2.6 = a = \dfrac{v_1^2 - v_0^2}{2(s_1 - s_0)}$, $v_1^2 = 2as_1 = 5.2(120) = 624$. Applying the formula again,

$-1.5 = a = \dfrac{v_2^2 - v_1^2}{2(s_2 - s_1)}$, $v_2^2 = v_1^2 - 3(s_2 - s_1)$, so

$s_2 = s_1 - (v_2^2 - v_1^2)/3 = 120 - (144 - 624)/3 = 280$ m.

33. The truck's velocity is $v_T = 50$ and its position is $s_T = 50t + 2500$. The car's acceleration is $a_C = 4$ ft/s^2, so $v_C = 4t$, $s_C = 2t^2$ (initial position and initial velocity of the car are both zero). $s_T = s_C$ when $50t + 2500 = 2t^2$, $2t^2 - 50t - 2500 = 2(t + 25)(t - 50) = 0$, $t = 50$ s and $s_C = s_T = 2t^2 = 5000$ ft.

35. $s = 0$ and $v = 112$ when $t = 0$ so $v(t) = -32t + 112$, $s(t) = -16t^2 + 112t$

(a) $v(3) = 16$ ft/s, $v(5) = -48$ ft/s

(b) $v = 0$ when the projectile is at its maximum height so $-32t + 112 = 0$, $t = 7/2$ s, $s(7/2) = -16(7/2)^2 + 112(7/2) = 196$ ft.

(c) $s = 0$ when it reaches the ground so $-16t^2 + 112t = 0$, $-16t(t - 7) = 0$, $t = 0, 7$ of which $t = 7$ is when it is at ground level on its way down. $v(7) = -112$, $|v| = 112$ ft/s.

37. **(a)** $s(t) = 0$ when it hits the ground, $s(t) = -16t^2 + 16t = -16t(t - 1) = 0$ when $t = 1$ s.

(b) The projectile moves upward until it gets to its highest point where $v(t) = 0$, $v(t) = -32t + 16 = 0$ when $t = 1/2$ s.

39. $s(t) = s_0 + v_0 t - \frac{1}{2}gt^2 = 60t - 4.9t^2$ m and $v(t) = v_0 - gt = 60 - 9.8t$ m/s

   (a) $v(t) = 0$ when $t = 60/9.8 \approx 6.12$ s

   (b) $s(60/9.8) \approx 183.67$ m

   (c) another 6.12 s; solve for t in $s(t) = 0$ to get this result, or use the symmetry of the parabola $s = 60t - 4.9t^2$ about the line $t = 6.12$ in the t-s plane

   (d) also 60 m/s, as seen from the symmetry of the parabola (or compute $v(6.12)$)

41. $s(t) = -4.9t^2 + 49t + 150$ and $v(t) = -9.8t + 49$

   (a) the projectile reaches its maximum height when $v(t) = 0$, $-9.8t + 49 = 0$, $t = 5$ s

   (b) $s(5) = -4.9(5)^2 + 49(5) + 150 = 272.5$ m

   (c) the projectile reaches its starting point when $s(t) = 150$, $-4.9t^2 + 49t + 150 = 150$, $-4.9t(t - 10) = 0$, $t = 10$ s

   (d) $v(10) = -9.8(10) + 49 = -49$ m/s

   (e) $s(t) = 0$ when the projectile hits the ground, $-4.9t^2 + 49t + 150 = 0$ when (use the quadratic formula) $t \approx 12.46$ s

   (f) $v(12.46) = -9.8(12.46) + 49 \approx -73.1$, the speed at impact is about 73.1 m/s

43. If $g = 32$ ft/s^2, $s_0 = 7$ and v_0 is unknown, then $s(t) = 7 + v_0 t - 16t^2$ and $v(t) = v_0 - 32t$; $s = s_{\max}$ when $v = 0$, or $t = v_0/32$; and $s_{\max} = 208$ yields $208 = s(v_0/32) = 7 + v_0(v_0/32) - 16(v_0/32)^2 = 7 + v_0^2/64$, so $v_0 = 8\sqrt{201} \approx 113.42$ ft/s.

EXERCISE SET 5.8

1. **(a)** $\dfrac{1}{2}\displaystyle\int_1^5 u^3\, du$ **(b)** $\dfrac{3}{2}\displaystyle\int_9^{25} \sqrt{u}\, du$

    (c) $\dfrac{1}{\pi}\displaystyle\int_{-\pi/2}^{\pi/2} \cos u\, du$ **(d)** $\displaystyle\int_1^2 (u+1)u^5\, du$

3. $u = 2x + 1$, $\dfrac{1}{2}\displaystyle\int_1^3 u^3\, du = \dfrac{1}{8}u^4 \Big]_1^3 = 10$ or $\dfrac{1}{8}(2x+1)^4 \Big]_0^1 = 10$

5. $u = 2x - 1$, $\dfrac{1}{2}\displaystyle\int_{-1}^1 u^3\, du = 0$, because u^3 is odd on $[-1, 1]$.

7. $u = 1 + x$, $\displaystyle\int_1^9 (u-1)u^{1/2}\, du = \int_1^9 (u^{3/2} - u^{1/2})\, du = \dfrac{2}{5}u^{5/2} - \dfrac{2}{3}u^{3/2} \Big]_1^9 = 1192/15,$

   or $\dfrac{2}{5}(1+x)^{5/2} - \dfrac{2}{3}(1+x)^{3/2} \Big]_0^8 = 1192/15$

9. $u = x/2$, $8\displaystyle\int_0^{\pi/4} \sin u\, du = -8\cos u \Big]_0^{\pi/4} = 8 - 4\sqrt{2}$, or $-8\cos(x/2) \Big]_0^{\pi/2} = 8 - 4\sqrt{2}$

11. $u = x^2 + 2$, $\dfrac{1}{2}\displaystyle\int_6^3 u^{-3}\, du = -\dfrac{1}{4u^2} \Big]_6^3 = -1/48$, or $-\dfrac{1}{4}\dfrac{1}{(x^2+2)^2} \Big]_{-2}^{-1} = -1/48$

13. $\dfrac{1}{3}\displaystyle\int_{-5}^5 \sqrt{25 - u^2}\, du = \dfrac{1}{3}\left[\dfrac{1}{2}\pi(5)^2\right] = \dfrac{25}{6}\pi$

15. $-\dfrac{1}{2} \displaystyle\int_1^0 \sqrt{1-u^2}\, du = \dfrac{1}{2}\displaystyle\int_0^1 \sqrt{1-u^2}\, du = \dfrac{1}{2}\cdot\dfrac{1}{4}[\pi(1)^2] = \pi/8$

17. $\displaystyle\int_0^1 \sin \pi x\, dx = -\dfrac{1}{\pi}\cos \pi x \Big]_0^1 = -\dfrac{1}{\pi}(-1-1) = 2/\pi$

19. $\displaystyle\int_{-1}^1 \dfrac{9}{(x+2)^2}\, dx = -9(x+2)^{-1}\Big]_{-1}^1 = -9\left[\dfrac{1}{3}-1\right] = 6$

21. $u = 2x-1,\ \dfrac{1}{2}\displaystyle\int_1^9 \dfrac{1}{\sqrt{u}}\, du = \sqrt{u}\Big]_1^9 = 2$
  23. $\dfrac{2}{3}(x^3+9)^{1/2}\Big]_{-1}^1 = \dfrac{2}{3}(\sqrt{10}-2\sqrt{2})$

25. $u = x^2+4x+7,\ \dfrac{1}{2}\displaystyle\int_{12}^{28} u^{-1/2}du = u^{1/2}\Big]_{12}^{28} = \sqrt{28}-\sqrt{12} = 2(\sqrt{7}-\sqrt{3})$

27. $2\sin^2 x\Big]_0^{\pi/4} = 1$
  29. $\dfrac{5}{2}\sin(x^2)\Big]_0^{\sqrt{\pi}} = 0$

31. $u = 3\theta,\ \dfrac{1}{3}\displaystyle\int_{\pi/4}^{\pi/3} \sec^2 u\, du = \dfrac{1}{3}\tan u\Big]_{\pi/4}^{\pi/3} = (\sqrt{3}-1)/3$

33. $u = 4-3y,\ y = \dfrac{1}{3}(4-u),\ dy = -\dfrac{1}{3}du$

$-\dfrac{1}{27}\displaystyle\int_4^1 \dfrac{16-8u+u^2}{u^{1/2}}\, du = \dfrac{1}{27}\displaystyle\int_1^4 (16u^{-1/2}-8u^{1/2}+u^{3/2})\, du$

$= \dfrac{1}{27}\left[32u^{1/2}-\dfrac{16}{3}u^{3/2}+\dfrac{2}{5}u^{5/2}\right]_1^4 = 106/405$

35. (b) $\displaystyle\int_0^{\pi/6} \sin^4 x(1-\sin^2 x)\cos x\, dx = \left(\dfrac{1}{5}\sin^5 x - \dfrac{1}{7}\sin^7 x\right)\Big|_0^{\pi/6} = \dfrac{1}{160}-\dfrac{1}{896} = \dfrac{23}{4480}$

37. (a) $u = 3x+1,\ \dfrac{1}{3}\displaystyle\int_1^4 f(u)\, du = 5/3$
  (b) $u = 3x,\ \dfrac{1}{3}\displaystyle\int_0^9 f(u)\, du = 5/3$

(c) $u = x^2,\ 1/2\displaystyle\int_4^0 f(u)\, du = -1/2\displaystyle\int_0^4 f(u)\, du = -1/2$

39. $\sin x = \cos(\pi/2 - x)$,

$\displaystyle\int_0^{\pi/2} \sin^n x\, dx = \displaystyle\int_0^{\pi/2} \cos^n(\pi/2-x)\, dx = -\displaystyle\int_{\pi/2}^0 \cos^n u\, du \quad (u = \pi/2 - x)$

$= \displaystyle\int_0^{\pi/2} \cos^n u\, du = \displaystyle\int_0^{\pi/2} \cos^n x\, dx \quad \text{(by replacing } u \text{ by } x)$

41. (a) $\displaystyle\int_0^1 \sin \pi x\, dx = 2/\pi$

43. (a) $I = -\displaystyle\int_a^0 \dfrac{f(a-u)}{f(a-u)+f(u)}\,du = \int_0^a \dfrac{f(a-u)+f(u)-f(u)}{f(a-u)+f(u)}\,du$

$ = \displaystyle\int_0^a du - \int_0^a \dfrac{f(u)}{f(a-u)+f(u)}\,du,\; I = a - I \text{ so } 2I = a,\, I = a/2$

(b) $3/2$ **(c)** $\pi/4$

45. (a) Let $u = -x$ then

$$\int_{-a}^a f(x)dx = -\int_a^{-a} f(-u)du = \int_{-a}^a f(-u)du = -\int_{-a}^a f(u)du$$

so, replacing u by x in the latter integral,

$$\int_{-a}^a f(x)dx = -\int_{-a}^a f(x)dx,\; 2\int_{-a}^a f(x)dx = 0,\; \int_{-a}^a f(x)dx = 0$$

The graph of f is symmetric about the origin so $\displaystyle\int_{-a}^0 f(x)dx$ is the negative of $\displaystyle\int_0^a f(x)dx$

thus $\displaystyle\int_{-a}^a f(x)dx = \int_{-a}^0 f(x)\,dx + \int_0^a f(x)dx = 0$

(b) $\displaystyle\int_{-a}^a f(x)dx = \int_{-a}^0 f(x)dx + \int_0^a f(x)dx$, let $u = -x$ in $\displaystyle\int_{-a}^0 f(x)dx$ to get

$$\int_{-a}^0 f(x)dx = -\int_a^0 f(-u)du = \int_0^a f(-u)du = \int_0^a f(u)du = \int_0^a f(x)dx$$

so $\displaystyle\int_{-a}^a f(x)dx = \int_0^a f(x)dx + \int_0^a f(x)dx = 2\int_0^a f(x)dx$

The graph of $f(x)$ is symmetric about the y-axis so there is as much signed area to the left of the y-axis as there is to the right.

REVIEW EXERCISE SET

3. $-\dfrac{1}{4x^2} + \dfrac{8}{3}x^{3/2} + C$ **5.** $-4\cos x + 2\sin x + C$

7. (a) $y(x) = 2\sqrt{x} - \dfrac{2}{3}x^{3/2} + C;\; y(1) = 0,\text{ so } C = -\dfrac{4}{3}, y(x) = 2\sqrt{x} - \dfrac{2}{3}x^{3/2} - \dfrac{4}{3}$

(b) $y(x) = \sin x - \dfrac{5}{2}x^2 + C, y(0) = 1 = C, y(x) = \sin x - \dfrac{5}{2}x^2 + 1$

9. (a) If $u = \sec x,\, du = \sec x \tan x\, dx,\, \displaystyle\int \sec^2 x \tan x\, dx = \int u\, du = u^2/2 + C_1 = (\sec^2 x)/2 + C_1$;

if $u = \tan x,\, du = \sec^2 x\, dx,\, \displaystyle\int \sec^2 x \tan x\, dx = \int u\, du = u^2/2 + C_2 = (\tan^2 x)/2 + C_2$.

(b) They are equal only if $\sec^2 x$ and $\tan^2 x$ differ by a constant, which is true.

11. $u = x^4 + 2, du = 4x^3\, dx,$

$\dfrac{1}{4}\displaystyle\int\left(\sqrt{u} - \dfrac{2}{\sqrt{u}}\right) + C = \dfrac{1}{6}u^{3/2} - 4\sqrt{u} + C = \dfrac{1}{6}(x^4 + 2)^{3/2} - 4\sqrt{x^4 + 2} + C$

13. $u = 5 + 2\sin 3x,\, du = 6\cos 3x\,dx;\, \displaystyle\int\dfrac{1}{6\sqrt{u}}du = \dfrac{1}{3}u^{1/2} + C = \dfrac{1}{3}\sqrt{5 + 2\sin 3x} + C$

15. $u = ax^3 + b$, $du = 3ax^2 dx$; $\displaystyle\int \frac{1}{3au^2} du = -\frac{1}{3au} + C = -\frac{1}{3a^2x^3 + 3ab} + C$

17. (a) $\displaystyle 1\cdot 2 + 2\cdot 3 + \cdots + n(n+1) = \sum_{k=1}^{n} k(k+1) = \sum_{k=1}^{n} k^2 + \sum_{k=1}^{n} k$

$$= \frac{1}{6}n(n+1)(2n+1) + \frac{1}{2}n(n+1) = \frac{1}{3}n(n+1)(n+2)$$

(b) $\displaystyle \sum_{k=1}^{n-1}\left(\frac{9}{n} - \frac{k}{n^2}\right) = \frac{9}{n}\sum_{k=1}^{n-1} 1 - \frac{1}{n^2}\sum_{k=1}^{n-1} k = \frac{9}{n}(n-1) - \frac{1}{n^2}\cdot\frac{1}{2}(n-1)(n) = \frac{17}{2}\left(\frac{n-1}{n}\right);$

$$\lim_{n\to+\infty} \frac{17}{2}\left(\frac{n-1}{n}\right) = \frac{17}{2}$$

(c) $\displaystyle \sum_{i=1}^{3}\left[\sum_{j=1}^{2} i + \sum_{j=1}^{2} j\right] = \sum_{i=1}^{3}\left[2i + \frac{1}{2}(2)(3)\right] = 2\sum_{i=1}^{3} i + \sum_{i=1}^{3} 3 = 2\cdot\frac{1}{2}(3)(4) + (3)(3) = 21$

19. For $1 \le k \le n$ the k-th L-shaped strip consists of the corner square, a strip above and a strip to the left for a combined area of $1 + (k-1) + (k-1) = 2k - 1$, so the total area is $\displaystyle\sum_{k=1}^{n}(2k-1) = n^2$.

21. left endpoints: $x_k^* = 1, 2, 3, 4$; $\displaystyle\sum_{k=1}^{4} f(x_k^*)\Delta x = (2 + 3 + 2 + 1)(1) = 8$

right endpoints: $x_k^* = 2, 3, 4, 5$; $\displaystyle\sum_{k=1}^{4} f(x_k^*)\Delta x = (3 + 2 + 1 + 2)(1) = 8$

23. $\displaystyle\lim_{n\to+\infty}\sum_{k=1}^{n}\left[4\frac{4k}{n} - \left(\frac{4k}{n}\right)^2\right]\frac{4}{n} = \lim_{n\to+\infty}\frac{64}{n^3}\sum_{k=1}^{n}(kn - k^2)$

$$= \lim_{n\to+\infty}\frac{64}{n^3}\left[\frac{n^2(n+1)}{2} - \frac{n(n+1)(2n+1)}{6}\right] = \lim_{n\to+\infty}\frac{64}{6n^3}[n^3 - n] = \frac{32}{3}$$

25. $0.7187714032, 0.6687714032, 0.6928353604$

27. $1.983523538, 1.983523538, 2.008248408$

29. (a) $\dfrac{1}{2} + \dfrac{1}{4} = \dfrac{3}{4}$ **(b)** $-1 - \dfrac{1}{2} = -\dfrac{3}{2}$

(c) $5\left(-1 - \dfrac{3}{4}\right) = -\dfrac{35}{4}$ **(d)** -2

(e) not enough information   **(f)** not enough information

31. (a) $\displaystyle\int_{-1}^{1} dx + \int_{-1}^{1}\sqrt{1 - x^2}\, dx = 2(1) + \pi(1)^2/2 = 2 + \pi/2$

(b) $\dfrac{1}{3}(x^2 + 1)^{3/2}\Big]_0^3 - \pi(3)^2/4 = \dfrac{1}{3}(10^{3/2} - 1) - 9\pi/4$

(c) $u = x^2$, $du = 2x\,dx$; $\dfrac{1}{2}\displaystyle\int_0^1 \sqrt{1 - u^2}\, du = \dfrac{1}{2}\pi(1)^2/4 = \pi/8$

33. The rectangle with vertices $(0,0)$, $(\pi,0)$, $(\pi,1)$ and $(0,1)$ has area π and is much too large; so is the triangle with vertices $(0,0)$, $(\pi,0)$ and $(\pi,1)$ which has area $\pi/2$; $1-\pi$ is negative; so the answer is $35\pi/128$.

35. **(a)** $\displaystyle\int_a^b \sum_{k=1}^n f_k(x)dx = \sum_{k=1}^n \int_a^b f_k(x)dx$

(b) yes; substitute $c_k f_k(x)$ for $f_k(x)$ in part (a), and then use $\displaystyle\int_a^b c_k f_k(x)dx = c_k \int_a^b f_k(x)dx$ from Theorem 5.5.4

37. $\displaystyle\int_1^9 \sqrt{x}\,dx = \frac{2}{3}x^{3/2}\Big]_1^9 = \frac{2}{3}(27-1) = 52/3$ **39.** $\displaystyle\left(\frac{1}{3}x^3 - 2x^2 + 7x\right)\Big]_{-3}^0 = 48$

41. $\displaystyle\int_1^3 x^{-2}dx = -\frac{1}{x}\Big]_1^3 = 2/3$ **43.** $\displaystyle\left(\frac{1}{2}x^2 - \sec x\right)\Big]_0^1 = 3/2 - \sec(1)$

45. $\displaystyle\int_0^{3/2}(3-2x)dx + \int_{3/2}^2 (2x-3)dx = (3x-x^2)\Big]_0^{3/2} + (x^2-3x)\Big]_{3/2}^2 = 9/4 + 1/4 = 5/2$

47. $\displaystyle A = \int_1^2 (-x^2 + 3x - 2)dx = \left(-\frac{1}{3}x^3 + \frac{3}{2}x^2 - 2x\right)\Big]_1^2 = 1/6$

49. **(a)** $x^3 + 1$ **(b)** $\displaystyle F(x) = \left(\frac{1}{4}t^4 + t\right)\Big]_1^x = \frac{1}{4}x^4 + x - \frac{5}{4}; F'(x) = x^3 + 1$

51. $\dfrac{1}{x^4 + 5}$ **53.** $|x - 1|$

57. $\displaystyle F'(x) = \frac{1}{1+x^2} + \frac{1}{1+(1/x)^2}(-1/x^2) = 0$ so F is constant on $(0, +\infty)$.

59. **(a)** The domain is $(-\infty, +\infty)$; $F(x)$ is 0 if $x = 1$, positive if $x > 1$, and negative if $x < 1$, because the integrand is positive, so the sign of the integral depends on the orientation (forwards or backwards).

(b) The domain is $[-2, 2]$; $F(x)$ is 0 if $x = -1$, positive if $-1 < x \le 2$, and negative if $-2 \le x < -1$; same reasons as in Part (a).

61. **(a)** $\displaystyle f_{ave} = \frac{1}{3}\int_0^3 x^{1/2}dx = 2\sqrt{3}/3; \sqrt{x^*} = 2\sqrt{3}/3, x^* = \frac{4}{3}$

(b) $\displaystyle f_{ave} = \frac{1}{2}\int_0^2 (2x - x^2)\,dx = \frac{1}{2}\left(x^2 - \frac{1}{3}x^3\right)\Big]_0^2 = \frac{2}{3}; 2x^* - (x^*)^2 = \frac{2}{3}, x^* = 1 \pm 1/\sqrt{3}$

63. If the acceleration $a = $ const, then $v(t) = at + v_0$, $s(t) = \frac{1}{2}at^2 + v_0 t + s_0$.

65. $\displaystyle s(t) = \int (t^3 - 2t^2 + 1)dt = \frac{1}{4}t^4 - \frac{2}{3}t^3 + t + C$,

$s(0) = \frac{1}{4}(0)^4 - \frac{2}{3}(0)^3 + 0 + C = 1, C = 1, s(t) = \frac{1}{4}t^4 - \frac{2}{3}t^3 + t + 1$

67. $\displaystyle s(t) = \int (2t - 3)dt = t^2 - 3t + C, s(1) = (1)^2 - 3(1) + C = 5, C = 7, s(t) = t^2 - 3t + 7$

69. displacement $= s(6) - s(0) = \int_0^6 (2t - 4)dt = (t^2 - 4t)\Big]_0^6 = 12$ m

distance $= \int_0^6 |2t - 4|dt = \int_0^2 (4 - 2t)dt + \int_2^6 (2t - 4)dt = (4t - t^2)\Big]_0^2 + (t^2 - 4t)\Big]_2^6 = 20$ m

71. displacement $= \int_1^3 \left(\frac{1}{2} - \frac{1}{t^2}\right) dt = 1/3$ m

distance $= \int_1^3 |v(t)|dt = -\int_1^{\sqrt{2}} v(t)dt + \int_{\sqrt{2}}^3 v(t)dt = 10/3 - 2\sqrt{2}$ m

73. $\qquad v(t) = -2t + 3$

displacement $= \int_1^4 (-2t + 3)dt = -6$ m

distance $= \int_1^4 |-2t + 3|dt = \int_1^{3/2} (-2t + 3)dt + \int_{3/2}^4 (2t - 3)dt = 13/2$ m

75. $A = A_1 + A_2 = \int_0^1 (1 - x^2)dx + \int_1^3 (x^2 - 1)dx = 2/3 + 20/3 = 22/3$

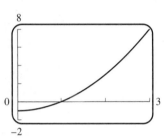

77. $A = A_1 + A_2 = \int_0^1 (x^3 - 4x^2 + 3x)\, dx - \int_1^3 (x^3 - 4x^2 + 3x)\, dx$

$= \dfrac{5}{12} + \dfrac{8}{3} = \dfrac{37}{12}$

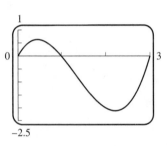

79. Take $t = 0$ when deceleration begins, then $a = -10$ so $v = -10t + C_1$, but $v = 88$ when $t = 0$ which gives $C_1 = 88$ thus $v = -10t + 88$, $t \geq 0$

(a) $v = 45$ mi/h $= 66$ ft/s, $66 = -10t + 88$, $t = 2.2$ s

(b) $v = 0$ (the car is stopped) when $t = 8.8$ s

$s = \int v\, dt = \int (-10t + 88)dt = -5t^2 + 88t + C_2$, and taking $s = 0$ when $t = 0$, $C_2 = 0$ so $s = -5t^2 + 88t$. At $t = 8.8$, $s = 387.2$. The car travels 387.2 ft before coming to a stop.

81. From the given data, $s(t) = -16t^2 + v_0 t + s_0$. The equation $s(t) = s_0$ has two solutions, $t = 0$ and $t = \dfrac{1}{16}v_0$. We want the average speed on the interval $[0, v_0/16]$, which is given by

speed$_{\text{ave}} = \dfrac{1}{v_0/16} \int_0^{v_0/16} |v(t)|\, dt = 2\dfrac{1}{v_0/16} \int_0^{v_0/32} (v_0 - 32t)\, dt = \dfrac{1}{2}v_0$

83. $u = 2x + 1$, $\dfrac{1}{2} \int_1^3 u^4 du = \dfrac{1}{10}u^5 \Big]_1^3 = 121/5$, or $\dfrac{1}{10}(2x + 1)^5 \Big]_0^1 = 121/5$

85. $\dfrac{2}{3}(3x+1)^{1/2}\Big]_0^1 = 2/3$

87. $\dfrac{1}{3\pi}\sin^3 \pi x\Big]_0^1 = 0$

89. Differentiate: $f(x) = -\dfrac{8}{(x+3)^2}$, so

$2 + \displaystyle\int_a^x f(t)dt = 2 + \int_a^x -\dfrac{8}{(t+3)^2}dt = 2 + \dfrac{8}{t+3}\Big]_a^x = 2 + \dfrac{8}{x+3} - \dfrac{8}{a+3}, 2 = \dfrac{8}{a+3}, 2a+6 = 8, a = 1.$

CHAPTER 6

Applications of the Definite Integral in Geometry, Science, and Engineering

EXERCISE SET 6.1

1. $A = \displaystyle\int_{-1}^{2} (x^2 + 1 - x)dx = (x^3/3 + x - x^2/2)\Big]_{-1}^{2} = 9/2$

3. $A = \displaystyle\int_{1}^{2} (y - 1/y^2)dy = (y^2/2 + 1/y)\Big]_{1}^{2} = 1$

5. **(a)** $A = \displaystyle\int_{0}^{4} (4x - x^2)dx = 32/3$ **(b)** $A = \displaystyle\int_{0}^{16} (\sqrt{y} - y/4)dy = 32/3$

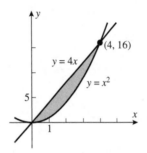

7. $A = \displaystyle\int_{1/4}^{1} (\sqrt{x} - x^2)dx = 49/192$

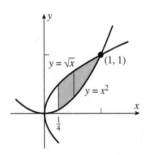

9. $A = \displaystyle\int_{\pi/4}^{\pi/2} (0 - \cos 2x)dx$

$= -\displaystyle\int_{\pi/4}^{\pi/2} \cos 2x\, dx = 1/2$

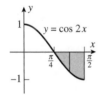

11. $A = \displaystyle\int_{\pi/4}^{3\pi/4} \sin y\, dy = \sqrt{2}$

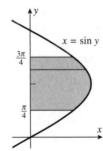

13. $y = 2 + |x - 1| = \begin{cases} 3 - x, & x \leq 1 \\ 1 + x, & x \geq 1 \end{cases}$,

$$A = \int_{-5}^{1} \left[\left(-\frac{1}{5}x + 7 \right) - (3 - x) \right] dx$$

$$+ \int_{1}^{5} \left[\left(-\frac{1}{5}x + 7 \right) - (1 + x) \right] dx$$

$$= \int_{-5}^{1} \left(\frac{4}{5}x + 4 \right) dx + \int_{1}^{5} \left(6 - \frac{6}{5}x \right) dx$$

$$= 72/5 + 48/5 = 24$$

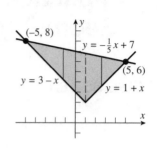

15. $A = \int_{0}^{1} (x^3 - 4x^2 + 3x) dx$

$$+ \int_{1}^{3} [-(x^3 - 4x^2 + 3x)] dx$$

$$= 5/12 + 32/12 = 37/12$$

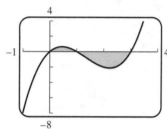

17. From the symmetry of the region

$$A = 2 \int_{\pi/4}^{5\pi/4} (\sin x - \cos x) dx = 4\sqrt{2}$$

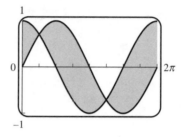

19. $A = \int_{-1}^{0} (y^3 - y) dy + \int_{0}^{1} -(y^3 - y) dy$

$$= 1/2$$

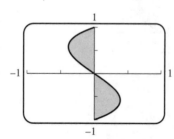

21. Solve $3 - 2x = x^6 + 2x^5 - 3x^4 + x^2$ to find the real roots $x = -3, 1$; from a plot it is seen that the line is above the polynomial when $-3 < x < 1$, so $A = \int_{-3}^{1} (3 - 2x - (x^6 + 2x^5 - 3x^4 + x^2)) \, dx = 9152/105$

23. $\displaystyle\int_0^k 2\sqrt{y}\,dy = \int_k^9 2\sqrt{y}\,dy$

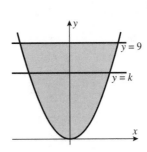

$$\int_0^k y^{1/2}\,dy = \int_k^9 y^{1/2}\,dy$$

$$\frac{2}{3}k^{3/2} = \frac{2}{3}(27 - k^{3/2})$$

$$k^{3/2} = 27/2$$

$$k = (27/2)^{2/3} = 9/\sqrt[3]{4}$$

25. **(a)** $\displaystyle A = \int_0^2 (2x - x^2)\,dx = 4/3$

(b) $y = mx$ intersects $y = 2x - x^2$ where $mx = 2x - x^2, x^2 + (m-2)x = 0, x(x + m - 2) = 0$ so $x = 0$ or $x = 2 - m$. The area below the curve and above the line is

$$\int_0^{2-m} (2x - x^2 - mx)\,dx = \int_0^{2-m} [(2-m)x - x^2]\,dx = \left[\frac{1}{2}(2-m)x^2 - \frac{1}{3}x^3\right]_0^{2-m} = \frac{1}{6}(2-m)^3$$

so $(2-m)^3/6 = (1/2)(4/3) = 2/3, (2-m)^3 = 4, m = 2 - \sqrt[3]{4}$.

27. The curves intersect at $x = 0$ and, by Newton's Method, at $x \approx 2.595739080 = b$, so

$$A \approx \int_0^b (\sin x - 0.2x)\,dx = -\left[\cos x + 0.1x^2\right]_0^b \approx 1.180898334$$

29. The x-coordinates of the points of intersection are $a \approx -0.423028$ and $b \approx 1.725171$; the area is

$$\int_a^b (2\sin x - x^2 + 1)\,dx \approx 2.542696.$$

31. $\displaystyle\int_0^{60} (v_2(t) - v_1(t))\,dt = s_2(60) - s_2(0) - (s_1(60) - s_1(0))$, but they are even at time $t = 60$, so $s_2(60) = s_1(60)$. Consequently the integral gives the difference $s_1(0) - s_2(0)$ of their starting points in meters.

33. **(a)** It gives the area of the region that is between f and g when $f(x) > g(x)$ <u>minus</u> the area of the region between f and g when $f(x) < g(x)$, for $a \le x \le b$.

(b) It gives the area of the region that is between f and g for $a \le x \le b$.

35. Solve $x^{1/2} + y^{1/2} = a^{1/2}$ for y to get

$$y = (a^{1/2} - x^{1/2})^2 = a - 2a^{1/2}x^{1/2} + x$$

$$A = \int_0^a (a - 2a^{1/2}x^{1/2} + x)\,dx = a^2/6$$

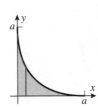

37. Let A be the area between the curve and the x-axis and A_R the area of the rectangle, then

$$A = \int_0^b kx^m\,dx = \frac{k}{m+1}x^{m+1}\Big]_0^b = \frac{kb^{m+1}}{m+1}, A_R = b(kb^m) = kb^{m+1}, \text{ so } A/A_R = 1/(m+1).$$

EXERCISE SET 6.2

1. $V = \pi \int_{-1}^{3} (3 - x)dx = 8\pi$

3. $V = \pi \int_{0}^{2} \frac{1}{4}(3 - y)^2 dy = 13\pi/6$

5. $V = \int_{0}^{2} x^4 dx = 32/5$

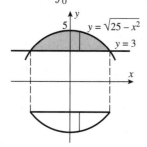

7. $V = \pi \int_{\pi/4}^{\pi/2} \cos x \, dx = (1 - \sqrt{2}/2)\pi$

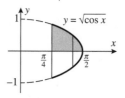

9. $V = \pi \int_{-4}^{4} [(25 - x^2) - 9]dx$

$= 2\pi \int_{0}^{4} (16 - x^2)dx = 256\pi/3$

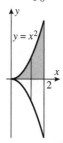

11. $V = \pi \int_{0}^{4} [(4x)^2 - (x^2)^2]dx$

$= \pi \int_{0}^{4} (16x^2 - x^4)dx = 2048\pi/15$

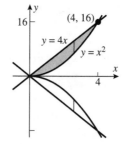

13. $V = \int_{0}^{1} \left(y^{1/3}\right)^2 dy = \frac{3}{5}$

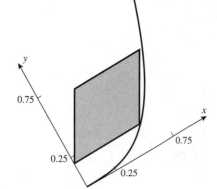

15. $V = \pi \int_{-1}^{3} (1 + y)dy = 8\pi$

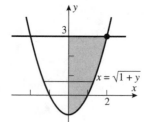

17. $V = \pi \int_{\pi/4}^{3\pi/4} \csc^2 y \, dy = 2\pi$

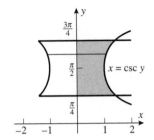

19. $V = \pi \int_{-1}^{2} [(y+2)^2 - y^4] dy = 72\pi/5$

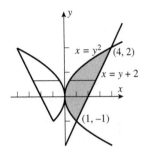

21. $V = \pi \int_{-a}^{a} \frac{b^2}{a^2}(a^2 - x^2) dx = 4\pi ab^2/3$

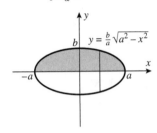

23. $V = \pi \int_{-1}^{0} (x+1) dx$

$\qquad + \pi \int_{0}^{1} [(x+1) - 2x] dx$

$\qquad = \pi/2 + \pi/2 = \pi$

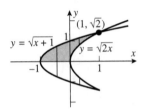

25. Partition the interval $[a, b]$ with $a = x_0 < x_1 < x_2 < \ldots < x_{n-1} < x_n = b$. Let x_k^* be an arbitrary point of $[x_{k-1}, x_k]$. The disk in question is obtained by revolving about the line $y = k$ the rectangle for which $x_{k-1} < x < x_k$ and y lies between $y = k$ and $y = f(x)$; the volume of this disk is $\Delta V_k = \pi(f(x_k^*) - k)^2 \Delta x_k$, and the total volume is given by $V = \pi \int_{a}^{b} (f(x) - k)^2 \, dx$.

27. (a) Intuitively, it seems that a line segment revolved about a line perpendicular to the line segment will generate a larger area, the farther it is from the line. This is because the average point on the line segment farther from the line will be revolved through a circle with a greater radius.

Consider the line segment which connects a point (x, y) on the curve $y = \sqrt{3 - x}$ to the point $(x, 0)$ beneath it. If this line segment is revolved around the x-axis we generate an area πy^2. If on the other hand the segment is revolved around the line $y = 2$ then the area of the resulting (infinitely thin) washer is $\pi[2^2 - (2-y)^2]$. So the question can be reduced to asking whether $y^2 \geq [2^2 - (2-y)^2]$, $y^2 \geq 4y - y^2$, or $y \geq 2$. In the present case the curve $y = \sqrt{3 - x}$ always satisfies $y \leq 2$, so V_2 has the larger volume.

(b) The volume of the solid generated by revolving the area around the x-axis is

$$V_1 = \pi \int_{-1}^{3} (3 - x) \, dx = 8\pi,$$ and the volume generated by revolving the area around the line

$y = 2$ is $V_2 = \pi \int_{-1}^{3} [2^2 - (2 - \sqrt{3 - x})^2] \, dx = \frac{40}{3}\pi$

29. $V = \pi \int_0^3 (9 - y^2)^2 dy$

$$= \pi \int_0^3 (81 - 18y^2 + y^4) dy$$

$$= 648\pi/5$$

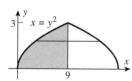

31. $V = \pi \int_0^1 [(\sqrt{x} + 1)^2 - (x + 1)^2] dx$

$$= \pi \int_0^1 (2\sqrt{x} - x - x^2) dx = \pi/2$$

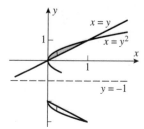

33. $A(x) = \pi(x^2/4)^2 = \pi x^4/16$,

$$V = \int_0^{20} (\pi x^4/16) dx = 40,000\pi \text{ ft}^3$$

35. $V = \int_0^1 (x - x^2)^2 dx$

$$= \int_0^1 (x^2 - 2x^3 + x^4) dx = 1/30$$

37. On the upper half of the circle, $y = \sqrt{1 - x^2}$, so:

(a) $A(x)$ is the area of a semicircle of radius y, so

$$A(x) = \pi y^2/2 = \pi(1 - x^2)/2; \quad V = \frac{\pi}{2} \int_{-1}^1 (1 - x^2)\, dx = \pi \int_0^1 (1 - x^2)\, dx = 2\pi/3$$

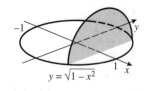

(b) $A(x)$ is the area of a square of side $2y$, so

$$A(x) = 4y^2 = 4(1 - x^2); \quad V = 4 \int_{-1}^1 (1 - x^2)\, dx = 8 \int_0^1 (1 - x^2)\, dx = 16/3$$

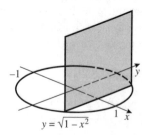

(c) $A(x)$ is the area of an equilateral triangle with sides $2y$, so

$$A(x) = \frac{\sqrt{3}}{4}(2y)^2 = \sqrt{3}y^2 = \sqrt{3}(1-x^2);$$

$$V = \int_{-1}^{1} \sqrt{3}(1-x^2)\,dx = 2\sqrt{3}\int_{0}^{1}(1-x^2)\,dx = 4\sqrt{3}/3$$

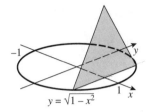

39. The two curves cross at $x = b \approx 1.403288534$, so

$$V = \pi \int_{0}^{b} ((2x/\pi)^2 - \sin^{16} x)\,dx + \pi \int_{b}^{\pi/2} (\sin^{16} x - (2x/\pi)^2)\,dx \approx 0.710172176.$$

41. (a) $V = \pi \displaystyle\int_{r-h}^{r} (r^2 - y^2)\,dy = \pi(rh^2 - h^3/3) = \frac{1}{3}\pi h^2(3r - h)$

(b) By the Pythagorean Theorem,

$$r^2 = (r-h)^2 + \rho^2,\ 2hr = h^2 + \rho^2;\ \text{from Part (a)},$$

$$V = \frac{\pi h}{3}(3hr - h^2) = \frac{\pi h}{3}\left(\frac{3}{2}(h^2 + \rho^2) - h^2\right)$$

$$= \frac{1}{6}\pi h(h^2 + 3\rho^2)$$

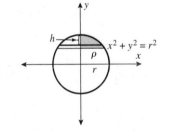

43. (b) $\Delta x = \dfrac{5}{10} = 0.5;\ \{y_0, y_1, \cdots, y_{10}\} = \{0, 2.00, 2.45, 2.45, 2.00, 1.46, 1.26, 1.25, 1.25, 1.25, 1.25\};$

$$\text{left} = \pi \sum_{i=0}^{9} \left(\frac{y_i}{2}\right)^2 \Delta x \approx 11.157;$$

$$\text{right} = \pi \sum_{i=1}^{10} \left(\frac{y_i}{2}\right)^2 \Delta x \approx 11.771;\ V \approx \text{average} = 11.464\ \text{cm}^3$$

45. (a)   **(b)**

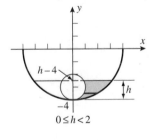

If the cherry is partially submerged then $0 \le h < 2$ as shown in Figure (a); if it is totally submerged then $2 \le h \le 4$ as shown in Figure (b). The radius of the glass is 4 cm and that of the cherry is 1 cm so points on the sections shown in the figures satisfy the equations $x^2 + y^2 = 16$ and $x^2 + (y+3)^2 = 1$. We will find the volumes of the solids that are generated when the shaded regions are revolved about the y-axis.

For $0 \leq h < 2$,

$$V = \pi \int_{-4}^{h-4} [(16 - y^2) - (1 - (y+3)^2)]dy = 6\pi \int_{-4}^{h-4} (y+4)dy = 3\pi h^2;$$

for $2 \leq h \leq 4$,

$$V = \pi \int_{-4}^{-2} [(16 - y^2) - (1 - (y+3)^2)]dy + \pi \int_{-2}^{h-4} (16 - y^2)dy$$

$$= 6\pi \int_{-4}^{-2} (y+4)dy + \pi \int_{-2}^{h-4} (16 - y^2)dy = 12\pi + \frac{1}{3}\pi(12h^2 - h^3 - 40)$$

$$= \frac{1}{3}\pi(12h^2 - h^3 - 4)$$

so

$$V = \begin{cases} 3\pi h^2 & \text{if } 0 \leq h < 2 \\ \dfrac{1}{3}\pi(12h^2 - h^3 - 4) & \text{if } 2 \leq h \leq 4 \end{cases}$$

47. $\tan\theta = h/x$ so $h = x\tan\theta$,

$A(y) = \dfrac{1}{2}hx = \dfrac{1}{2}x^2\tan\theta = \dfrac{1}{2}(r^2 - y^2)\tan\theta$

because $x^2 = r^2 - y^2$,

$V = \dfrac{1}{2}\tan\theta \int_{-r}^{r} (r^2 - y^2)dy$

$= \tan\theta \int_{0}^{r} (r^2 - y^2)dy = \dfrac{2}{3}r^3\tan\theta$

49. Each cross section perpendicular to the y-axis is a square so

$A(y) = x^2 = r^2 - y^2$,

$\dfrac{1}{8}V = \int_{0}^{r} (r^2 - y^2)dy$

$V = 8(2r^3/3) = 16r^3/3$

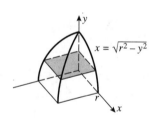

EXERCISE SET 6.3

1. $V = \int_{1}^{2} 2\pi x(x^2)dx = 2\pi \int_{1}^{2} x^3 dx = 15\pi/2$

3. $V = \int_{0}^{1} 2\pi y(2y - 2y^2)dy = 4\pi \int_{0}^{1} (y^2 - y^3)dy = \pi/3$

5. $V = \int_0^1 2\pi(x)(x^3)dx$

$= 2\pi \int_0^1 x^4 dx = 2\pi/5$

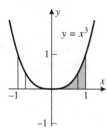

7. $V = \int_1^3 2\pi x(1/x)dx = 2\pi \int_1^3 dx = 4\pi$

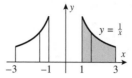

9. $V = \int_1^2 2\pi x[(2x - 1) - (-2x + 3)]dx$

$= 8\pi \int_1^2 (x^2 - x)dx = 20\pi/3$

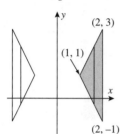

11. $V = \int_0^1 2\pi y^3 dy = \pi/2$

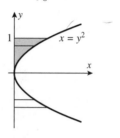

13. $V = \int_0^1 2\pi y(1 - \sqrt{y})dy$

$= 2\pi \int_0^1 (y - y^{3/2})dy = \pi/5$

15. $V = 2\pi \int_0^\pi x \sin x \, dx = 2\pi(-x \cos x + \sin x)\Big]_0^\pi = 2\pi^2$

17. The volume is given by $2\pi \int_0^k x \sin x \, dx = 2\pi(\sin k - k \cos k) = 8$; solve for k to get $k = 1.736796$.

19. **(a)** $V = \int_0^1 2\pi x(x^3 - 3x^2 + 2x)dx = 7\pi/30$

(b) much easier; the method of slicing would require that x be expressed in terms of y.

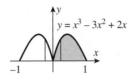

21. $V = \int_1^2 2\pi(x+1)(1/x^3)dx$

$= 2\pi \int_1^2 (x^{-2} + x^{-3})dx = 7\pi/4$

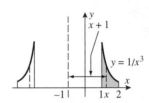

23. $x = \dfrac{h}{r}(r-y)$ is an equation of the line

through $(0,r)$ and $(h,0)$ so

$V = \int_0^r 2\pi y \left[\dfrac{h}{r}(r-y)\right] dy$

$= \dfrac{2\pi h}{r} \int_0^r (ry - y^2)dy = \pi r^2 h/3$

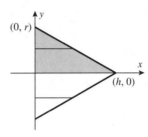

25. Let the sphere have radius R, the hole radius r. By the Pythagorean Theorem, $r^2 + (L/2)^2 = R^2$. Use cylindrical shells to calculate the volume of the solid obtained by rotating about the y-axis the region $r < x < R$, $-\sqrt{R^2 - x^2} < y < \sqrt{R^2 - x^2}$:

$V = \int_r^R (2\pi x)2\sqrt{R^2 - x^2}\,dx = -\dfrac{4}{3}\pi(R^2 - x^2)^{3/2}\bigg]_r^R = \dfrac{4}{3}\pi(L/2)^3,$

so the volume is independent of R.

27. $V_x = \pi \int_{1/2}^b \dfrac{1}{x^2}dx = \pi(2 - 1/b)$, $V_y = 2\pi \int_{1/2}^b dx = \pi(2b - 1)$;

$V_x = V_y$ if $2 - 1/b = 2b - 1$, $2b^2 - 3b + 1 = 0$, solve to get $b = 1/2$ (reject) or $b = 1$.

EXERCISE SET 6.4

1. (a) $\dfrac{dy}{dx} = 2$, $L = \int_1^2 \sqrt{1 + 4}dx = \sqrt{5}$

(b) $\dfrac{dx}{dy} = \dfrac{1}{2}$, $L = \int_2^4 \sqrt{1 + 1/4}\,dy = 2\sqrt{5}/2 = \sqrt{5}$

3. $f'(x) = \dfrac{9}{2}x^{1/2}$, $1 + [f'(x)]^2 = 1 + \dfrac{81}{4}x$,

$L = \int_0^1 \sqrt{1 + 81x/4}\,dx = \dfrac{8}{243}\left(1 + \dfrac{81}{4}x\right)^{3/2}\bigg]_0^1 = (85\sqrt{85} - 8)/243$

5. $\dfrac{dy}{dx} = \dfrac{2}{3}x^{-1/3},\ 1 + \left(\dfrac{dy}{dx}\right)^2 = 1 + \dfrac{4}{9}x^{-2/3} = \dfrac{9x^{2/3} + 4}{9x^{2/3}},$

$$L = \int_1^8 \dfrac{\sqrt{9x^{2/3} + 4}}{3x^{1/3}}\,dx = \dfrac{1}{18}\int_{13}^{40} u^{1/2}du,\ \ u = 9x^{2/3} + 4$$

$$= \dfrac{1}{27}u^{3/2}\Big]_{13}^{40} = \dfrac{1}{27}(40\sqrt{40} - 13\sqrt{13}) = \dfrac{1}{27}(80\sqrt{10} - 13\sqrt{13})$$

or (alternate solution)

$$x = y^{3/2},\ \dfrac{dx}{dy} = \dfrac{3}{2}y^{1/2},\ 1 + \left(\dfrac{dx}{dy}\right)^2 = 1 + \dfrac{9}{4}y = \dfrac{4 + 9y}{4},$$

$$L = \dfrac{1}{2}\int_1^4 \sqrt{4 + 9y}\,dy = \dfrac{1}{18}\int_{13}^{40} u^{1/2}du = \dfrac{1}{27}(80\sqrt{10} - 13\sqrt{13})$$

7. $x = g(y) = \dfrac{1}{24}y^3 + 2y^{-1},\ g'(y) = \dfrac{1}{8}y^2 - 2y^{-2},$

$$1 + [g'(y)]^2 = 1 + \left(\dfrac{1}{64}y^4 - \dfrac{1}{2} + 4y^{-4}\right) = \dfrac{1}{64}y^4 + \dfrac{1}{2} + 4y^{-4} = \left(\dfrac{1}{8}y^2 + 2y^{-2}\right)^2,$$

$$L = \int_2^4 \left(\dfrac{1}{8}y^2 + 2y^{-2}\right)dy = 17/6$$

9. $(dx/dt)^2 + (dy/dt)^2 = (t^2)^2 + (t)^2 = t^2(t^2 + 1),\ L = \int_0^1 t(t^2 + 1)^{1/2}dt = (2\sqrt{2} - 1)/3$

11. $(dx/dt)^2 + (dy/dt)^2 = (-2\sin 2t)^2 + (2\cos 2t)^2 = 4,\ L = \int_0^{\pi/2} 2\,dt = \pi$

13. (a)

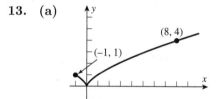

(b) dy/dx does not exist at $x = 0$.

(c) $x = g(y) = y^{3/2},\ g'(y) = \dfrac{3}{2}\,y^{1/2},$

$$L = \int_0^1 \sqrt{1 + 9y/4}\,dy \quad \text{(portion for } -1 \le x \le 0\text{)}$$

$$+ \int_0^4 \sqrt{1 + 9y/4}\,dy \quad \text{(portion for } 0 \le x \le 8\text{)}$$

$$= \dfrac{8}{27}\left(\dfrac{13}{8}\sqrt{13} - 1\right) + \dfrac{8}{27}(10\sqrt{10} - 1) = (13\sqrt{13} + 80\sqrt{10} - 16)/27$$

15. (a) The function $y = f(x) = x^2$ is inverse to the
function $x = g(y) = \sqrt{y} : f(g(y)) = y$ for $1/4 \le y \le 4$,
and $g(f(x)) = x$ for $1/2 \le x \le 2$. Geometrically this means
that the graphs of $y = f(x)$ and $x = g(y)$ are symmetric to
each other with respect to the line $y = x$ and hence have the
same arc length.

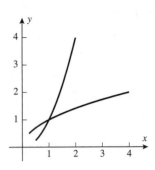

(b) $L_1 = \displaystyle\int_{1/2}^{2} \sqrt{1 + (2x)^2}\, dx \quad$ and $\quad L_2 = \displaystyle\int_{1/4}^{4} \sqrt{1 + \left(\dfrac{1}{2\sqrt{x}}\right)^2}\, dx$

Make the change of variables $x = \sqrt{y}$ in the first integral to obtain

$$L_1 = \int_{1/4}^{4} \sqrt{1 + (2\sqrt{y})^2}\,\frac{1}{2\sqrt{y}}\, dy = \int_{1/4}^{4} \sqrt{\left(\frac{1}{2\sqrt{y}}\right)^2 + 1}\, dy = L_2$$

(c) $L_1 = \displaystyle\int_{1/2}^{2} \sqrt{1 + (2y)^2}\, dy, \qquad L_2 = \displaystyle\int_{1/4}^{4} \sqrt{1 + \left(\dfrac{1}{2\sqrt{y}}\right)^2}\, dy$

(d) For L_1, $\Delta x = \dfrac{3}{20}$, $x_k = \dfrac{1}{2} + k\dfrac{3}{20} = \dfrac{3k + 10}{20}$, and thus

$$L_1 \approx \sum_{k=1}^{10} \sqrt{(\Delta x)^2 + [f(x_k) - (f(x_{k-1})]^2}$$

$$= \sum_{k=1}^{10} \sqrt{\left(\frac{3}{20}\right)^2 + \left(\frac{(3k+10)^2 - (3k+7)^2}{400}\right)^2} \approx 4.072396336$$

For L_2, $\Delta x = \dfrac{15}{40} = \dfrac{3}{8}$, $x_k = \dfrac{1}{4} + \dfrac{3k}{8} = \dfrac{3k + 2}{8}$, and thus

$$L_2 \approx \sum_{k=1}^{10} \sqrt{\left(\frac{3}{8}\right)^2 + \left[\sqrt{\frac{3k+2}{8}} - \sqrt{\frac{3k-1}{8}}\right]^2} \approx 4.071626502$$

(e) The expression for L_1 is better, perhaps because L_1 has in general a smaller slope and so approximations of the true slope are better.

(f) For L_1, $\Delta x = \dfrac{3}{20}$, the midpoint is $x_k^* = \dfrac{1}{2} + \left(k - \dfrac{1}{2}\right)\dfrac{3}{20} = \dfrac{6k + 17}{40}$, and thus

$$L_1 \approx \sum_{k=1}^{10} \frac{3}{20}\sqrt{1 + \left(2\frac{6k+17}{40}\right)^2} \approx 4.072396336.$$

For L_2, $\Delta x = \dfrac{15}{40}$, and the midpoint is $x_k^* = \dfrac{1}{4} + \left(k - \dfrac{1}{2}\right)\dfrac{15}{40} = \dfrac{6k + 1}{16}$, and thus

$$L_2 \approx \sum_{k=1}^{10} \frac{15}{40}\sqrt{1 + \left(4\frac{6k+1}{16}\right)^{-1}} \approx 4.066160149$$

(g) $L_1 = \displaystyle\int_{1/2}^{2} \sqrt{1 + (2x)^2}\, dx \approx 4.0729, \; L_2 = \displaystyle\int_{1/4}^{4} \sqrt{1 + \left(\dfrac{1}{2\sqrt{x}}\right)^2}\, dx \approx 4.0729$

17. (a) The function $y = f(x) = 1 + 1/x$ is inverse to the function $x = g(y) = 1/(y-1)$: $f(g(y)) = y$ for $4/3 \le y \le 2$, and $g(f(x)) = x$ for $1 \le x \le 3$. Geometrically this means that the graphs of $y = f(x)$ and $x = g(y)$ are symmetric to each other with respect to the line $y = x$.

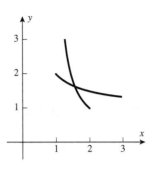

(b) $L_1 = \int_1^3 \sqrt{1 + 1/x^4}\, dx, \qquad L_2 = \int_{4/3}^2 \sqrt{1 + \dfrac{1}{(x-1)^4}}\, dx;$

In the expression for L_1 make the change of variable $y = 1 + 1/x$ to transform the first integral into the second:

$$x = \frac{1}{y-1}, dx = -\frac{dy}{(y-1)^2},$$

$$L_1 = -\int_2^{4/3} \frac{1}{(y-1)^2}\sqrt{1 + (y-1)^4}\, dy = \int_{4/3}^2 \sqrt{\frac{1}{(y-1)^4} + 1}\, dy = L_2$$

(c) $L_1 = \int_1^3 \sqrt{1 + \dfrac{1}{y^4}}\, dy, \qquad L_2 = \int_{4/3}^2 \sqrt{1 + \dfrac{1}{(y-1)^4}}\, dy.$

(d) For L_1, $\Delta x_k = \dfrac{1}{5}, x[k] = 1 + \dfrac{k}{5}$, and thus $L_1 \approx 2.145900021$

For L_2, $\Delta x_k = \dfrac{4}{15}, x[k] = \dfrac{4}{3} + k\dfrac{4}{15}$, and thus $L_2 \approx 2.146262783$

(e) $L \approx 2.146622210$, so the expression for L_2 is slightly more accurate. The slope of $1 + 1/x$ is on average greater than the slope of $1 + 1/(x-1)$ as indicated in the graph in Part (a).

(f) For L_1, $\Delta x_k = \dfrac{2}{5}$, the midpoint is $x_k^* = 1 + \left(k - \dfrac{1}{2}\right)\dfrac{1}{5}$, and thus
$L_1 \approx 2.144326862$.

For L_2, $\Delta x_k = \dfrac{4}{15}$, and the midpoint is $x_k^* = \dfrac{4}{3} + \left(k - \dfrac{1}{2}\right)\dfrac{4}{15}$, and thus
$L_2 \approx 2.137080471$

(g) $L_1 = \int_1^3 \sqrt{1 + \dfrac{1}{x^4}}\, dx \approx 2.1443$

$L_2 = \int_2^{4/3} \sqrt{1 + \dfrac{1}{(x-1)^4}}\, dx \approx 2.1466$

19. $f'(x) = \sec x \tan x, 0 \le \sec x \tan x \le 2\sqrt{3}$ for $0 \le x \le \pi/3$ so $\dfrac{\pi}{3} \le L \le \dfrac{\pi}{3}\sqrt{13}$.

21. $L = \int_0^\pi \sqrt{1 + (k\cos x)^2}\, dx$

k	1	2	1.84	1.83	1.832
L	3.8202	5.2704	5.0135	4.9977	5.0008

Experimentation yields the values in the table, which by the Intermediate-Value Theorem show that the true solution k to $L = 5$ lies between $k = 1.83$ and $k = 1.832$, so $k = 1.83$ to two decimal places.

23. $y = 0$ at $x = b = 30.585$; distance $= \int_0^b \sqrt{1 + (12.54 - 0.82x)^2}\, dx = 196.306$ yd

25. (a) Use the interval $0 \le \phi < 2\pi$.

(b) $(dx/d\phi)^2 + (dy/d\phi)^2 = (-3a\cos^2\phi\sin\phi)^2 + (3a\sin^2\phi\cos\phi)^2$

$$= 9a^2\cos^2\phi\sin^2\phi(\cos^2\phi + \sin^2\phi) = (9a^2/4)\sin^2 2\phi, \text{ so}$$

$$L = (3a/2)\int_0^{2\pi} |\sin 2\phi|\,d\phi = 6a\int_0^{\pi/2} \sin 2\phi\,d\phi = -3a\cos 2\phi\Big]_0^{\pi/2} = 6a$$

27. (a) $(dx/dt)^2 + (dy/dt)^2 = 4\sin^2 t + \cos^2 t = 4\sin^2 t + (1 - \sin^2 t) = 1 + 3\sin^2 t,$

$$L = \int_0^{2\pi} \sqrt{1 + 3\sin^2 t}\,dt = 4\int_0^{\pi/2} \sqrt{1 + 3\sin^2 t}\,dt$$

(b) 9.69

(c) distance traveled $= \displaystyle\int_{1.5}^{4.8} \sqrt{1 + 3\sin^2 t}\,dt \approx 5.16$ cm

EXERCISE SET 6.5

1. $S = \displaystyle\int_0^1 2\pi(7x)\sqrt{1 + 49}dx = 70\pi\sqrt{2}\int_0^1 x\,dx = 35\pi\sqrt{2}$

3. $f'(x) = -x/\sqrt{4 - x^2}$, $1 + [f'(x)]^2 = 1 + \dfrac{x^2}{4 - x^2} = \dfrac{4}{4 - x^2}$,

$$S = \int_{-1}^1 2\pi\sqrt{4 - x^2}(2/\sqrt{4 - x^2})dx = 4\pi\int_{-1}^1 dx = 8\pi$$

5. $S = \displaystyle\int_0^2 2\pi(9y + 1)\sqrt{82}dy = 2\pi\sqrt{82}\int_0^2 (9y + 1)dy = 40\pi\sqrt{82}$

7. $g'(y) = -y/\sqrt{9 - y^2}$, $1 + [g'(y)]^2 = \dfrac{9}{9 - y^2}$, $S = \displaystyle\int_{-2}^2 2\pi\sqrt{9 - y^2}\cdot\dfrac{3}{\sqrt{9 - y^2}}dy = 6\pi\int_{-2}^2 dy = 24\pi$

9. $f'(x) = \dfrac{1}{2}x^{-1/2} - \dfrac{1}{2}x^{1/2}$, $1 + [f'(x)]^2 = 1 + \dfrac{1}{4}x^{-1} - \dfrac{1}{2} + \dfrac{1}{4}x = \left(\dfrac{1}{2}x^{-1/2} + \dfrac{1}{2}x^{1/2}\right)^2$,

$$S = \int_1^3 2\pi\left(x^{1/2} - \dfrac{1}{3}x^{3/2}\right)\left(\dfrac{1}{2}x^{-1/2} + \dfrac{1}{2}x^{1/2}\right)dx = \dfrac{\pi}{3}\int_1^3 (3 + 2x - x^2)dx = 16\pi/9$$

11. $x = g(y) = \dfrac{1}{4}y^4 + \dfrac{1}{8}y^{-2}$, $g'(y) = y^3 - \dfrac{1}{4}y^{-3}$,

$$1 + [g'(y)]^2 = 1 + \left(y^6 - \dfrac{1}{2} + \dfrac{1}{16}y^{-6}\right) = \left(y^3 + \dfrac{1}{4}y^{-3}\right)^2,$$

$$S = \int_1^2 2\pi\left(\dfrac{1}{4}y^4 + \dfrac{1}{8}y^{-2}\right)\left(y^3 + \dfrac{1}{4}y^{-3}\right)dy = \dfrac{\pi}{16}\int_1^2 (8y^7 + 6y + y^{-5})dy = 16{,}911\pi/1024$$

13. $f'(x) = \cos x$, $1 + [f'(x)]^2 = 1 + \cos^2 x$,
$$S = \int_0^\pi 2\pi\sin x\sqrt{1 + \cos^2 x}\,dx = 2\pi(\sqrt{2} + \ln(\sqrt{2} + 1)) \approx 14.42$$

15. $n = 20, a = 0, b = \pi, \Delta x = (b - a)/20 = \pi/20, x_k = k\pi/20$,
$$S \approx \pi\sum_{k=1}^{20} [\sin(k - 1)\pi/20 + \sin k\pi/20]\sqrt{(\pi/20)^2 + [\sin(k - 1)\pi/20 - \sin k\pi/20]^2} \approx 14.39394496$$

17. Revolve the line segment joining the points $(0,0)$ and (h,r) about the x-axis. An equation of the line segment is $y = (r/h)x$ for $0 \le x \le h$ so

$$S = \int_0^h 2\pi(r/h)x\sqrt{1 + r^2/h^2}\,dx = \frac{2\pi r}{h^2}\sqrt{r^2 + h^2}\int_0^h x\,dx = \pi r\sqrt{r^2 + h^2}$$

19. $g(y) = \sqrt{r^2 - y^2}$, $g'(y) = -y/\sqrt{r^2 - y^2}$, $1 + [g'(y)]^2 = r^2/(r^2 - y^2)$,

(a) $S = \displaystyle\int_{r-h}^r 2\pi\sqrt{r^2 - y^2}\sqrt{r^2/(r^2 - y^2)}\,dy = 2\pi r \int_{r-h}^r dy = 2\pi rh$

(b) From Part (a), the surface area common to two polar caps of height $h_1 > h_2$ is
$2\pi rh_1 - 2\pi rh_2 = 2\pi r(h_1 - h_2)$.

21. $S = \displaystyle\int_a^b 2\pi[f(x) + k]\sqrt{1 + [f'(x)]^2}\,dx$

23. Note that $1 \le \sec x \le 2$ for $0 \le x \le \pi/3$. Let L be the arc length of the curve $y = \tan x$ for $0 < x < \pi/3$. Then $L = \displaystyle\int_0^{\pi/3} \sqrt{1 + \sec^2 x}\,dx$, and by Exercise 24, and the inequaities above, $2\pi L \le S \le 4\pi L$. But from the inequalities for $\sec x$ above, we can show that $\sqrt{2}\pi/3 \le L \le \sqrt{5}\pi/3$. Hence, combining the two sets of inequalities, $2\pi(\sqrt{2}\pi/3) \le 2\pi L \le S \le 4\pi L \le 4\pi\sqrt{5}\pi/3$. To obtain the inequalities in the text, observe that

$$\frac{2\pi^2}{3} < 2\pi\frac{\sqrt{2}\pi}{3} \le 2\pi L \le S \le 4\pi L \le 4\pi\frac{\sqrt{5}\pi}{3} < \frac{4\pi^2}{3}\sqrt{13}.$$

25. Let $a = t_0 < t_1 < \ldots < t_{n-1} < t_n = b$ be a partition of $[a,b]$. Then the lateral area of the frustum of slant height $\ell = \sqrt{\Delta x_k^2 + \Delta y_k^2}$ and radii $y(t_1)$ and $y(t_2)$ is $\pi(y(t_k) + y(t_{k-1}))\ell$. Thus the area of the frustum S_k is given by $S_k = \pi(y(t_{k-1}) + y(t_k))\sqrt{[x(t_k) - x(t_{k-1})]^2 + [y(t_k) - y(t_{k-1})]^2}$ with the limit as $\max \Delta t_k \to 0$ of $S = \displaystyle\int_a^b 2\pi y(t)\sqrt{[x'(t)]^2 + y'(t)^2}\,dt$

27. $x' = 2t$, $y' = 2$, $(x')^2 + (y')^2 = 4t^2 + 4$

$$S = 2\pi\int_0^4 (2t)\sqrt{4t^2 + 4}\,dt = 8\pi\int_0^4 t\sqrt{t^2 + 1}\,dt = \frac{8\pi}{3}(17\sqrt{17} - 1)$$

29. $x' = 1$, $y' = 4t$, $(x')^2 + (y')^2 = 1 + 16t^2$, $S = 2\pi\displaystyle\int_0^1 t\sqrt{1 + 16t^2}\,dt = \frac{\pi}{24}(17\sqrt{17} - 1)$

31. $x' = -r\sin t$, $y' = r\cos t$, $(x')^2 + (y')^2 = r^2$,

$$S = 2\pi\int_0^\pi r\sin t\sqrt{r^2}\,dt = 2\pi r^2\int_0^\pi \sin t\,dt = 4\pi r^2$$

33. For (4), express the curve $y = f(x)$ in the parametric form $x = t, y = f(t)$ so $dx/dt = 1$ and $dy/dt = f'(t) = f'(x) = dy/dx$. For (5), express $x = g(y)$ as $x = g(t), y = t$ so $dx/dt = g'(t) = g'(y) = dx/dy$ and $dy/dt = 1$.

EXERCISE SET 6.6

1. (a) $f_{\text{ave}} = \dfrac{1}{4 - 0}\displaystyle\int_0^4 2x\,dx = 4$ (b) $2x^* = 4$, $x^* = 2$

(c)

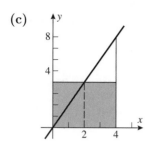

3. $f_{\text{ave}} = \dfrac{1}{3-1} \displaystyle\int_1^3 3x\, dx = \dfrac{3}{4}x^2 \Big]_1^3 = 6$

5. $f_{\text{ave}} = \dfrac{1}{\pi} \displaystyle\int_0^\pi \sin x\, dx = -\dfrac{1}{\pi}\cos x \Big]_0^\pi = \dfrac{2}{\pi}$

7. $\dfrac{1}{2-0}\displaystyle\int_0^2 \dfrac{x}{(5x^2+1)^2}\, dx = -\dfrac{1}{2}\dfrac{1}{10}\dfrac{1}{5x^2+1}\Big|_0^2 = \dfrac{1}{21}$

9. **(a)** $\frac{1}{5}[f(0.4)+f(0.8)+f(1.2)+f(1.6)+f(2.0)] = \frac{1}{5}[0.48+1.92+4.32+7.68+12.00] = 5.28$

   (b) $\dfrac{1}{20}3[(0.1)^2+(0.2)^2+\ldots+(1.9)^2+(2.0)^2] = \dfrac{861}{200} = 4.305$

   (c) $f_{\text{ave}} = \dfrac{1}{2}\displaystyle\int_0^2 3x^2\, dx = \dfrac{1}{2}x^3 \Big]_0^2 = 4$

   (d) Parts (a) and (b) can be interpreted as being two Riemann sums (n = 5, n = 20) for the average, using right endpoints. Since f is increasing, these sums overestimate the integral.

11. **(a)** $\displaystyle\int_0^3 v(t)\, dt = \int_0^2 (1-t)\, dt + \int_2^3 (t-3)\, dt = -\dfrac{1}{2}$, so $v_{\text{ave}} = -\dfrac{1}{6}$

   (b) $\displaystyle\int_0^3 v(t)\, dt = \int_0^1 t\, dt + \int_1^2 dt + \int_2^3 (-2t+5)\, dt = \dfrac{1}{2}+1+0 = \dfrac{3}{2}$, so $v_{\text{ave}} = \dfrac{1}{2}$

13. Linear means $f(\alpha x_1 + \beta x_2) = \alpha f(x_1) + \beta f(x_2)$, so $f\left(\dfrac{a+b}{2}\right) = \dfrac{1}{2}f(a) + \dfrac{1}{2}f(b) = \dfrac{f(a)+f(b)}{2}$.

15. **(a)** $v_{\text{ave}} = \dfrac{1}{4-1}\displaystyle\int_1^4 (3t^3+2)dt = \dfrac{1}{3}\dfrac{789}{4} = \dfrac{263}{4}$

   (b) $v_{\text{ave}} = \dfrac{s(4)-s(1)}{4-1} = \dfrac{100-7}{3} = 31$

17. time to fill tank = (volume of tank)/(rate of filling) = $[\pi(3)^2 5]/(1) = 45\pi$, weight of water in tank at time $t = (62.4)$ (rate of filling)(time) = $62.4t$,

   $\text{weight}_{\text{ave}} = \dfrac{1}{45\pi}\displaystyle\int_0^{45\pi} 62.4t\, dt = 1404\pi$ lb

19. $\displaystyle\int_0^{30} 100(1-0.0001t^2)dt = 2910$ cars, so an average of $\dfrac{2910}{30} = 97$ cars/min.

21. **(a)** $\dfrac{1}{7}[0.74+0.65+0.56+0.45+0.35+0.25+0.16] = 0.4514285714$

   (b) $\dfrac{1}{7}\displaystyle\int_0^7 [0.5+0.5\sin(0.213x+2.481)]\, dx \approx 0.4614$

23. Solve for k: $\displaystyle\int_0^k \sqrt{3x}\, dx = 6k$, so $\sqrt{3}\dfrac{2}{3}x^{3/2}\Big]_0^k = \dfrac{2}{3}\sqrt{3}k^{3/2} = 6k$, $k = (3\sqrt{3})^2 = 27$

25. (a) $V_{\text{rms}}^2 = \dfrac{1}{1/f - 0} \displaystyle\int_0^{1/f} V_p^2 \sin^2(2\pi ft)\,dt = \dfrac{1}{2}fV_p^2 \displaystyle\int_0^{1/f} [1 - \cos(4\pi ft)]\,dt$

$= \dfrac{1}{2}fV_p^2 \left[t - \dfrac{1}{4\pi f}\sin(4\pi ft) \right]\Big]_0^{1/f} = \dfrac{1}{2}V_p^2$, so $V_{\text{rms}} = V_p/\sqrt{2}$

(b) $V_p/\sqrt{2} = 120, V_p = 120\sqrt{2} \approx 169.7$ V

EXERCISE SET 6.7

1. (a) $W = F \cdot d = 30(7) = 210$ ft·lb

(b) $W = \displaystyle\int_1^6 F(x)\,dx = \int_1^6 x^{-2}\,dx = -\dfrac{1}{x}\Big]_1^6 = 5/6$ ft·lb

3. Since $W = \displaystyle\int_a^b F(x)\,dx =$ the area under the curve, it follows that $d < 2.5$ since the area increases

faster under the left part of the curve. In fact, $W_d = \displaystyle\int_0^d F(x)\,dx = 40d$, and

$W = \displaystyle\int_0^5 F(x)\,dx = 140$, so $d = 7/4$.

5. The calculus book has displacement zero, so no work is done holding it.

7. distance traveled $= \displaystyle\int_0^5 v(t)\,dt = \int_0^5 \dfrac{4t}{5}\,dt = \dfrac{2}{5}t^2\Big]_0^5 = 10$ ft. The force is a constant 10 lb, so the
work done is $10 \cdot 10 = 100$ ft·lb.

9. $F(x) = kx$, $F(0.2) = 0.2k = 100$, $k = 500$ N/m, $W = \displaystyle\int_0^{0.8} 500x\,dx = 160$ J

11. $W = \displaystyle\int_0^1 kx\,dx = k/2 = 10$, $k = 20$ lb/ft   13. $W = \displaystyle\int_0^6 (9 - x)\rho(25\pi)\,dx = 900\pi\rho$ ft·lb

15. $w/4 = x/3, w = 4x/3$,

$W = \displaystyle\int_0^2 (3 - x)(9810)(4x/3)(6)\,dx$

$= 78480 \displaystyle\int_0^2 (3x - x^2)\,dx$

$= 261,600$ J

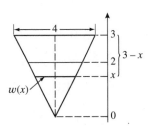

17. (a) $W = \displaystyle\int_0^9 (10 - x)62.4(300)\,dx$

$= 18,720 \displaystyle\int_0^9 (10 - x)\,dx$

$= 926,640$ ft·lb

(b) to empty the pool in one hour would require
$926,640/3600 = 257.4$ ft·lb of work per second
so hp of motor $= 257.4/550 = 0.468$

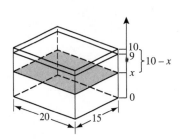

19. $W = \int_0^{100} 15(100 - x)dx$

$= 75,000 \text{ ft·lb}$

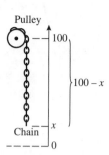
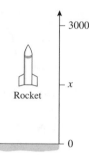

21. When the rocket is x ft above the ground

total weight = weight of rocket + weight of fuel

$= 3 + [40 - 2(x/1000)]$

$= 43 - x/500 \text{ tons},$

$W = \int_0^{3000} (43 - x/500)dx = 120,000 \text{ ft·tons}$

23. **(a)** $150 = k/(4000)^2$, $k = 2.4 \times 10^9$, $w(x) = k/x^2 = 2,400,000,000/x^2 \text{ lb}$

(b) $6000 = k/(4000)^2$, $k = 9.6 \times 10^{10}$, $w(x) = \left(9.6 \times 10^{10}\right)/(x + 4000)^2 \text{ lb}$

(c) $W = \int_{4000}^{5000} 9.6(10^{10})x^{-2}dx = 4,800,000 \text{ mi·lb} = 2.5344 \times 10^{10} \text{ ft·lb}$

25. $W = \frac{1}{2}mv_f^2 - \frac{1}{2}mv_i^2 = \frac{1}{2}4.00 \times 10^5(v_f^2 - 20^2)$. But $W = F \cdot d = (6.40 \times 10^5) \cdot (3.00 \times 10^3)$, so
$19.2 \times 10^8 = 2.00 \times 10^5 v_f^2 - 8.00 \times 10^7$, $19200 = 2v_f^2 - 800$, $v_f = 100 \text{ m/s}$.

27. **(a)** The kinetic energy would have decreased by $\frac{1}{2}mv^2 = \frac{1}{2}4 \cdot 10^6(15000)^2 = 4.5 \times 10^{14} \text{ J}$

(b) $(4.5 \times 10^{14})/(4.2 \times 10^{15}) \approx 0.107$ **(c)** $\frac{1000}{13}(0.107) \approx 8.24 \text{ bombs}$

EXERCISE SET 6.8

1. **(a)** $F = \rho hA = 62.4(5)(100) = 31,200 \text{ lb}$ **(b)** $F = \rho hA = 9810(10)(25) = 2,452,500 \text{ N}$

$P = \rho h = 62.4(5) = 312 \text{ lb/ft}^2$ $P = \rho h = 9810(10) = 98.1 \text{ kPa}$

3. $F = \int_0^2 62.4x(4)dx$

$= 249.6 \int_0^2 x\,dx = 499.2\,\text{lb}$

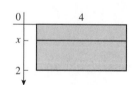

5. $F = \int_0^5 9810x(2\sqrt{25 - x^2})\,dx$

$= 19{,}620 \int_0^5 x(25 - x^2)^{1/2}\,dx$

$= 8.175 \times 10^5 \text{ N}$

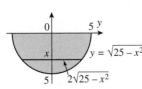

7. By similar triangles

$\dfrac{w(x)}{6} = \dfrac{10 - x}{8}$

$w(x) = \dfrac{3}{4}(10 - x),$

$F = \int_2^{10} 9810x\left[\dfrac{3}{4}(10 - x)\right]dx$

$= 7357.5 \int_2^{10} (10x - x^2)\,dx = 1{,}098{,}720 \text{ N}$

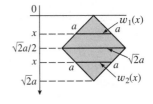

9. Yes: if $\rho_2 = 2\rho_1$ then $F_2 = \int_a^b \rho_2 h(x)w(x)\,dx = \int_a^b 2\rho_1 h(x)w(x)\,dx = 2\int_a^b \rho_1 h(x)w(x)\,dx = 2F_1.$

11. Find the forces on the upper and lower halves and add them:

$\dfrac{w_1(x)}{\sqrt{2}a} = \dfrac{x}{\sqrt{2}a/2}, \; w_1(x) = 2x$

$F_1 = \int_0^{\sqrt{2}a/2} \rho x(2x)\,dx = 2\rho \int_0^{\sqrt{2}a/2} x^2\,dx = \sqrt{2}\rho a^3/6,$

$\dfrac{w_2(x)}{\sqrt{2}a} = \dfrac{\sqrt{2}a - x}{\sqrt{2}a/2}, \; w_2(x) = 2(\sqrt{2}a - x)$

$F_2 = \int_{\sqrt{2}a/2}^{\sqrt{2}a} \rho x[2(\sqrt{2}a - x)]\,dx = 2\rho \int_{\sqrt{2}a/2}^{\sqrt{2}a} (\sqrt{2}ax - x^2)\,dx = \sqrt{2}\rho a^3/3,$

$F = F_1 + F_2 = \sqrt{2}\rho a^3/6 + \sqrt{2}\rho a^3/3 = \rho a^3/\sqrt{2} \text{ lb}$

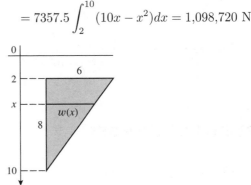

13. $\sqrt{16^2 + 4^2} = \sqrt{272} = 4\sqrt{17}$ is the
other dimension of the bottom.
$(h(x) - 4)/4 = x/(4\sqrt{17})$
$h(x) = x/\sqrt{17} + 4,$
$\sec\theta = 4\sqrt{17}/16 = \sqrt{17}/4$

$F = \int_0^{4\sqrt{17}} 62.4(x/\sqrt{17} + 4)10(\sqrt{17}/4)\,dx$

$= 156\sqrt{17} \int_0^{4\sqrt{17}} (x/\sqrt{17} + 4)\,dx$

$= 63{,}648 \text{ lb}$

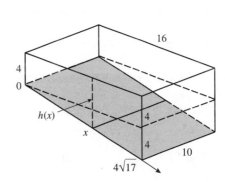

15. $h(x) = x \sin 60° = \sqrt{3}x/2$,

$\theta = 30°$, $\sec\theta = 2/\sqrt{3}$,

$$F = \int_0^{100} 9810(\sqrt{3}x/2)(200)(2/\sqrt{3})\, dx$$

$$= 200 \cdot 9810 \int_0^{100} x\, dx$$

$$= 9810 \cdot 100^3 = 9.81 \times 10^9 \text{ N}$$

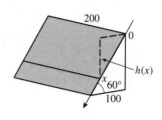

17. (a) From Exercise 16, $F = 4\rho_0(h+1)$ so (assuming that ρ_0 is constant) $dF/dt = 4\rho_0(dh/dt)$ which is a positive constant if dh/dt is a positive constant.

(b) If $dh/dt = 20$ then $dF/dt = 80\rho_0$ lb/min from Part (a).

REVIEW EXERCISE SET

7. (a) $A = \int_a^b (f(x) - g(x))\, dx + \int_b^c (g(x) - f(x))\, dx + \int_c^d (f(x) - g(x))\, dx$

(b) $A = \int_{-1}^0 (x^3 - x)\, dx + \int_0^1 (x - x^3)\, dx + \int_1^2 (x^3 - x)\, dx = \dfrac{1}{4} + \dfrac{1}{4} + \dfrac{9}{4} = \dfrac{11}{4}$

9. Find where the curves cross: set $x^3 = 4x^2$, by observation $x = 2$ is a solution. Then

$$V = \pi \int_0^2 [(x^2 + 4)^2 - (x^3)^2]\, dx = \frac{4352}{105}\pi.$$

11. By implicit differentiation $\dfrac{dy}{dx} = -\left(\dfrac{y}{x}\right)^{1/3}$, so $1 + \left(\dfrac{dy}{dx}\right)^2 = 1 + \left(\dfrac{y}{x}\right)^{2/3} = \dfrac{x^{2/3} + y^{2/3}}{x^{2/3}} = \dfrac{4}{x^{2/3}}$,

$$L = \int_{-8}^{-1} \frac{2}{(-x)^{1/3}}\, dx = 9.$$

13. $A = 2\pi \displaystyle\int_9^{16} \sqrt{25 - x}\sqrt{4 + \dfrac{1}{25 - x}}\, dx = \left(65^{3/2} - 37^{3/2}\right)\dfrac{\pi}{6}$

15. For $0 < x < 3$ the area between the curve and the x-axis consists of two triangles of equal area but of opposite signs, hence 0. For $3 < x < 5$ the area is a rectangle of width 2 and height 3. For $5 < x < 7$ the area consists of two triangles of equal area but opposite sign, hence 0; and for $7 < x < 10$ the curve is given by $y = (4t - 37)/3$ and $\displaystyle\int_7^{10} (4t - 37)/3\, dt = -3$. Thus the desired average is $\dfrac{1}{10}(0 + 6 + 0 - 3) = 0.3$.

17. A cross section of the solid, perpendicular to the x-axis, has area equal to $\pi(\sec x)^2$, and the average of these cross sectional areas is given by $A = \dfrac{1}{\pi/3} \displaystyle\int_0^{\pi/3} \pi(\sec x)^2\, dx = \dfrac{3}{\pi}\pi \tan x \Big]_0^{\pi/3} = 3\sqrt{3}$

19. (a) $F = kx$, $\dfrac{1}{2} = k\dfrac{1}{4}$, $k = 2$, $W = \displaystyle\int_0^{1/4} kx\, dx = 1/16$ J

(b) $25 = \displaystyle\int_0^L kx\, dx = kL^2/2$, $L = 5$ m

21. **(a)** $F = \displaystyle\int_0^1 \rho x 3 \, dx$ N

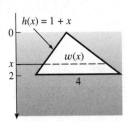

(b) By similar triangles $\dfrac{w(x)}{4} = \dfrac{x}{2}$, $w(x) = 2x$, so

$F = \displaystyle\int_1^4 \rho(1+x)2x \, dx \ \text{lb/ft}^2$.

(c) A formula for the parabola is $y = \dfrac{8}{125}x^2 - 10$, so $F = \displaystyle\int_{-10}^0 9810|y|2\sqrt{\dfrac{125}{8}(y+10)} \, dy$ N.

CHAPTER 7

Exponential, Logarithmic, and Inverse Trigonometric Functions

EXERCISE SET 7.1

1. (a) -4 (b) 4 (c) $1/4$

3. (a) 2.9690 (b) 0.0341

5. (a) $\log_2 16 = \log_2(2^4) = 4$ (b) $\log_2\left(\dfrac{1}{32}\right) = \log_2(2^{-5}) = -5$

    (c) $\log_4 4 = 1$ (d) $\log_9 3 = \log_9(9^{1/2}) = 1/2$

7. (a) 1.3655 (b) -0.3011

9. (a) $2\ln a + \dfrac{1}{2}\ln b + \dfrac{1}{2}\ln c = 2r + s/2 + t/2$ (b) $\ln b - 3\ln a - \ln c = s - 3r - t$

11. (a) $1 + \log x + \dfrac{1}{2}\log(x - 3)$ (b) $2\ln|x| + 3\ln\sin x - \dfrac{1}{2}\ln(x^2 + 1)$

13. $\log\dfrac{2^4(16)}{3} = \log(256/3)$ 15. $\ln\dfrac{\sqrt[3]{x}(x + 1)^2}{\cos x}$

17. $\sqrt{x} = 10^{-1} = 0.1,\ x = 0.01$ 19. $1/x = e^{-2},\ x = e^2$

21. $2x = 8,\ x = 4$ 23. $\log_{10} x = 5,\ x = 10^5$

25. $\ln 2x^2 = \ln 3,\ 2x^2 = 3,\ x^2 = 3/2,\ x = \sqrt{3/2}$ (we discard $-\sqrt{3/2}$ because it does not satisfy the original equation)

27. $\ln 5^{-2x} = \ln 3,\ -2x\ln 5 = \ln 3,\ x = -\dfrac{\ln 3}{2\ln 5}$

29. $e^{3x} = 7/2,\ 3x = \ln(7/2),\ x = \dfrac{1}{3}\ln(7/2)$

31. $e^{-x}(x + 2) = 0$ so $e^{-x} = 0$ (impossible) or $x + 2 = 0,\ x = -2$

33. $e^{-2x} - 3e^{-x} + 2 = (e^{-x} - 2)(e^{-x} - 1) = 0$ so $e^{-x} = 2,\ x = -\ln 2$ or $e^{-x} = 1,\ x = 0$

35. (a) 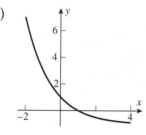   (b)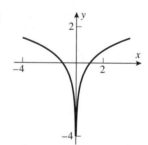

37. $\log_2 7.35 = (\log 7.35)/(\log 2) = (\ln 7.35)/(\ln 2) \approx 2.8777$;
    $\log_5 0.6 = (\log 0.6)/(\log 5) = (\ln 0.6)/(\ln 5) \approx -0.3174$

39.

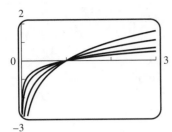

41. $x = y \approx 1.4710, x = y \approx 7.8571$

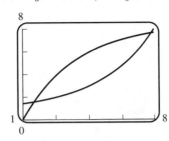

43. (a) no, the curve passes through the origin   **(b)** $y = 2^{x/4}$

(c) $y = 2^{-x}$ **(d)** $y = (\sqrt{5})^x$

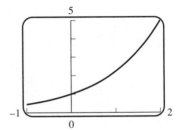

45. $\log(1/2) < 0$ so $3\log(1/2) < 2\log(1/2)$

47. $75e^{-t/125} = 15, t = -125\ln(1/5) = 125\ln 5 \approx 201$ days.

49. (a) 7.4; basic   **(b)** 4.2; acidic   **(c)** 6.4; acidic   **(d)** 5.9; acidic

51. (a) 140 dB; damage   **(b)** 120 dB; damage
     (c) 80 dB; no damage   **(d)** 75 dB; no damage

53. Let I_A and I_B be the intensities of the automobile and blender, respectively. Then $\log_{10} I_A/I_0 = 7$ and $\log_{10} I_B/I_0 = 9.3$, $I_A = 10^7 I_0$ and $I_B = 10^{9.3} I_0$, so $I_B/I_A = 10^{2.3} \approx 200$.

55. (a) $\log E = 4.4 + 1.5(8.2) = 16.7, E = 10^{16.7} \approx 5 \times 10^{16}$ J

     (b) Let M_1 and M_2 be the magnitudes of earthquakes with energies of E and $10E$, respectively. Then $1.5(M_2 - M_1) = \log(10E) - \log E = \log 10 = 1$, $M_2 - M_1 = 1/1.5 = 2/3 \approx 0.67$.

EXERCISE SET 7.2

1. $\dfrac{1}{5x}(5) = \dfrac{1}{x}$ **3.** $\dfrac{1}{1+x}$ **5.** $\dfrac{1}{1-x^2}(2x) = \dfrac{2x}{1-x^2}$

7. $\dfrac{1}{x/(1+x^2)}\left[\dfrac{(1+x^2)(1) - x(2x)}{(1+x^2)^2}\right] = \dfrac{1-x^2}{x(1+x^2)}$

9. $\dfrac{d}{dx}(2\ln x) = 2\dfrac{d}{dx}\ln x = \dfrac{2}{x}$ **11.** $\dfrac{1}{2}(\ln x)^{-1/2}\left(\dfrac{1}{x}\right) = \dfrac{1}{2x\sqrt{\ln x}}$

13. $\ln x + x\dfrac{1}{x} = 1 + \ln x$

15. $2x \log_2(3 - 2x) + \dfrac{-2}{(\ln 2)(3 - 2x)}$

17. $\dfrac{2x(1 + \log x) - x/(\ln 10)}{(1 + \log x)^2}$

19. $\dfrac{1}{\ln x}\left(\dfrac{1}{x}\right) = \dfrac{1}{x \ln x}$

21. $\dfrac{1}{\tan x}(\sec^2 x) = \dfrac{\sec^2 x}{\tan x}$

23. $-\dfrac{1}{x}\sin(\ln x)$

25. $\dfrac{1}{(\ln 10)\sin^2 x}(2\sin x \cos x) = 2\dfrac{\cot x}{\ln 10}$

27. $\dfrac{d}{dx}\left[3\ln(x - 1) + 4\ln(x^2 + 1)\right] = \dfrac{3}{x - 1} + \dfrac{8x}{x^2 + 1} = \dfrac{11x^2 - 8x + 3}{(x - 1)(x^2 + 1)}$

29. $\dfrac{d}{dx}\left[\ln \cos x - \dfrac{1}{2}\ln(4 - 3x^2)\right] = -\tan x + \dfrac{3x}{4 - 3x^2}$

31. $\ln |y| = \ln |x| + \dfrac{1}{3}\ln |1 + x^2|,\ \dfrac{dy}{dx} = x\sqrt[3]{1 + x^2}\left[\dfrac{1}{x} + \dfrac{2x}{3(1 + x^2)}\right]$

33. $\ln |y| = \dfrac{1}{3}\ln |x^2 - 8| + \dfrac{1}{2}\ln |x^3 + 1| - \ln |x^6 - 7x + 5|$

$\dfrac{dy}{dx} = \dfrac{(x^2 - 8)^{1/3}\sqrt{x^3 + 1}}{x^6 - 7x + 5}\left[\dfrac{2x}{3(x^2 - 8)} + \dfrac{3x^2}{2(x^3 + 1)} - \dfrac{6x^5 - 7}{x^6 - 7x + 5}\right]$

35. $f'(x) = ex^{e-1}$

37. **(a)** $\log_x e = \dfrac{\ln e}{\ln x} = \dfrac{1}{\ln x},\ \dfrac{d}{dx}[\log_x e] = -\dfrac{1}{x(\ln x)^2}$

(b) $\log_x 2 = \dfrac{\ln 2}{\ln x},\ \dfrac{d}{dx}[\log_x 2] = -\dfrac{\ln 2}{x(\ln x)^2}$

39. $f(x_0) = \ln e^{-1} = -1,\quad f'(x) = \dfrac{1}{x},\quad f'(x_0) = \dfrac{1}{e^{-1}} = e$

$y = 1 + e\left(x - e^{-1}\right) = ex$

41. $f(x_0) = f(-e) = 1,\ f'(x)\Big|_{x=-e} = -\dfrac{1}{e},$

$y - 1 = -\dfrac{1}{e}(x + e),\ y = -\dfrac{1}{e}x$

43. Let the equation of the tangent line be $y = mx$ and suppose that it meets the curve at (x_0, y_0). Then $m = \dfrac{1}{x}\Big|_{x=x_0} = \dfrac{1}{x_0}$ and $y_0 = mx_0 = \ln x_0$. So $m = \dfrac{1}{x_0} = \dfrac{\ln x_0}{x_0}$ and $\ln x_0 = 1, x_0 = e, m = \dfrac{1}{e}$ and the equation of the tangent line is $y = \dfrac{1}{e}x$.

45. The area of the triangle PQR, given by $|PQ||QR|/2$ is required. $|PQ| = w$, and, by Exercise 44, $|QR| = 1$, so area $= w/2$.

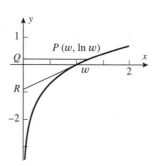

47. If $x = 0$ then $y = \ln e = 1$, and $\dfrac{dy}{dx} = \dfrac{1}{x+e}$. But $e^y = x + e$, so $\dfrac{dy}{dx} = \dfrac{1}{e^y} = e^{-y}$.

49. Let $y = \ln(x+a)$. Following Exercise 47 we get $\dfrac{dy}{dx} = \dfrac{1}{x+a} = e^{-y}$, and when $x = 0, y = \ln(a) = 0$ if $a = 1$, so let $a = 1$, then $y = \ln(x+1)$.

51. **(a)** $f(x) = \ln x; f'(e^2) = \lim\limits_{\Delta x \to 0} \dfrac{\ln(e^2 + \Delta x) - 2}{\Delta x} = \dfrac{d}{dx}(\ln x)\Big|_{x=e^2} = \dfrac{1}{x}\Big|_{x=e^2} = e^{-2}$

   (b) $f(w) = \ln w; f'(1) = \lim\limits_{h \to 0} \dfrac{\ln(1+h) - \ln 1}{h} = \lim\limits_{h \to 0} \dfrac{\ln(1+h)}{h} = \dfrac{1}{w}\Big|_{w=1} = 1$

53. $2\ln x - 3\cos x + C$

55. $\displaystyle\int \dfrac{1}{u}\,du = \ln|u| + C = \ln|\ln x| + C$

57. $u = 1 + x^5, du = 5x^4\,dx, \displaystyle\int \dfrac{du}{5u} = \dfrac{1}{5}\ln u + C = \dfrac{1}{5}\ln(1+x^5) + C$

59. $\displaystyle\int \left(1 + \dfrac{1}{t}\right)\,dt = t + \ln|t| + C$

61. Let $F(x) = \dfrac{3}{2}\ln(1+x^2)$; then $F(2) - F(0) = \dfrac{3}{2}\ln 5$

63. $\displaystyle\int_1^2 u\,du = \dfrac{1}{2}u^2\Big]_1^2 = \dfrac{3}{2}$

65. $\dfrac{1}{2}\ln(2x+e)\Big]_0^e = \dfrac{1}{2}(\ln(3e) - \ln e) = \dfrac{\ln 3}{2}$

67. $y(t) = \displaystyle\int t^{-1}\,dt = \ln|t| + C, y(-1) = C = 5, C = 5; y(t) = \ln|t| + 5$

EXERCISE SET 7.3

1. **(a)** $f'(x) = 5x^4 + 3x^2 + 1 \geq 1$ so f is one-to-one on $-\infty < x < +\infty$.

   (b) $f(1) = 3$ so $1 = f^{-1}(3); \dfrac{d}{dx}f^{-1}(x) = \dfrac{1}{f'(f^{-1}(x))}, (f^{-1})'(3) = \dfrac{1}{f'(1)} = \dfrac{1}{9}$

3. $f^{-1}(x) = \dfrac{2}{x} - 3$, so directly $\dfrac{d}{dx}f^{-1}(x) = -\dfrac{2}{x^2}$. Using Formula (1),

   $f'(x) = \dfrac{-2}{(x+3)^2}$, so $\dfrac{1}{f'(f^{-1}(x))} = -(1/2)(f^{-1}(x)+3)^2$,

   $\dfrac{d}{dx}f^{-1}(x) = -(1/2)\left(\dfrac{2}{x}\right)^2 = -\dfrac{2}{x^2}$

5. **(a)** $f'(x) = 2x + 8$; $f' < 0$ on $(-\infty, -4)$ and $f' > 0$ on $(-4, +\infty)$; not enough information. By inspection, $f(1) = 10 = f(-9)$, so not one-to-one

   (b) $f'(x) = 10x^4 + 3x^2 + 3 \geq 3 > 0$; $f'(x)$ is positive for all x, so f is one-to-one

   (c) $f'(x) = 2 + \cos x \geq 1 > 0$ for all x, so f is one-to-one

   (d) $f'(x) = -(\ln 2) \left(\frac{1}{2}\right)^x < 0$ because $\ln 2 > 0$, so f is one-to-one for all x.

7. $y = f^{-1}(x)$, $x = f(y) = 5y^3 + y - 7$, $\dfrac{dx}{dy} = 15y^2 + 1$, $\dfrac{dy}{dx} = \dfrac{1}{15y^2 + 1}$;

   check: $1 = 15y^2 \dfrac{dy}{dx} + \dfrac{dy}{dx}$, $\dfrac{dy}{dx} = \dfrac{1}{15y^2 + 1}$

9. $y = f^{-1}(x)$, $x = f(y) = 2y^5 + y^3 + 1$, $\dfrac{dx}{dy} = 10y^4 + 3y^2$, $\dfrac{dy}{dx} = \dfrac{1}{10y^4 + 3y^2}$;

   check: $1 = 10y^4 \dfrac{dy}{dx} + 3y^2 \dfrac{dy}{dx}$, $\dfrac{dy}{dx} = \dfrac{1}{10y^4 + 3y^2}$

11. $7e^{7x}$

13. $x^3 e^x + 3x^2 e^x = x^2 e^x (x + 3)$

15. $\dfrac{dy}{dx} = \dfrac{(e^x + e^{-x})(e^x + e^{-x}) - (e^x - e^{-x})(e^x - e^{-x})}{(e^x + e^{-x})^2}$

   $= \dfrac{(e^{2x} + 2 + e^{-2x}) - (e^{2x} - 2 + e^{-2x})}{(e^x + e^{-x})^2} = 4/(e^x + e^{-x})^2$

17. $(x \sec^2 x + \tan x)e^{x \tan x}$

19. $(1 - 3e^{3x})e^{(x - e^{3x})}$

21. $\dfrac{(x - 1)e^{-x}}{1 - xe^{-x}} = \dfrac{x - 1}{e^x - x}$

23. $f'(x) = 2^x \ln 2$; $y = 2^x$, $\ln y = x \ln 2$, $\dfrac{1}{y}y' = \ln 2$, $y' = y \ln 2 = 2^x \ln 2$

25. $f'(x) = \pi^{\sin x}(\ln \pi) \cos x$;

   $y = \pi^{\sin x}$, $\ln y = (\sin x) \ln \pi$, $\dfrac{1}{y}y' = (\ln \pi) \cos x$, $y' = \pi^{\sin x}(\ln \pi) \cos x$

27. $\ln y = (\ln x) \ln(x^3 - 2x)$, $\dfrac{1}{y}\dfrac{dy}{dx} = \dfrac{3x^2 - 2}{x^3 - 2x} \ln x + \dfrac{1}{x} \ln(x^3 - 2x)$,

   $\dfrac{dy}{dx} = (x^3 - 2x)^{\ln x} \left[\dfrac{3x^2 - 2}{x^3 - 2x} \ln x + \dfrac{1}{x} \ln(x^3 - 2x) \right]$

29. $\ln y = (\tan x) \ln(\ln x)$, $\dfrac{1}{y}\dfrac{dy}{dx} = \dfrac{1}{x \ln x} \tan x + (\sec^2 x) \ln(\ln x)$,

   $\dfrac{dy}{dx} = (\ln x)^{\tan x} \left[\dfrac{\tan x}{x \ln x} + (\sec^2 x) \ln(\ln x) \right]$

31. $f'(x) = ex^{e-1}$

33. **(a)** $f(x) = x^3 - 3x^2 + 2x = x(x - 1)(x - 2)$ so $f(0) = f(1) = f(2) = 0$ thus f is not one-to-one.

(b) $f'(x) = 3x^2 - 6x + 2$, $f'(x) = 0$ when $x = \dfrac{6 \pm \sqrt{36 - 24}}{6} = 1 \pm \sqrt{3}/3$. $f'(x) > 0$ (f is increasing) if $x < 1 - \sqrt{3}/3$, $f'(x) < 0$ (f is decreasing) if $1 - \sqrt{3}/3 < x < 1 + \sqrt{3}/3$, so $f(x)$ takes on values less than $f(1 - \sqrt{3}/3)$ on both sides of $1 - \sqrt{3}/3$ thus $1 - \sqrt{3}/3$ is the largest value of k.

35. (a) $f'(x) = 4x^3 + 3x^2 = (4x + 3)x^2 = 0$ only at $x = 0$. But on $[0, 2]$, f' has no sign change, so f is one-to-one.

(b) $F'(x) = 2f'(2g(x))g'(x)$ so $F'(3) = 2f'(2g(3))g'(3)$. By inspection $f(1) = 3$, so $g(3) = f^{-1}(3) = 1$ and $g'(3) = (f^{-1})'(3) = 1/f'(f^{-1}(3)) = 1/f'(1) = 1/7$ because $f'(x) = 4x^3 + 3x^2$. Thus $F'(3) = 2f'(2)(1/7) = 2(44)(1/7) = 88/7$.
$F(3) = f(2g(3)) = f(2 \cdot 1) = f(2) = 25$, so the line tangent to $F(x)$ at $(3, 25)$ has the equation $y - 25 = (88/7)(x - 3)$, $y = (88/7)x - 89/7$.

37. (a) $f'(x) = ke^{kx}$, $f''(x) = k^2 e^{kx}$, $f'''(x) = k^3 e^{kx}, \ldots, f^{(n)}(x) = k^n e^{kx}$
(b) $g'(x) = -ke^{-kx}$, $g''(x) = k^2 e^{-kx}$, $g'''(x) = -k^3 e^{-kx}, \ldots, g^{(n)}(x) = (-1)^n k^n e^{-kx}$

39. $f'(x) = \dfrac{1}{\sqrt{2\pi}\sigma} \exp\left[-\dfrac{1}{2}\left(\dfrac{x - \mu}{\sigma}\right)^2\right] \dfrac{d}{dx}\left[-\dfrac{1}{2}\left(\dfrac{x - \mu}{\sigma}\right)^2\right]$

$= \dfrac{1}{\sqrt{2\pi}\sigma} \exp\left[-\dfrac{1}{2}\left(\dfrac{x - \mu}{\sigma}\right)^2\right]\left[-\left(\dfrac{x - \mu}{\sigma}\right)\left(\dfrac{1}{\sigma}\right)\right]$

$= -\dfrac{1}{\sqrt{2\pi}\sigma^3}(x - \mu)\exp\left[-\dfrac{1}{2}\left(\dfrac{x - \mu}{\sigma}\right)^2\right]$

41. $y = Ae^{2x} + Be^{-4x}$, $y' = 2Ae^{2x} - 4Be^{-4x}$, $y'' = 4Ae^{2x} + 16Be^{-4x}$ so
$y'' + 2y' - 8y = (4Ae^{2x} + 16Be^{-4x}) + 2(2Ae^{2x} - 4Be^{-4x}) - 8(Ae^{2x} + Be^{-4x}) = 0$

43. $\dfrac{dy}{dx} = 100(-0.2)e^{-0.2x} = -20y$, $k = -0.2$

45. $\ln y = \ln 60 - \ln(5 + 7e^{-t})$, $\dfrac{y'}{y} = \dfrac{7e^{-t}}{5 + 7e^{-t}} = \dfrac{7e^{-t} + 5 - 5}{5 + 7e^{-t}} = 1 - \dfrac{1}{12}y$, so
$\dfrac{dy}{dt} = r\left(1 - \dfrac{y}{K}\right)y$, with $r = 1$, $K = 12$.

47. $\lim\limits_{h \to 0} \dfrac{10^h - 1}{h} = \dfrac{d}{dx}10^x\bigg|_{x=0} = \dfrac{d}{dx}e^{x \ln 10}\bigg|_{x=0} = \ln 10$

49. $\displaystyle\int\left[\dfrac{2}{x} + 3e^x\right]dx = 2\ln|x| + 3e^x + C$ **51.** $-\dfrac{1}{5}\displaystyle\int e^u\, du = -\dfrac{1}{5}e^u + C = -\dfrac{1}{5}e^{-5x} + C$

53. $u = 2x$, $du = 2dx$; $\dfrac{1}{2}\displaystyle\int e^u\, du = \dfrac{1}{2}e^u + C = \dfrac{1}{2}e^{2x} + C$

55. $u = \sin x$, $du = \cos x\, dx$; $\displaystyle\int e^u\, du = e^u + C = e^{\sin x} + C$

57. $u = -2x^3$, $du = -6x^2$, $-\dfrac{1}{6}\displaystyle\int e^u du = -\dfrac{1}{6}e^u + C = -\dfrac{1}{6}e^{-2x^3} + C$

59. $\int e^{-x}dx; u = -x, du = -dx; -\int e^u du = -e^u + C = -e^{-x} + C$

61. $\ln(e^x) + \ln(e^{-x}) = \ln(e^x e^{-x}) = \ln 1 = 0$ so $\int [\ln(e^x) + \ln(e^{-x})]dx = C$

63. $5e^x \Big]_{\ln 2}^3 = 5e^3 - 5(2) = 5e^3 - 10$

65. $\dfrac{1}{2}\int_{-1}^1 e^u\, du = \dfrac{1}{2}\left(e - e^{-1}\right)$

67. $u = e^x + 4, du = e^x dx, u = e^{-\ln 3} + 4 = \dfrac{1}{3} + 4 = \dfrac{13}{3}$ when $x = -\ln 3$

$u = e^{\ln 3} + 4 = 3 + 4 = 7$ when $x = \ln 3$, $\int_{13/3}^7 \dfrac{1}{u}du = \ln u \Big]_{13/3}^7 = \ln(7) - \ln(13/3) = \ln(21/13)$, or

$\ln(e^x + 4)\Big]_{-\ln 3}^{\ln 3} = \ln 7 - \ln(13/3) = \ln 21/13$

69. $y(t) = (802.137)\int e^{1.528t}dt = 524.959e^{1.528t} + C; y(0) = 750 = 524.959 + C, C = 225.041,$

$y(t) = 524959e^{1.528t} + 225.041, y(12) = 48{,}233{,}500{,}000$

71. $\int_0^k e^{2x}dx = 3, \dfrac{1}{2}e^{2x}\Big]_0^k = 3, \dfrac{1}{2}(e^{2k} - 1) = 3, e^{2k} = 7, k = \dfrac{1}{2}\ln 7$

EXERCISE SET 7.4

1. critical points $x = 0$: f':
 $x = 0$: relative minimum;

3. critical points $x = -1, 1$: f':
 $x = -1$: relative minimum;
 $x = 1$: relative maximum

5. $f'(x) = x^2(3 - 2x)e^{-2x}$, $f'(x) = 0$ for x in $[1, 4]$ when $x = 3/2$;
 if $x = 1, 3/2, 4$, then $f(x) = e^{-2}, \dfrac{27}{8}e^{-3}, 64e^{-8}$;
 critical point at $x = 3/2$; absolute maximum of $\dfrac{27}{8}e^{-3}$ at $x = 3/2$,
 absolute minimum of $64e^{-8}$ at $x = 4$

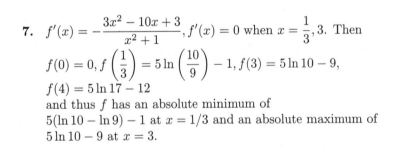

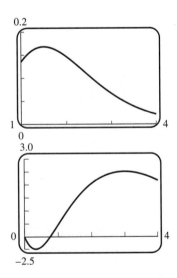

7. $f'(x) = -\dfrac{3x^2 - 10x + 3}{x^2 + 1}, f'(x) = 0$ when $x = \dfrac{1}{3}, 3$. Then
 $f(0) = 0, f\left(\dfrac{1}{3}\right) = 5\ln\left(\dfrac{10}{9}\right) - 1, f(3) = 5\ln 10 - 9,$
 $f(4) = 5\ln 17 - 12$
 and thus f has an absolute minimum of
 $5(\ln 10 - \ln 9) - 1$ at $x = 1/3$ and an absolute maximum of
 $5\ln 10 - 9$ at $x = 3$.

9. (a) $\displaystyle\lim_{x\to+\infty} xe^x = +\infty$, $\displaystyle\lim_{x\to-\infty} xe^x = 0$

(b) $y = xe^x$;
$y' = (x+1)e^x$;
$y'' = (x+2)e^x$

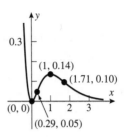

11. (a) $\displaystyle\lim_{x\to+\infty} \frac{x^2}{e^{2x}} = 0$, $\displaystyle\lim_{x\to-\infty} \frac{x^2}{e^{2x}} = +\infty$

(b) $y = x^2/e^{2x} = x^2e^{-2x}$;
$y' = 2x(1-x)e^{-2x}$;
$y'' = 2(2x^2 - 4x + 1)e^{-2x}$;
$y'' = 0$ if $2x^2 - 4x + 1 = 0$, when
$$x = \frac{4 \pm \sqrt{16-8}}{4} = 1 \pm \sqrt{2}/2 \approx 0.29, 1.71$$

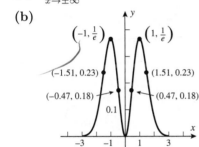

13. (a) $\displaystyle\lim_{x\to\pm\infty} x^2e^{-x^2} = 0$

(b)

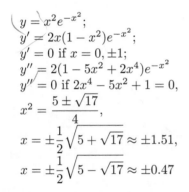

$y = x^2e^{-x^2}$;
$y' = 2x(1-x^2)e^{-x^2}$;
$y' = 0$ if $x = 0, \pm 1$;
$y'' = 2(1 - 5x^2 + 2x^4)e^{-x^2}$
$y'' = 0$ if $2x^4 - 5x^2 + 1 = 0$,
$x^2 = \dfrac{5 \pm \sqrt{17}}{4}$,
$x = \pm\dfrac{1}{2}\sqrt{5 + \sqrt{17}} \approx \pm 1.51$,
$x = \pm\dfrac{1}{2}\sqrt{5 - \sqrt{17}} \approx \pm 0.47$

15. (a) $\displaystyle\lim_{x\to-\infty} f(x) = 0$, $\displaystyle\lim_{x\to+\infty} f(x) = -\infty$

(b) $f'(x) = -\dfrac{e^x(x-2)}{(x-1)^2}$ so
$f'(x) = 0$ when $x = 2$
$f''(x) = -\dfrac{e^x(x^2-4x+5)}{(x-1)^3}$ so
$f''(x) \neq 0$ always
relative maximum when $x = 2$, no point of inflection
asymptote $x = 1$

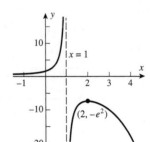

17. $\lim\limits_{x\to+\infty} f(x) = 0$, $\lim\limits_{x\to-\infty} f(x) = +\infty$
$f'(x) = x(2-x)e^{1-x}$, $f''(x) = (x^2 - 4x + 2)e^{1-x}$
critical points at $x = 0, 2$;
relative minimum at $x = 0$,
relative maximum at $x = 2$
points of inflection at $x = 2 \pm \sqrt{2}$
horizontal asymptote $y = 0$ as $x \to +\infty$

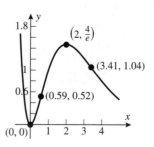

19. (a) $\lim\limits_{x\to 0^+} y = \lim\limits_{x\to 0^+} x\ln x = \lim\limits_{x\to 0^+} \dfrac{\ln x}{1/x} = \lim\limits_{x\to 0^+} \dfrac{1/x}{-1/x^2} = 0;$
$\lim\limits_{x\to+\infty} y = +\infty$

(b) $y = x\ln x$,
$y' = 1 + \ln x$,
$y'' = 1/x$,
$y' = 0$ when $x = e^{-1}$

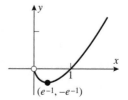

21. (a) $\lim\limits_{x\to 0^+} x^2\ln x = 0$ by the rule given, $\lim\limits_{x\to+\infty} x^2\ln x = +\infty$ by inspection, and $f(x)$ not defined for $x < 0$

(b) $y = x^2\ln 2x$, $y' = 2x\ln 2x + x$
$y'' = 2\ln 2x + 3$
$y' = 0$ if $x = 1/(2\sqrt{e})$,
$y'' = 0$ if $x = 1/(2e^{3/2})$

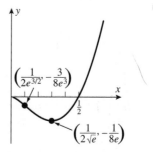

23. (a) $\lim\limits_{x\to+\infty} f(x) = +\infty$, $\lim\limits_{x\to 0^+} f(x) = 0$

(b) $y = x^{2/3}\ln x$
$y' = \dfrac{2\ln x + 3}{3x^{1/3}}$
$y' = 0$ when $\ln x = -\dfrac{3}{2}, x = e^{-3/2}$
$y'' = \dfrac{-3 + 2\ln x}{9x^{4/3}}$,
$y'' = 0$ when $\ln x = \dfrac{3}{2}, x = e^{3/2}$

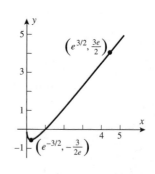

25. (a)

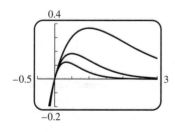

(b) $y' = (1 - bx)e^{-bx}$, $y'' = b^2(x - 2/b)e^{-bx}$;
relative maximum at $x = 1/b$, $y = 1/be$;
point of inflection at $x = 2/b$, $y = 2/be^2$.
Increasing b moves the relative maximum
and the point of inflection to the left and
down, i.e. towards the origin.

27. (a) The oscillations of $e^x \cos x$ about zero increase as $x \to +\infty$ so the limit does not exist, and $\lim\limits_{x \to -\infty} e^x \cos x = 0$.

(b)

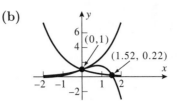

(c) The curve $y = e^{ax} \cos bx$ oscillates between $y = e^{ax}$ and $y = -e^{ax}$. The frequency of oscillation increases when b increases.

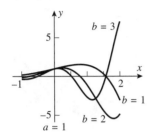 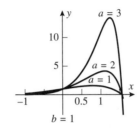

29. (a) $y'(t) = \dfrac{LAke^{-kt}}{(1 + Ae^{-kt})^2} S$, so $y'(0) = \dfrac{LAk}{(1 + A)^2}$

(b) The rate of growth increases to its maximum, which occurs when y is halfway between 0 and L, or when $t = \dfrac{1}{k} \ln A$; it then decreases back towards zero.

(c) From (2) one sees that $\dfrac{dy}{dt}$ is maximized when y lies half way between 0 and L, i.e. $y = L/2$. This follows since the right side of (2) is a parabola (with y as independent variable) with y-intercepts $y = 0, L$. The value $y = L/2$ corresponds to $t = \dfrac{1}{k} \ln A$, from (4).

31. $t = 7.67$

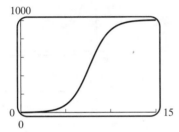

33. (a) $\dfrac{dN}{dt} = 250(20 - t)e^{-t/20} = 0$ at $t = 20$, $N(0) = 125{,}000$, $N(20) \approx 161{,}788$, and $N(100) \approx 128{,}369$; the absolute maximum is $N = 161{,}788$ at $t = 20$, the absolute minimum is $N = 125{,}000$ at $t = 0$.

(b) The absolute minimum of $\dfrac{dN}{dt}$ occurs when $\dfrac{d^2N}{dt^2} = 12.5(t - 40)e^{-t/20} = 0$, $t = 40$.

35. $\dfrac{dk}{dT} = k_0 \exp\left(-\dfrac{q(T - T_0)}{2T_0 T}\right)\left[-\dfrac{d}{dT}\left(\dfrac{q(T - T_0)}{2T_0 T}\right)\right] = -\dfrac{k_0 q}{2T^2} \exp\left(-\dfrac{q(T - T_0)}{2T_0 T}\right)$

37. $\displaystyle\int_0^{\ln 2} e^{2x}\, dx = \dfrac{1}{2} e^{2x}\bigg]_0^{\ln 2} = \dfrac{1}{2}(4 - 1) = \dfrac{3}{2}$

39. $A = \displaystyle\int_0^{\ln 2} \left(e^{2x} - e^x\right)\,dx$

$= \left(\dfrac{1}{2}e^{2x} - e^x\right)\Bigg]_0^{\ln 2} = 1/2$

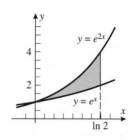

41. $\text{Area} = -\displaystyle\int_{-1}^0 (e^x - 1)\,dx + \int_0^1 (e^x - 1)\,dx = 1/e + e - 2$

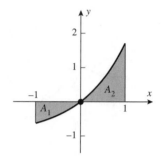

43. $f_{\text{ave}} = \dfrac{1}{e - 1}\displaystyle\int_1^e \dfrac{1}{x}\,dx = \dfrac{1}{e - 1}(\ln e - \ln 1) = \dfrac{1}{e - 1}$

45. $f_{\text{ave}} = \dfrac{1}{4}\displaystyle\int_0^4 e^{-2x}\,dx = -\dfrac{1}{8}e^{-2x}\Bigg]_0^4 = \dfrac{1}{8}\left(1 - e^{-8}\right)$

47. **(a)** the displacement is positive on $(0, 5)$

    (b) For $0 < T < 5$ the displacement is

       $\text{disp} = \dfrac{1}{2}T + (T + 1)e^{-T} - 1$

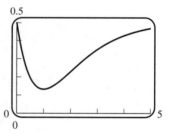

49. The graphs of $y = 1$ and $y = e^x \sin x$ intersect near the points $x = 1$ and $x = 3$. Let $f(x) = 1 - e^x \sin x$, $f'(x) = -e^x(\cos x + \sin x)$, and

$x_{n+1} = x_n + \dfrac{1 - e^x \sin x}{e^x(\cos x + \sin x)}$.   If $x_1 = 1$ then $x_2 = 0.65725814$, $x_3 = 0.59118311, \ldots, x_5 = x_6 = 0.58853274$, and if $x_1 = 3$ then $x_2 = 3.10759324$, $x_3 = 3.09649396, \ldots, x_5 = x_6 = 3.09636393$.

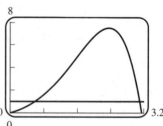

51. A graphing utility shows that there are two inflection points at $x \approx 0.25, -1.25$. These points are the zeros of $f''(x) = (x^4 + 4x^3 + 8x^2 + 4x - 1)\dfrac{e^{-x}}{(x^2 + 1)^3}$. It is equivalent to find the zeros of $g(x) = x^4 + 4x^3 + 8x^2 + 4x - 1$. One root is $x = -1$ by inspection. Since $g'(x) = 4x^3 + 12x^2 + 16x + 4$, Newton's Method becomes

$x_{n+1} = x_n - \dfrac{x_n^4 + 4x_n^3 + 8x_n^2 + 4x_n - 1}{4x_n^3 + 12x_n^2 + 16x_n + 4}$

With $x_0 = 0.25$, $x_1 = 0.18572695$, $x_2 = 0.179563312$, $x_3 = 0.179509029$, $x_4 = x_5 = 0.179509025$. So the points of inflection are at $x \approx 0.18, x = -1$.

53. $V = \pi \displaystyle\int_0^{\ln 3} e^{2x}\,dx = \dfrac{\pi}{2}e^{2x}\Big]_0^{\ln 3} = 4\pi$

55. $V = 2\pi \displaystyle\int_0^1 \dfrac{x}{x^2+1}\,dx$

$\qquad = \pi \ln(x^2+1)\Big]_0^1 = \pi \ln 2$

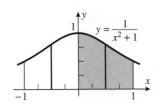

57. $(dx/dt)^2 + (dy/dt)^2 = [e^t(\cos t - \sin t)]^2 + [e^t(\cos t + \sin t)]^2 = 2e^{2t}$,

$\qquad L = \displaystyle\int_0^{\pi/2} \sqrt{2}\,e^t\,dt = \sqrt{2}(e^{\pi/2} - 1)$

59. $dy/dx = \dfrac{\sec x \tan x}{\sec x} = \tan x,\ \sqrt{1+(y')^2} = \sqrt{1 + \tan^2 x} = \sec x$ when $0 < x < \pi/4$, so

$\qquad L = \displaystyle\int_0^{\pi/4} \sec x\,dx = \ln(1 + \sqrt{2})$

61. $f'(x) = e^x,\quad 1 + [f'(x)]^2 = 1 + e^{2x},\ S = \displaystyle\int_0^1 2\pi e^x \sqrt{1 + e^{2x}}\,dx \approx 22.94$

63. $x' = e^t(\cos t - \sin t), y' = e^t(\cos t + \sin t), (x')^2 + (y')^2 = 2e^{2t}$

$\qquad S = 2\pi \displaystyle\int_0^{\pi/2} (e^t \sin t)\sqrt{2e^{2t}}\,dt = 2\sqrt{2}\pi \int_0^{\pi/2} e^{2t}\sin t\,dt$

$\qquad = 2\sqrt{2}\pi \left[\dfrac{1}{5}e^{2t}(2\sin t - \cos t)\right]_0^{\pi/2} = \dfrac{2\sqrt{2}}{5}\pi(2e^\pi + 1)$

EXERCISE SET 7.5

1. **(a)** $\displaystyle\lim_{x\to 2} \dfrac{x^2-4}{x^2+2x-8} = \lim_{x\to 2} \dfrac{(x-2)(x+2)}{(x+4)(x-2)} = \lim_{x\to 2} \dfrac{x+2}{x+4} = \dfrac{2}{3}$

    (b) $\displaystyle\lim_{x\to+\infty} \dfrac{2x-5}{3x+7} = \dfrac{2 - \lim\limits_{x\to+\infty}\dfrac{5}{x}}{3 + \lim\limits_{x\to+\infty}\dfrac{7}{x}} = \dfrac{2}{3}$

3. **(a)** As $x \to -1, f(x) \to 0,\ g(x) \to 0$, so the limit is an indeterminate of form $0/0$.

    (b) $T_f(x) = f(-1) + f'(-1)(x+1) = 0 - 2(x+1) = -2x - 2$;

       $T_g(x) = 0 - 3(x+1) = -3x - 3$, or $f(x) = -2(x+1) + (x+1)\epsilon_1(x)$,

       $g(x) = -3(x+1) + (x+1)\epsilon_2(x)$, where $\epsilon_i(x) \to 0$ as $x \to -1, i = 1, 2$.

    (c) $\displaystyle\lim_{x\to-1} \dfrac{f(x)}{g(x)} = \lim_{x\to-1} \dfrac{-2(x+1)+(x+1)\epsilon_1(x)}{-3(x+1)+(x+1)\epsilon_2(x)} = \lim_{x\to-1} \dfrac{-2+\epsilon_1(x)}{-3+\epsilon_2(x)} = \dfrac{2}{3}$

5. $\displaystyle\lim_{x\to 0}$

7. $\displaystyle\lim_{\theta\to 0} \dfrac{\sec^2 \theta}{1} = 1$

9. $\displaystyle\lim_{x\to\pi^+} \dfrac{\cos x}{1} = -1$

11. $\displaystyle\lim_{x\to+\infty} \dfrac{1/x}{1} = 0$

13. $\displaystyle\lim_{x\to 0^+} \frac{-\csc^2 x}{1/x} = \lim_{x\to 0^+} \frac{-x}{\sin^2 x} = \lim_{x\to 0^+} \frac{-1}{2\sin x \cos x} = -\infty$

15. $\displaystyle\lim_{x\to +\infty} \frac{100x^{99}}{e^x} = \lim_{x\to +\infty} \frac{(100)(99)x^{98}}{e^x} = \cdots = \lim_{x\to +\infty} \frac{(100)(99)(98)\cdots(1)}{e^x} = 0$

17. $\displaystyle\lim_{x\to 0} \frac{2}{\sqrt{1-4x^2}} = 2$

19. $\displaystyle\lim_{x\to +\infty} xe^{-x} = \lim_{x\to +\infty} \frac{x}{e^x} = \lim_{x\to +\infty} \frac{1}{e^x} = 0$

21. $\displaystyle\lim_{x\to +\infty} x\sin(\pi/x) = \lim_{x\to +\infty} \frac{\sin(\pi/x)}{1/x} = \lim_{x\to +\infty} \frac{(-\pi/x^2)\cos(\pi/x)}{-1/x^2} = \lim_{x\to +\infty} \pi\cos(\pi/x) = \pi$

23. $\displaystyle\lim_{x\to(\pi/2)^-} \sec 3x\cos 5x = \lim_{x\to(\pi/2)^-} \frac{\cos 5x}{\cos 3x} = \lim_{x\to(\pi/2)^-} \frac{-5\sin 5x}{-3\sin 3x} = \frac{-5(+1)}{(-3)(-1)} = -\frac{5}{3}$

25. $y = (1-3/x)^x$, $\displaystyle\lim_{x\to +\infty} \ln y = \lim_{x\to +\infty} \frac{\ln(1-3/x)}{1/x} = \lim_{x\to +\infty} \frac{-3}{1-3/x} = -3$, $\displaystyle\lim_{x\to +\infty} y = e^{-3}$

27. $y = (e^x + x)^{1/x}$, $\displaystyle\lim_{x\to 0} \ln y = \lim_{x\to 0} \frac{\ln(e^x + x)}{x} = \lim_{x\to 0} \frac{e^x + 1}{e^x + x} = 2$, $\displaystyle\lim_{x\to 0} y = e^2$

29. $y = (2-x)^{\tan(\pi x/2)}$, $\displaystyle\lim_{x\to 1} \ln y = \lim_{x\to 1} \frac{\ln(2-x)}{\cot(\pi x/2)} = \lim_{x\to 1} \frac{2\sin^2(\pi x/2)}{\pi(2-x)} = 2/\pi$, $\displaystyle\lim_{x\to 1} y = e^{2/\pi}$

31. $\displaystyle\lim_{x\to 0} \left(\frac{1}{\sin x} - \frac{1}{x}\right) = \lim_{x\to 0} \frac{x-\sin x}{x\sin x} = \lim_{x\to 0} \frac{1-\cos x}{x\cos x + \sin x} = \lim_{x\to 0} \frac{\sin x}{2\cos x - x\sin x} = 0$

33. $\displaystyle\lim_{x\to +\infty} \frac{(x^2+x) - x^2}{\sqrt{x^2+x}+x} = \lim_{x\to +\infty} \frac{x}{\sqrt{x^2+x}+x} = \lim_{x\to +\infty} \frac{1}{\sqrt{1+1/x}+1} = 1/2$

35. $\displaystyle\lim_{x\to +\infty} [x - \ln(x^2+1)] = \lim_{x\to +\infty} [\ln e^x - \ln(x^2+1)] = \lim_{x\to +\infty} \ln \frac{e^x}{x^2+1}$,

$\displaystyle\lim_{x\to +\infty} \frac{e^x}{x^2+1} = \lim_{x\to +\infty} \frac{e^x}{2x} = \lim_{x\to +\infty} \frac{e^x}{2} = +\infty$ so $\displaystyle\lim_{x\to +\infty} [x-\ln(x^2+1)] = +\infty$

39. **(a)** L'Hôpital's Rule does not apply to the problem $\displaystyle\lim_{x\to 1} \frac{3x^2 - 2x + 1}{3x^2 - 2x}$ because it is not a $\dfrac{0}{0}$ form.

(b) $\displaystyle\lim_{x\to 1} \frac{3x^2 - 2x + 1}{3x^2 - 2x} = 2$

41. $\displaystyle\lim_{x\to +\infty} \frac{1/(x\ln x)}{1/(2\sqrt{x})} = \lim_{x\to +\infty} \frac{2}{\sqrt{x}\ln x} = 0$

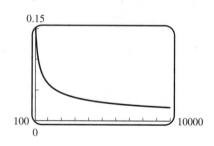

43. $y = (\sin x)^{3/\ln x}$,

$$\lim_{x \to 0^+} \ln y = \lim_{x \to 0^+} \frac{3\ln \sin x}{\ln x} = \lim_{x \to 0^+} (3\cos x)\frac{x}{\sin x} = 3,$$

$$\lim_{x \to 0^+} y = e^3$$

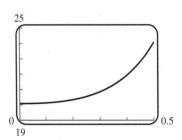

45. $\ln x - e^x = \ln x - \dfrac{1}{e^{-x}} = \dfrac{e^{-x}\ln x - 1}{e^{-x}}$;

$$\lim_{x \to +\infty} e^{-x}\ln x = \lim_{x \to +\infty} \frac{\ln x}{e^x} = \lim_{x \to +\infty} \frac{1/x}{e^x} = 0 \text{ by L'Hôpital's Rule,}$$

$$\text{so } \lim_{x \to +\infty} [\ln x - e^x] = \lim_{x \to +\infty} \frac{e^{-x}\ln x - 1}{e^{-x}} = -\infty$$

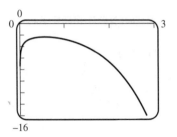

47. $y = (\ln x)^{1/x}$,

$$\lim_{x \to +\infty} \ln y = \lim_{x \to +\infty} \frac{\ln(\ln x)}{x} = \lim_{x \to +\infty} \frac{1}{x\ln x} = 0;$$

$$\lim_{x \to +\infty} y = 1, \ y = 1 \text{ is the horizontal asymptote}$$

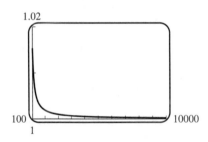

49. (a) 0 **(b)** $+\infty$ **(c)** 0 **(d)** $-\infty$ **(e)** $+\infty$ **(f)** $-\infty$

51. $\displaystyle\lim_{x \to +\infty} \frac{1 + 2\cos 2x}{1}$ does not exist, nor is it $\pm\infty$; $\displaystyle\lim_{x \to +\infty} \frac{x + \sin 2x}{x} = \lim_{x \to +\infty} \left(1 + \frac{\sin 2x}{x}\right) = 1$

53. $\displaystyle\lim_{x \to +\infty} (2 + x\cos 2x + \sin 2x)$ does not exist, nor is it $\pm\infty$; $\displaystyle\lim_{x \to +\infty} \frac{x(2 + \sin 2x)}{x + 1} = \lim_{x \to +\infty} \frac{2 + \sin 2x}{1 + 1/x}$,
which does not exist because $\sin 2x$ oscillates between -1 and 1 as $x \to +\infty$

55. $\displaystyle\lim_{R \to 0^+} \frac{\frac{Vt}{L}e^{-Rt/L}}{1} = \frac{Vt}{L}$

57. (b) $\displaystyle\lim_{x \to +\infty} x(k^{1/x} - 1) = \lim_{t \to 0^+} \frac{k^t - 1}{t} = \lim_{t \to 0^+} \frac{(\ln k)k^t}{1} = \ln k$

  (c) $\ln 0.3 = -1.20397$, $1024\left(\sqrt[1024]{0.3} - 1\right) = -1.20327$;

  $\ln 2 = 0.69315$, $1024\left(\sqrt[1024]{2} - 1\right) = 0.69338$

59. (a) No; $\sin(1/x)$ oscillates as $x \to 0$. **(b)**

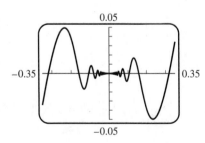

(c)   For the limit as $x \to 0^+$ use the Squeezing Theorem together with the inequalities $-x^2 \le x^2 \sin(1/x) \le x^2$. For $x \to 0^-$ do the same; thus $\lim\limits_{x \to 0} f(x) = 0$.

61. $\lim\limits_{x \to 0^+} \dfrac{\sin(1/x)}{(\sin x)/x}$, $\lim\limits_{x \to 0^+} \dfrac{\sin x}{x} = 1$ but $\lim\limits_{x \to 0^+} \sin(1/x)$ does not exist because $\sin(1/x)$ oscillates between

-1 and 1 as $x \to +\infty$, so $\lim\limits_{x \to 0^+} \dfrac{x \sin(1/x)}{\sin x}$ does not exist.

EXERCISE SET 7.6

1.   (a) 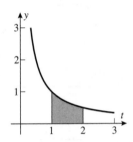   **(b)** 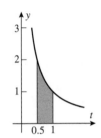   **(c)**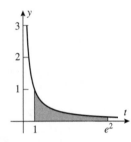

3.   (a) $\ln t\Big]_1^{ac} = \ln(ac) = \ln a + \ln c = 7$ **(b)** $\ln t\Big]_1^{1/c} = \ln(1/c) = -5$

(c) $\ln t\Big]_1^{a/c} = \ln(a/c) = 2 - 5 = -3$ **(d)** $\ln t\Big]_1^{a^3} = \ln a^3 = 3\ln a = 6$

5. $\ln 5 \approx 1.603210678$; $\ln 5 = 1.609437912$; magnitude of error is < 0.0063

7.   (a) x^{-1}, $x > 0$ **(b)** x^2, $x \ne 0$
   (c) $-x^2$, $-\infty < x < +\infty$ **(d)** $-x$, $-\infty < x < +\infty$
   (e) x^3, $x > 0$ **(f)** $\ln x + x$, $x > 0$
   (g) $x - \sqrt[3]{x}$, $-\infty < x < +\infty$ **(h)** $\dfrac{e^x}{x}$, $x > 0$

9.   (a) $3^\pi = e^{\pi \ln 3}$ **(b)** $2^{\sqrt{2}} = e^{\sqrt{2} \ln 2}$

11.   (a) $y = 2x$, $\lim\limits_{x \to +\infty} \left(1 + \dfrac{1}{2x}\right)^x = \lim\limits_{x \to +\infty} \left[\left(1 + \dfrac{1}{2x}\right)^{2x}\right]^{1/2} = \lim\limits_{y \to +\infty} \left[\left(1 + \dfrac{1}{y}\right)^y\right]^{1/2} = e^{1/2}$

(b) $y = 2x$, $\lim\limits_{y \to 0} (1 + y)^{2/y} = \lim\limits_{y \to 0} \left[(1 + y)^{1/y}\right]^2 = e^2$

13. $g'(x) = x^2 - x$

15.   (a) $\dfrac{1}{x^3}(3x^2) = \dfrac{3}{x}$ **(b)** $e^{\ln x}\dfrac{1}{x} = 1$

17. $F'(x) = \dfrac{\sin x}{x^2 + 1}$, $F''(x) = \dfrac{(x^2 + 1)\cos x - 2x\sin x}{(x^2 + 1)^2}$
   (a) 0 **(b)** 0 **(c)** 1

19.   (a) $\dfrac{d}{dx} \int_1^{x^2} t\sqrt{1 + t}\,dt = x^2\sqrt{1 + x^2}(2x) = 2x^3\sqrt{1 + x^2}$

(b) $\displaystyle\int_{1}^{x^2} t\sqrt{1+t}\,dt = -\frac{2}{3}(x^2+1)^{3/2} + \frac{2}{5}(x^2+1)^{5/2} - \frac{4\sqrt{2}}{15}$

21. **(a)** $-\cos x^3$

(b) $-\dfrac{\tan^2 x}{1+\tan^2 x}\sec^2 x = -\tan^2 x$

23. $-3\dfrac{3x-1}{9x^2+1} + 2x\dfrac{x^2-1}{x^4+1}$

25. **(a)** $\sin^2(x^3)(3x^2) - \sin^2(x^2)(2x) = 3x^2\sin^2(x^3) - 2x\sin^2(x^2)$

(b) $\dfrac{1}{1+x}(1) - \dfrac{1}{1-x}(-1) = \dfrac{2}{1-x^2}$

27. from geometry, $\displaystyle\int_{0}^{3} f(t)\,dt = 0,\ \int_{3}^{5} f(t)\,dt = 6,\ \int_{5}^{7} f(t)\,dt = 0;$ and $\displaystyle\int_{7}^{10} f(t)\,dt$

$$= \int_{7}^{10} (4t-37)/3\,dt = -3$$

(a) $F(0) = 0,\ F(3) = 0,\ F(5) = 6,\ F(7) = 6,\ F(10) = 3$

(b) F is increasing where $F' = f$ is positive, so on $[3/2, 6]$ and $[37/4, 10]$, decreasing on $[0, 3/2]$ and $[6, 37/4]$

(c) critical points when $F'(x) = f(x) = 0$, so $x = 3/2, 6, 37/4$; maximum $15/2$ at $x = 6$, minimum $-9/4$ at $x = 3/2$

(d)

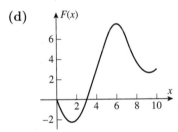

29. $x < 0 : F(x) = \displaystyle\int_{-1}^{x} (-t)\,dt = -\frac{1}{2}t^2\Big]_{-1}^{x} = \frac{1}{2}(1 - x^2),$

$x \geq 0 : F(x) = \displaystyle\int_{-1}^{0} (-t)\,dt + \int_{0}^{x} t\,dt = \frac{1}{2} + \frac{1}{2}x^2;\ F(x) = \begin{cases} (1-x^2)/2, & x < 0 \\ (1+x^2)/2, & x \geq 0 \end{cases}$

31. $y(x) = 2 + \displaystyle\int_{1}^{x} \frac{2t^2+1}{t}\,dt = 2 + (t^2 + \ln t)\Big]_{1}^{x} = x^2 + \ln x + 1$

33. $y(x) = 1 + \displaystyle\int_{\pi/4}^{x} (\sec^2 t - \sin t)\,dt = \tan x + \cos x - \sqrt{2}/2$

35. $P(x) = P_0 + \displaystyle\int_{0}^{x} r(t)\,dt$ individuals

37. II has a minimum at $x = 12$, and I has a zero there, so I could be the derivative of II; on the other hand I has a minimum near $x = 1/3$, but II is not zero there, so II could not be the derivative of I, so I is the graph of $f(x)$ and II is the graph of $\int_{0}^{x} f(t)\,dt$.

39. **(a)** where $f(t) = 0$; by the First Derivative Test, at $t = 3$

(b) where $f(t) = 0$; by the First Derivative Test, at $t = 1, 5$

(c) at $t = 0, 1$ or 5; from the graph it is evident that it is at $t = 5$

(d) at $t = 0, 3$ or 5; from the graph it is evident that it is at $t = 3$

(e) F is concave up when $F'' = f'$ is positive, i.e. where f is increasing, so on $(0, 1/2)$ and $(2, 4)$; it is concave down on $(1/2, 2)$ and $(4, 5)$

(f)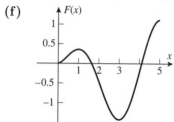

41. $C'(x) = \cos(\pi x^2/2)$, $C''(x) = -\pi x \sin(\pi x^2/2)$

(a) $\cos t$ goes from negative to positive at $2k\pi - \pi/2$, and from positive to negative at $t = 2k\pi + \pi/2$, so $C(x)$ has relative minima when $\pi x^2/2 = 2k\pi - \pi/2$, $x = \pm\sqrt{4k - 1}$, $k = 1, 2, \ldots$, and $C(x)$ has relative maxima when $\pi x^2/2 = (4k + 1)\pi/2$, $x = \pm\sqrt{4k + 1}$, $k = 0, 1, \ldots$.

(b) $\sin t$ changes sign at $t = k\pi$, so $C(x)$ has inflection points at $\pi x^2/2 = k\pi$, $x = \pm\sqrt{2k}$, $k = 1, 2, \ldots$; the case $k = 0$ is distinct due to the factor of x in $C''(x)$, but x changes sign at $x = 0$ and $\sin(\pi x^2/2)$ does not, so there is also a point of inflection at $x = 0$

43. Differentiate: $f(x) = 2e^{2x}$, so $4 + \int_a^x f(t)\,dt = 4 + \int_a^x 2e^{2t}\,dt = 4 + e^{2t}\Big]_a^x = 4 + e^{2x} - e^{2a} = e^{2x}$ provided $e^{2a} = 4$, $a = (\ln 4)/2$.

45. From Exercise 44(d) $\left| e - \left(1 + \dfrac{1}{50}\right)^{50} \right| < y(50)$, and from the graph $y(50) < 0.06$

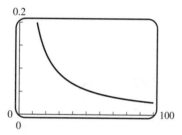

EXERCISE SET 7.7

1. $\tan \theta = 4/3$, $0 < \theta < \pi/2$; use the triangle shown to get $\sin \theta = 4/5$, $\cos \theta = 3/5$, $\cot \theta = 3/4$, $\sec \theta = 5/3$, $\csc \theta = 5/4$

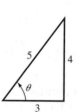

3. (a) $0 \le x \le \pi$ **(b)** $-1 \le x \le 1$

(c) $-\pi/2 < x < \pi/2$ **(d)** $-\infty < x < +\infty$

5. Let $\theta = \cos^{-1}(3/5)$,
$\sin 2\theta = 2 \sin\theta \cos\theta = 2(4/5)(3/5) = 24/25$

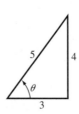

7. **(a)** $\cos(\tan^{-1} x) = \dfrac{1}{\sqrt{1 + x^2}}$

(b) $\tan(\cos^{-1} x) = \dfrac{\sqrt{1 - x^2}}{x}$

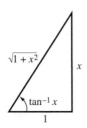

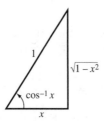

(c) $\sin(\sec^{-1} x) = \dfrac{\sqrt{x^2 - 1}}{x}$

(d) $\cot(\sec^{-1} x) = \dfrac{1}{\sqrt{x^2 - 1}}$

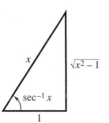

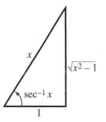

9. **(a)**

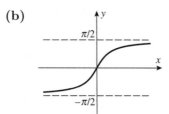

(b)

11. **(a)** $x = \pi + \cos^{-1}(0.85) \approx 3.6964$ rad

(b) $\theta = -\cos^{-1}(0.23) \approx -76.7°$

13. **(a)** $\sin^{-1} 0.9 > 1$, so it is not in the domain of $\sin^{-1} x$

(b) $-1 \le \sin^{-1} x \le 1$ is necessary, or $-0.841471 \le x \le 0.841471$

15. **(a)**

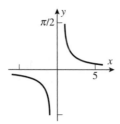

(b) The domain of $\cot^{-1} x$ is $(-\infty, +\infty)$, the range is $(0, \pi)$; the domain of $\csc^{-1} x$ is $(-\infty, -1] \cup [1, +\infty)$, the range is $[-\pi/2, 0) \cup (0, \pi/2]$.

17. $\dfrac{3}{\sqrt{1-(3x)^2}} = \dfrac{3}{\sqrt{1-9x^2}}$

19. $\dfrac{1}{\sqrt{1-1/x^2}}(-1/x^2) = -\dfrac{1}{|x|\sqrt{x^2-1}}$

21. $\dfrac{3x^2}{1+(x^3)^2} = \dfrac{3x^2}{1+x^6}$

23. $y = 1/\tan x = \cot x,\ dy/dx = -\csc^2 x$

25. $\dfrac{e^x}{|x|\sqrt{x^2-1}} + e^x \sec^{-1} x$ **27.** 0 **29.** 0

31. $-\dfrac{1}{1+x}\left(\dfrac{1}{2}x^{-1/2}\right) = -\dfrac{1}{2(1+x)\sqrt{x}}$

33. (a) Let $x = f(y) = \cot y,\ 0 < y < \pi,\ -\infty < x < +\infty$. Then f is differentiable and one-to-one and $f'(f^{-1}(x)) = -\csc^2(\cot^{-1} x) = -x^2 - 1 \neq 0$, and

$$\frac{d}{dx}[\cot^{-1} x]\Big|_{x=0} = \lim_{x\to 0}\frac{1}{f'(f^{-1}(x))} = -\lim_{x\to 0}\frac{1}{x^2+1} = -1.$$

(b) If $x \neq 0$ then, from Exercise 50(a) of Section 1.5,

$$\frac{d}{dx}\cot^{-1} x = \frac{d}{dx}\tan^{-1}\frac{1}{x} = -\frac{1}{x^2}\frac{1}{1+(1/x)^2} = -\frac{1}{x^2+1}.\ \text{For } x = 0,\ \text{Part (a) shows the same};$$

thus for $-\infty < x < +\infty$, $\dfrac{d}{dx}[\cot^{-1} x] = -\dfrac{1}{x^2+1}$.

(c) For $-\infty < u < +\infty$, by the chain rule it follows that $\dfrac{d}{dx}[\cot^{-1} u] = -\dfrac{1}{u^2+1}\dfrac{du}{dx}$.

35. $x^3 + x\tan^{-1} y = e^y,\ 3x^2 + \dfrac{x}{1+y^2}y' + \tan^{-1} y = e^y y',\ y' = \dfrac{(3x^2 + \tan^{-1} y)(1+y^2)}{(1+y^2)e^y - x}$

37. $\displaystyle\int\left[\frac{1}{2\sqrt{1-x^2}} - \frac{3}{1+x^2}\right]dx = \frac{1}{2}\sin^{-1} x - 3\tan^{-1} x + C$

39. $u = 2x,\ \dfrac{1}{2}\displaystyle\int\frac{1}{\sqrt{1-u^2}}du = \frac{1}{2}\sin^{-1}(2x) + C$

41. $u = e^x,\ \displaystyle\int\frac{1}{1+u^2}du = \tan^{-1}(e^x) + C$ **43.** $u = \tan x,\ \displaystyle\int\frac{1}{\sqrt{1-u^2}}du = \sin^{-1}(\tan x) + C$

45. $\sin^{-1} x\Big]_0^{1/\sqrt{2}} = \sin^{-1}(1/\sqrt{2}) - \sin^{-1} 0 = \pi/4$

47. $\sec^{-1} x\Big]_{\sqrt{2}}^{2} = \sec^{-1} 2 - \sec^{-1}\sqrt{2} = \pi/3 - \pi/4 = \pi/12$

49. $\displaystyle\int_{\pi/4}^{\pi/3}\sqrt{u}\,du = -\frac{1}{12}\pi^{3/2} + \frac{2}{27}\sqrt{3}\pi^{3/2}$

51. $u = \sqrt{x},\ 2\displaystyle\int_1^{\sqrt{3}}\frac{1}{u^2+1}du = 2\tan^{-1} u\Big]_1^{\sqrt{3}} = 2(\tan^{-1}\sqrt{3} - \tan^{-1} 1) = 2(\pi/3 - \pi/4) = \pi/6$ or

$2\tan^{-1}\sqrt{x}\Big]_1^3 = \pi/6$

53. $u = \sqrt{3}x^2$, $\dfrac{1}{2\sqrt{3}} \displaystyle\int_0^{\sqrt{3}} \dfrac{1}{\sqrt{4 - u^2}}\,du = \dfrac{1}{2\sqrt{3}}\sin^{-1}\dfrac{u}{2}\Big]_0^{\sqrt{3}} = \dfrac{1}{2\sqrt{3}}\left(\dfrac{\pi}{3}\right) = \dfrac{\pi}{6\sqrt{3}}$

55. $u = 3x$, $\dfrac{1}{3}\displaystyle\int_0^{\sqrt{3}} \dfrac{1}{1 + u^2}\,du = \dfrac{1}{3}\tan^{-1}u\Big]_0^{\sqrt{3}} = \dfrac{1}{3}\dfrac{\pi}{3} = \dfrac{\pi}{9}$

57. **(a)** $\sin^{-1}(x/3) + C$ **(b)** $(1/\sqrt{5})\tan^{-1}(x/\sqrt{5}) + C$
     (c) $(1/\sqrt{\pi})\sec^{-1}(x/\sqrt{\pi}) + C$

59. **(a)** $55.0°$ **(b)** $33.6°$ **(c)** $25.8°$

61. **(a)** If $\gamma = 90°$, then $\sin\gamma = 1$, $\sqrt{1 - \sin^2\phi\sin^2\gamma} = \sqrt{1 - \sin^2\phi} = \cos\phi$,
      $D = \tan\phi\tan\lambda = (\tan 23.45°)(\tan 65°) \approx 0.93023374$ so $h \approx 21.1$ hours.
   (b) If $\gamma = 270°$, then $\sin\gamma = -1$, $D = -\tan\phi\tan\lambda \approx -0.93023374$ so $h \approx 2.9$ hours.

63. $y = 0$ when $x^2 = 6000v^2/g$, $x = 10v\sqrt{60/g} = 1000\sqrt{30}$ for $v = 400$ and $g = 32$;
    $\tan\theta = 3000/x = 3/\sqrt{30}$, $\theta = \tan^{-1}(3/\sqrt{30}) \approx 29°$.

65. $A = \displaystyle\int_{-1}^1 \left(\dfrac{2}{1 + x^2} - |x|\right)\,dx$

$\qquad = 2\displaystyle\int_0^1 \left(\dfrac{2}{1 + x^2} - x\right)\,dx$

$\qquad = 4\tan^{-1}x - x^2\Big]_0^1 = \pi - 1$

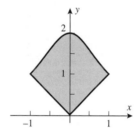

67. The area is given by $\displaystyle\int_0^k (1/\sqrt{1 - x^2} - x)\,dx = \sin^{-1}k - k^2/2 = 1$; solve for k to get
    $k = 0.997301$.

69. $V = \displaystyle\int_{-2}^2 \pi\dfrac{1}{4 + x^2}\,dx = \dfrac{\pi}{2}\tan^{-1}(x/2)\Big]_{-2}^2 = \pi^2/4$

71. Disks: $V = \pi\displaystyle\int_0^1 y^2\,dx = \pi\displaystyle\int_0^1 \left[\sin^{-1}x\right]^2\,dx = \dfrac{1}{4}\pi^3 - 2\pi$
    Cylindrical shells: $V = 2\pi\displaystyle\int_0^{\pi/2} (x_2 - x_1)y\,dy = 2\pi\displaystyle\int_0^{\pi/2} (1 - \sin y)y\,dy = \dfrac{1}{4}\pi^3 - 2\pi$

73. $f_{\text{ave}} = \dfrac{1}{\sqrt{3} - 1}\displaystyle\int_1^{\sqrt{3}} \dfrac{dx}{1 + x^2} = \dfrac{1}{\sqrt{3} - 1}\tan^{-1}x\Big]_1^{\sqrt{3}} = \dfrac{1}{\sqrt{3} - 1}\left(\dfrac{\pi}{3} - \dfrac{\pi}{4}\right) = \dfrac{1}{\sqrt{3} - 1}\dfrac{\pi}{12}$

75. By the Mean-Value Theorem on the interval $[0, x]$,
    $\dfrac{\tan^{-1}x - \tan^{-1}0}{x - 0} = \dfrac{\tan^{-1}x}{x} = \dfrac{1}{1 + c^2}$ for c in $(0, x)$, but

    $\dfrac{1}{1 + x^2} < \dfrac{1}{1 + c^2} < 1$ for c in $(0, x)$ so $\dfrac{1}{1 + x^2} < \dfrac{\tan^{-1}x}{x} < 1$, $\dfrac{x}{1 + x^2} < \tan^{-1}x < x$.

77. $y = \displaystyle\int \dfrac{3}{\sqrt{1 - t^2}}\,dt = 3\sin^{-1}t + C$, $y\left(\dfrac{\sqrt{3}}{2}\right) = 0 = \pi + C$, $C = -\pi$, $y = 3\sin^{-1}t - \pi$

79. $y = \int \dfrac{1}{25 + 9t^2}\, dt = \dfrac{1}{15}\tan^{-1}\left(\dfrac{3}{5}t\right) + C,\; \dfrac{\pi}{30} = y\left(-\dfrac{5}{3}\right) = -\dfrac{1}{15}\dfrac{\pi}{4} + C,$

$C = \dfrac{\pi}{60}, y = \dfrac{1}{15}\tan^{-1}\left(\dfrac{3}{5}t\right) + \dfrac{\pi}{60}$

81. (a) Let $\theta = \sin^{-1}(-x)$ then $\sin\theta = -x$, $-\pi/2 \le \theta \le \pi/2$. But $\sin(-\theta) = -\sin\theta$ and $-\pi/2 \le -\theta \le \pi/2$ so $\sin(-\theta) = -(-x) = x$, $-\theta = \sin^{-1} x$, $\theta = -\sin^{-1} x$.

(b) proof is similar to that in Part (a)

83. (a) $\sin^{-1} x = \tan^{-1}\dfrac{x}{\sqrt{1 - x^2}}$ (see figure)

(b) $\sin^{-1} x + \cos^{-1} x = \pi/2;$

$\cos^{-1} x = \pi/2 - \sin^{-1} x = \pi/2 - \tan^{-1}\dfrac{x}{\sqrt{1 - x^2}}$

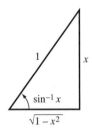

85. (a) $\tan^{-1}\dfrac{1}{2} + \tan^{-1}\dfrac{1}{3} = \tan^{-1}\dfrac{1/2 + 1/3}{1 - (1/2)(1/3)} = \tan^{-1} 1 = \pi/4$

(b) $2\tan^{-1}\dfrac{1}{3} = \tan^{-1}\dfrac{1}{3} + \tan^{-1}\dfrac{1}{3} = \tan^{-1}\dfrac{1/3 + 1/3}{1 - (1/3)(1/3)} = \tan^{-1}\dfrac{3}{4},$

$2\tan^{-1}\dfrac{1}{3} + \tan^{-1}\dfrac{1}{7} = \tan^{-1}\dfrac{3}{4} + \tan^{-1}\dfrac{1}{7} = \tan^{-1}\dfrac{3/4 + 1/7}{1 - (3/4)(1/7)} = \tan^{-1} 1 = \pi/4$

EXERCISE SET 7.8

1. (a) $\sinh 3 \approx 10.0179$

(b) $\cosh(-2) \approx 3.7622$

(c) $\tanh(\ln 4) = 15/17 \approx 0.8824$

(d) $\sinh^{-1}(-2) \approx -1.4436$

(e) $\cosh^{-1} 3 \approx 1.7627$

(f) $\tanh^{-1}\dfrac{3}{4} \approx 0.9730$

3. (a) $\sinh(\ln 3) = \dfrac{1}{2}(e^{\ln 3} - e^{-\ln 3}) = \dfrac{1}{2}\left(3 - \dfrac{1}{3}\right) = \dfrac{4}{3}$

(b) $\cosh(-\ln 2) = \dfrac{1}{2}(e^{-\ln 2} + e^{\ln 2}) = \dfrac{1}{2}\left(\dfrac{1}{2} + 2\right) = \dfrac{5}{4}$

(c) $\tanh(2\ln 5) = \dfrac{e^{2\ln 5} - e^{-2\ln 5}}{e^{2\ln 5} + e^{-2\ln 5}} = \dfrac{25 - 1/25}{25 + 1/25} = \dfrac{312}{313}$

(d) $\sinh(-3\ln 2) = \dfrac{1}{2}(e^{-3\ln 2} - e^{3\ln 2}) = \dfrac{1}{2}\left(\dfrac{1}{8} - 8\right) = -\dfrac{63}{16}$

5.

	$\sinh x_0$	$\cosh x_0$	$\tanh x_0$	$\coth x_0$	$\operatorname{sech} x_0$	$\operatorname{csch} x_0$
(a)	2	$\sqrt{5}$	$2/\sqrt{5}$	$\sqrt{5}/2$	$1/\sqrt{5}$	$1/2$
(b)	$3/4$	$5/4$	$3/5$	$5/3$	$4/5$	$4/3$
(c)	$4/3$	$5/3$	$4/5$	$5/4$	$3/5$	$3/4$

(a) $\cosh^2 x_0 = 1 + \sinh^2 x_0 = 1 + (2)^2 = 5$, $\cosh x_0 = \sqrt{5}$

(b) $\sinh^2 x_0 = \cosh^2 x_0 - 1 = \dfrac{25}{16} - 1 = \dfrac{9}{16}$, $\sinh x_0 = \dfrac{3}{4}$ (because $x_0 > 0$)

(c) $\text{sech}^2 x_0 = 1 - \tanh^2 x_0 = 1 - \left(\dfrac{4}{5}\right)^2 = 1 - \dfrac{16}{25} = \dfrac{9}{25}$, $\text{sech}\, x_0 = \dfrac{3}{5}$,

$\cosh x_0 = \dfrac{1}{\text{sech}\, x_0} = \dfrac{5}{3}$, from $\dfrac{\sinh x_0}{\cosh x_0} = \tanh x_0$ we get $\sinh x_0 = \left(\dfrac{5}{3}\right)\left(\dfrac{4}{5}\right) = \dfrac{4}{3}$

7. (a) $y = \sinh^{-1} x$ if and only if $x = \sinh y$; $1 = \dfrac{dy}{dx}\dfrac{dx}{dy} = \dfrac{dy}{dx}\cosh y$; so

$$\frac{d}{dx}[\sinh^{-1} x] = \frac{dy}{dx} = \frac{1}{\cosh y} = \frac{1}{\sqrt{1 + \sinh^2 y}} = \frac{1}{\sqrt{1 + x^2}} \text{ for all } x.$$

(b) Let $x \geq 1$. Then $y = \cosh^{-1} x$ if and only if $x = \cosh y$; $1 = \dfrac{dy}{dx}\dfrac{dx}{dy} = \dfrac{dy}{dx}\sinh y$, so

$$\frac{d}{dx}[\cosh^{-1} x] = \frac{dy}{dx} = \frac{1}{\sinh y} = \frac{1}{\sqrt{\cosh^2 y - 1}} = \frac{1}{x^2 - 1} \text{ for } x \geq 1.$$

(c) Let $-1 < x < 1$. Then $y = \tanh^{-1} x$ if and only if $x = \tanh y$; thus

$$1 = \frac{dy}{dx}\frac{dx}{dy} = \frac{dy}{dx}\text{sech}^2 y = \frac{dy}{dx}(1 - \tanh^2 y) = 1 - x^2, \text{ so } \frac{d}{dx}[\tanh^{-1} x] = \frac{dy}{dx} = \frac{1}{1 - x^2}.$$

9. $4\cosh(4x - 8)$

11. $-\dfrac{1}{x}\text{csch}^2(\ln x)$

13. $\dfrac{1}{x^2}\text{csch}(1/x)\coth(1/x)$

15. $\dfrac{2 + 5\cosh(5x)\sinh(5x)}{\sqrt{4x + \cosh^2(5x)}}$

17. $x^{5/2}\tanh(\sqrt{x})\text{sech}^2(\sqrt{x}) + 3x^2\tanh^2(\sqrt{x})$

19. $\dfrac{1}{\sqrt{1 + x^2/9}}\left(\dfrac{1}{3}\right) = 1/\sqrt{9 + x^2}$

21. $1/\left[(\cosh^{-1} x)\sqrt{x^2 - 1}\right]$

23. $-(\tanh^{-1} x)^{-2}/(1 - x^2)$

25. $\dfrac{\sinh x}{\sqrt{\cosh^2 x - 1}} = \dfrac{\sinh x}{|\sinh x|} = \begin{cases} 1, & x > 0 \\ -1, & x < 0 \end{cases}$

27. $-\dfrac{e^x}{2x\sqrt{1 - x}} + e^x\,\text{sech}^{-1}x$

29. $\dfrac{1}{7}\sinh^7 x + C$

33. $\dfrac{2}{3}(\tanh x)^{3/2} + C$

35. $\ln(\cosh x) + C$

37. $-\dfrac{1}{3}\text{sech}^3 x\Big]_{\ln 2}^{\ln 3} = 37/375$

39. $u = 3x, \dfrac{1}{3}\displaystyle\int \dfrac{1}{\sqrt{1 + u^2}}\,du = \dfrac{1}{3}\sinh^{-1} 3x + C$

41. $u = e^x, \displaystyle\int \dfrac{1}{u\sqrt{1 - u^2}}\,du = -\text{sech}^{-1}(e^x) + C$

43. $u = 2x, \displaystyle\int \dfrac{du}{u\sqrt{1 + u^2}} = -\text{csch}^{-1}|u| + C = -\text{csch}^{-1}|2x| + C$

45. $\tanh^{-1} x \Big]_0^{1/2} = \tanh^{-1}(1/2) - \tanh^{-1}(0) = \dfrac{1}{2} \ln \dfrac{1 + 1/2}{1 - 1/2} = \dfrac{1}{2} \ln 3$

49. $A = \displaystyle\int_0^{\ln 3} \sinh 2x \, dx = \dfrac{1}{2} \cosh 2x \Big]_0^{\ln 3} = \dfrac{1}{2}[\cosh(2 \ln 3) - 1],$

but $\cosh(2 \ln 3) = \cosh(\ln 9) = \dfrac{1}{2}(e^{\ln 9} + e^{-\ln 9}) = \dfrac{1}{2}(9 + 1/9) = 41/9$ so $A = \dfrac{1}{2}[41/9 - 1] = 16/9.$

51. $V = \pi \displaystyle\int_0^5 (\cosh^2 2x - \sinh^2 2x) dx = \pi \displaystyle\int_0^5 dx = 5\pi$

53. $y' = \sinh x, \ 1 + (y')^2 = 1 + \sinh^2 x = \cosh^2 x$

$L = \displaystyle\int_0^{\ln 2} \cosh x \, dx = \sinh x \Big]_0^{\ln 2} = \sinh(\ln 2) = \dfrac{1}{2}(e^{\ln 2} - e^{-\ln 2}) = \dfrac{1}{2}\left(2 - \dfrac{1}{2}\right) = \dfrac{3}{4}$

55. **(a)** $\displaystyle\lim_{x \to +\infty} \sinh x = \lim_{x \to +\infty} \dfrac{1}{2}(e^x - e^{-x}) = +\infty - 0 = +\infty$

(b) $\displaystyle\lim_{x \to -\infty} \sinh x = \lim_{x \to -\infty} \dfrac{1}{2}(e^x - e^{-x}) = 0 - \infty = -\infty$

(c) $\displaystyle\lim_{x \to +\infty} \tanh x = \lim_{x \to +\infty} \dfrac{e^x - e^{-x}}{e^x + e^{-x}} = 1$

(d) $\displaystyle\lim_{x \to -\infty} \tanh x = \lim_{x \to -\infty} \dfrac{e^x - e^{-x}}{e^x + e^{-x}} = -1$

(e) $\displaystyle\lim_{x \to +\infty} \sinh^{-1} x = \lim_{x \to +\infty} \ln(x + \sqrt{x^2 + 1}) = +\infty$

(f) $\displaystyle\lim_{x \to 1^-} \tanh^{-1} x = \lim_{x \to 1^-} \dfrac{1}{2}[\ln(1 + x) - \ln(1 - x)] = +\infty$

57. $\sinh(-x) = \dfrac{1}{2}(e^{-x} - e^x) = -\dfrac{1}{2}(e^x - e^{-x}) = -\sinh x$

$\cosh(-x) = \dfrac{1}{2}(e^{-x} + e^x) = \dfrac{1}{2}(e^x + e^{-x}) = \cosh x$

59. **(a)** Divide $\cosh^2 x - \sinh^2 x = 1$ by $\cosh^2 x.$

(b) $\tanh(x + y) = \dfrac{\sinh x \cosh y + \cosh x \sinh y}{\cosh x \cosh y + \sinh x \sinh y} = \dfrac{\dfrac{\sinh x}{\cosh x} + \dfrac{\sinh y}{\cosh y}}{1 + \dfrac{\sinh x \sinh y}{\cosh x \cosh y}} = \dfrac{\tanh x + \tanh y}{1 + \tanh x \tanh y}$

(c) Let $y = x$ in Part (b).

61. **(a)** $\dfrac{d}{dx}(\cosh^{-1} x) = \dfrac{1 + x/\sqrt{x^2 - 1}}{x + \sqrt{x^2 - 1}} = 1/\sqrt{x^2 - 1}$

(b) $\dfrac{d}{dx}(\tanh^{-1} x) = \dfrac{d}{dx}\left[\dfrac{1}{2}(\ln(1 + x) - \ln(1 - x))\right] = \dfrac{1}{2}\left(\dfrac{1}{1 + x} + \dfrac{1}{1 - x}\right) = 1/(1 - x^2)$

63. If $|u| < 1$ then, by Theorem 7.8.6, $\displaystyle\int \dfrac{du}{1 - u^2} = \tanh^{-1} u + C.$

For $|u| > 1, \displaystyle\int \dfrac{du}{1 - u^2} = \coth^{-1} u + C = \tanh^{-1}(1/u) + C.$

65. (a) $\lim\limits_{x \to +\infty} (\cosh^{-1} x - \ln x) = \lim\limits_{x \to +\infty} [\ln(x + \sqrt{x^2 - 1}) - \ln x]$

$$= \lim\limits_{x \to +\infty} \ln \frac{x + \sqrt{x^2 - 1}}{x} = \lim\limits_{x \to +\infty} \ln(1 + \sqrt{1 - 1/x^2}) = \ln 2$$

(b) $\lim\limits_{x \to +\infty} \dfrac{\cosh x}{e^x} = \lim\limits_{x \to +\infty} \dfrac{e^x + e^{-x}}{2e^x} = \lim\limits_{x \to +\infty} \dfrac{1}{2}(1 + e^{-2x}) = 1/2$

67. Let $x = -u/a$, $\displaystyle\int \frac{1}{\sqrt{u^2 - a^2}}\, du = -\int \frac{a}{a\sqrt{x^2 - 1}}\, dx = -\cosh^{-1} x + C = -\cosh^{-1}(-u/a) + C.$

$$-\cosh^{-1}(-u/a) = -\ln(-u/a + \sqrt{u^2/a^2 - 1}) = \ln\left[\frac{a}{-u + \sqrt{u^2 - a^2}} \frac{u + \sqrt{u^2 - a^2}}{u + \sqrt{u^2 - a^2}}\right]$$

$$= \ln\left|u + \sqrt{u^2 - a^2}\right| - \ln a = \ln\left|u + \sqrt{u^2 - a^2}\right| + C_1$$

so $\displaystyle\int \frac{1}{\sqrt{u^2 - a^2}}\, du = \ln\left|u + \sqrt{u^2 - a^2}\right| + C_2.$

69. $\displaystyle\int_{-a}^{a} e^{tx}\, dx = \frac{1}{t} e^{tx}\Big]_{-a}^{a} = \frac{1}{t}(e^{at} - e^{-at}) = \frac{2 \sinh at}{t}$ for $t \neq 0$.

71. From Part (b) of Exercise 70, $S = a \cosh(b/a) - a$ so $30 = a \cosh(200/a) - a$. Let $u = 200/a$, then $a = 200/u$ so $30 = (200/u)[\cosh u - 1]$, $\cosh u - 1 = 0.15u$. If $f(u) = \cosh u - 0.15u - 1$, then $u_{n+1} = u_n - \dfrac{\cosh u_n - 0.15u_n - 1}{\sinh u_n - 0.15}$; $u_1 = 0.3, \ldots, u_4 = u_5 = 0.297792782 \approx 200/a$ so $a \approx 671.6079505$. From Part (a), $L = 2a \sinh(b/a) \approx 2(671.6079505) \sinh(0.297792782) \approx 405.9\,\text{ft}$.

73. Set $a = 68.7672$, $b = 0.0100333$, $c = 693.8597$, $d = 299.2239$.

(a)

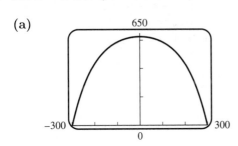

(b) $L = 2\displaystyle\int_0^d \sqrt{1 + a^2 b^2 \sinh^2 bx}\, dx$

$$= 1480.2798\,\text{ft}$$

(c) $x = 283.6249$ ft

(d) $82°$

75. (a) When the bow of the boat is at the point (x, y) and the person has walked a distance D, then the person is located at the point $(0, D)$, the line segment connecting $(0, D)$ and (x, y) has length a; thus $a^2 = x^2 + (D - y)^2$, $D = y + \sqrt{a^2 - x^2} = a\,\text{sech}^{-1}(x/a)$.

(b) Find D when $a = 15$, $x = 10$: $D = 15\,\text{sech}^{-1}(10/15) = 15\ln\left(\dfrac{1 + \sqrt{5/9}}{2/3}\right) \approx 14.44$ m.

(c) $dy/dx = -\dfrac{a^2}{x\sqrt{a^2 - x^2}} + \dfrac{x}{\sqrt{a^2 - x^2}} = \dfrac{1}{\sqrt{a^2 - x^2}}\left[-\dfrac{a^2}{x} + x\right] = -\dfrac{1}{x}\sqrt{a^2 - x^2},$

$1 + [y']^2 = 1 + \dfrac{a^2 - x^2}{x^2} = \dfrac{a^2}{x^2}$; with $a = 15$ and $x = 5$, $L = \displaystyle\int_5^{15} \frac{225}{x^2}\, dx = -\dfrac{225}{x}\Big]_5^{15} = 30$ m.

REVIEW EXERCISES CHAPTER 7

1. **(a)** $x = f(y) = (e^y)^2 + 1$; $y = f^{-1}(x) = \ln\sqrt{x-1} = \frac{1}{2}\ln(x-1)$

   (b) $x = f(y) = \sin\left(\dfrac{1-2y}{y}\right)$; $y = \dfrac{1}{2 + \sin^{-1} x}$

   (c) $x = \dfrac{1}{1 + 3\tan^{-1} y}$; $y = \tan\left(\dfrac{1-x}{3x}\right)$

3. Draw equilateral triangles of sides 5, 12, 13, and 3, 4, 5. Then $\sin[\cos^{-1}(4/5)] = 3/5$, $\sin[\cos^{-1}(5/13)] = 12/13$, $\cos[\sin^{-1}(4/5)] = 3/5$, $\cos[\sin^{-1}(5/13)] = 12/13$

   (a) $\cos[\cos^{-1}(4/5) + \sin^{-1}(5/13)] = \cos(\cos^{-1}(4/5))\cos(\sin^{-1}(5/13))$
   $$- \sin(\cos^{-1}(4/5))\sin(\sin^{-1}(5/13))$$
   $$= \frac{4}{5}\frac{12}{13} - \frac{3}{5}\frac{5}{13} = \frac{33}{65}.$$

   (b) $\sin[\sin^{-1}(4/5) + \cos^{-1}(5/13)] = \sin(\sin^{-1}(4/5))\cos(\cos^{-1}(5/13))$
   $$+ \cos(\sin^{-1}(4/5))\sin(\cos^{-1}(5/13))$$
   $$= \frac{4}{5}\frac{5}{13} + \frac{3}{5}\frac{12}{13} = \frac{56}{65}.$$

5. $y = 5$ ft $= 60$ in, so $60 = \log x$, $x = 10^{60}$ in $\approx 1.58 \times 10^{55}$ mi.

7. $3\ln\left(e^{2x}(e^x)^3\right) + 2\exp(\ln 1) = 3\ln e^{2x} + 3\ln(e^x)^3 + 2\cdot 1 = 3(2x) + (3\cdot 3)x + 2 = 15x + 2$

9. **(a)**
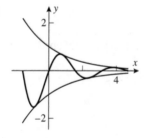

   (b) The curve $y = e^{-x/2}\sin 2x$ has x−intercepts at $x = -\pi/2, 0, \pi/2, \pi, 3\pi/2$. It intersects the curve $y = e^{-x/2}$ at $x = \pi/4, 5\pi/4$ and it intersects the curve $y = -e^{-x/2}$ at $x = -\pi/4, 3\pi/4$.

11. **(a)**
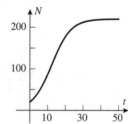

   (b) $N = 80$ when $t = 9.35$ yrs
   (c) 220 sheep

13. **(a)** The function $\ln x - x^{0.2}$ is negative at $x = 1$ and positive at $x = 4$, so it is reasonable to expect it to be zero somewhere in between. (This will be established later in this book.)
   (b) $x = 3.654$

15. $+\infty$

17. $\left(1 + \dfrac{3}{x}\right)^{-x} = \left[\left(1 + \dfrac{3}{x}\right)^{x/3}\right]^{(-3)}$ so the limit is e^{-3}

19. $y = \ln(x+1) + 2\ln(x+2) - 3\ln(x+3) - 4\ln(x+4)$, $dy/dx = \dfrac{1}{x+1} + \dfrac{2}{x+2} - \dfrac{3}{x+3} - \dfrac{4}{x+4}$

21. $\dfrac{1}{2x}(2) = 1/x$

23. $\dfrac{1}{3x(\ln x + 1)^{2/3}}$

25. $\log_{10}\ln x = \dfrac{\ln\ln x}{\ln 10}$, $y' = \dfrac{1}{(\ln 10)(x\ln x)}$

27. $y = \dfrac{3}{2}\ln x + \dfrac{1}{2}\ln(1+x^4)$, $y' = \dfrac{3}{2x} + \dfrac{2x^3}{(1+x^4)}$

29. $y = x^2 + 1$ so $y' = 2x$.

31. $y' = 2e^{\sqrt{x}} + 2xe^{\sqrt{x}}\dfrac{d}{dx}\sqrt{x} = 2e^{\sqrt{x}} + \sqrt{x}e^{\sqrt{x}}$

33. $y' = \dfrac{2}{\pi(1 + 4x^2)}$

35. $\ln y = e^x\ln x$, $\dfrac{y'}{y} = e^x\left(\dfrac{1}{x} + \ln x\right)$, $\dfrac{dy}{dx} = x^{e^x}e^x\left(\dfrac{1}{x} + \ln x\right) = e^x\left[x^{e^x - 1} + x^{e^x}\ln x\right]$

37. $y' = \dfrac{2}{|2x+1|\sqrt{(2x+1)^2 - 1}}$

39. $\ln y = 3\ln x - \dfrac{1}{2}\ln(x^2+1)$, $y'/y = \dfrac{3}{x} - \dfrac{x}{x^2+1}$, $y = \dfrac{3x^2}{\sqrt{x^2+1}} - \dfrac{x^4}{(x^2+1)^{3/2}}$

41. (b)

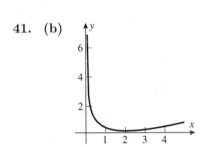

(c) $\dfrac{dy}{dx} = \dfrac{1}{2} - \dfrac{1}{x}$ so $\dfrac{dy}{dx} < 0$ at $x = 1$ and $\dfrac{dy}{dx} > 0$ at $x = e$

(d) The slope is a continuous function which goes from a negative value to a positive value; therefore it must take the value zero between, by the Intermediate Value Theorem.

(e) $\dfrac{dy}{dx} = 0$ when $x = 2$

43. Solve $\dfrac{dy}{dt} = 3\dfrac{dx}{dt}$ given $y = x\ln x$. Then $\dfrac{dy}{dt} = \dfrac{dy}{dx}\dfrac{dx}{dt} = (1 + \ln x)\dfrac{dx}{dt}$, so $1 + \ln x = 3$, $\ln x = 2$, $x = e^2$.

45. Set $y = \log_b x$ and solve $y' = 1$: $y' = \dfrac{1}{x\ln b} = 1$ so $x = \dfrac{1}{\ln b}$.
The curves intersect when (x, x) lies on the graph of $y = \log_b x$,
so $x = \log_b x$. From Formula (8), Section 1.6, $\log_b x = \dfrac{\ln x}{\ln b}$
from which $\ln x = 1$, $x = e$, $\ln b = 1/e$, $b = e^{1/e} \approx 1.4447$.

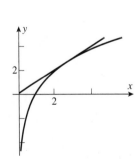

47. Yes, g must be differentiable (where $f' \neq 0$); this can be inferred from the graphs. Note that if $f' = 0$ at a point then g' cannot exist (infinite slope).

49. Let $P(x_0, y_0)$ be a point on $y = e^{3x}$ then $y_0 = e^{3x_0}$. $dy/dx = 3e^{3x}$ so $m_{\tan} = 3e^{3x_0}$ at P and an equation of the tangent line at P is $y - y_0 = 3e^{3x_0}(x - x_0)$, $y - e^{3x_0} = 3e^{3x_0}(x - x_0)$. If the line passes through the origin then $(0,0)$ must satisfy the equation so $-e^{3x_0} = -3x_0 e^{3x_0}$ which gives $x_0 = 1/3$ and thus $y_0 = e$. The point is $(1/3, e)$.

51. $y' = ae^{ax}\sin bx + be^{ax}\cos bx$ and $y'' = (a^2 - b^2)e^{ax}\sin bx + 2abe^{ax}\cos bx$, so $y'' - 2ay' + (a^2 + b^2)y$
$= (a^2 - b^2)e^{ax}\sin bx + 2abe^{ax}\cos bx - 2a(ae^{ax}\sin bx + be^{ax}\cos bx) + (a^2 + b^2)e^{ax}\sin bx = 0$.

53. **(a)**

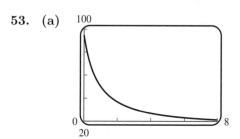

(b) as t tends to $+\infty$, the population tends to 19
$$\lim_{t \to +\infty} P(t) = \lim_{t \to +\infty} \frac{95}{5 - 4e^{-t/4}} = \frac{95}{5 - 4\lim\limits_{t \to +\infty} e^{-t/4}} = \frac{95}{5} = 19$$

(c) the rate of population growth tends to zero

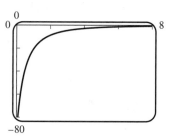

55. In the case $+\infty - (-\infty)$ the limit is $+\infty$; in the case $-\infty - (+\infty)$ the limit is $-\infty$, because large positive (negative) quantities are added to large positive (negative) quantities. The cases $+\infty - (+\infty)$ and $-\infty - (-\infty)$ are indeterminate; large numbers of opposite sign are subtracted, and more information about the sizes is needed.

57. $\lim\limits_{x \to +\infty} (e^x - x^2) = \lim\limits_{x \to +\infty} x^2(e^x/x^2 - 1)$, but $\lim\limits_{x \to +\infty} \dfrac{e^x}{x^2} = \lim\limits_{x \to +\infty} \dfrac{e^x}{2x} = \lim\limits_{x \to +\infty} \dfrac{e^x}{2} = +\infty$

so $\lim\limits_{x \to +\infty} (e^x/x^2 - 1) = +\infty$ and thus $\lim\limits_{x \to +\infty} x^2(e^x/x^2 - 1) = +\infty$

59. $= \lim\limits_{x \to 0} \dfrac{(x^2 + 2x)e^x}{6\sin 3x \cos 3x} = \lim\limits_{x \to 0} \dfrac{(x^2 + 2x)e^x}{3\sin 6x} = \lim\limits_{x \to 0} \dfrac{(x^2 + 4x + 2)e^x}{18\cos 6x} = \dfrac{1}{9}$

61. $f'(x) = -\dfrac{2x}{e^{x^2}}$ **(a)** $(-\infty, 0]$ **(b)** $[0, +\infty)$

$f''(x) = \dfrac{2(2x^2 - 1)}{e^{x^2}}$ **(c)** $(-\infty, -\sqrt{2}/2), (\sqrt{2}/2, +\infty)$ **(d)** $(-\sqrt{2}/2, \sqrt{2}/2)$

(e) $-\sqrt{2}/2, \sqrt{2}/2$

63. $f'(x) = 2x/(1 + x^2)$; critical point at $x = 0$; relative minimum of 0 at $x = 0$ (first derivative test)

65. $\lim\limits_{x \to 0^+} f(x) = \lim\limits_{x \to +\infty} f(x) = +\infty$ and $f'(x) = \dfrac{e^x(x - 2)}{x^3}$, stationary point at $x = 2$; by Theorem 5.4.4 $f(x)$ has an absolute minimum at $x = 2$, and $m = e^2/4$.

67. $f'(x) = 1/2 + 2x/(x^2 + 1)$,

$f'(x) = 0$ on $[-4, 0]$ for $x = -2 \pm \sqrt{3}$

if $x = -2 - \sqrt{3}, -2 + \sqrt{3}$ then

$f(x) = -1 - \sqrt{3}/2 + \ln 4 + \ln(2 + \sqrt{3}) \approx 0.84$,

$-1 + \sqrt{3}/2 + \ln 4 + \ln(2 - \sqrt{3}) \approx -0.06$,

absolute maximum at $x = -2 - \sqrt{3}$,

absolute minimum at $x = -2 + \sqrt{3}$

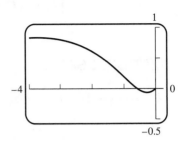

69. $3x^{1/3} - 5e^x + C$

71. $\tan^{-1} x + 2\sin^{-1} x + C$

73. $0.351220577, 0.420535296, 0.386502483$

75. Since $y = e^x$ and $y = \ln x$ are inverse functions, their graphs
are symmetric with respect to the line $y = x$; consequently the
areas A_1 and A_3 are equal (see figure). But $A_1 + A_2 = e$, so

$$\int_1^e \ln x \, dx + \int_0^1 e^x \, dx = A_2 + A_3 = A_2 + A_1 = e$$

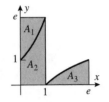

77. $f(x) = e^x, [a, b] = [0, 1], \Delta x = \dfrac{1}{n}; \displaystyle\lim_{n \to +\infty} \sum_{k=1}^n f(x_k^*)\frac{1}{n} = \int_0^1 e^x \, dx = e - 1$

79. $\displaystyle\int_1^3 e^x \, dx = e^x \Big]_1^3 = e^3 - e$

81. **(a)** $dy = x^{1/3} \, dx, y(x) = \displaystyle\int x^{1/3} \, dx = \frac{3}{4}x^{4/3} + C, y(1) = 2 = \frac{3}{4} + C, C = \frac{5}{4}, y(x) = \frac{3}{4}x^{4/3} + \frac{5}{4}$

(b) $dy = \displaystyle\int xe^{x^2} \, dx = \frac{1}{2}e^{x^2} + C, y(0) = 0, y(x) = \frac{1}{2}e^{x^2} - \frac{1}{2}$

83. $\displaystyle\int_0^1 e^{-x/2} dx = 2(1 - 1/\sqrt{e})$

85. $V = \displaystyle\int_1^4 \left(\sqrt{x} - \frac{1}{\sqrt{x}}\right)^2 dx = 2\ln 2 + \frac{3}{2}$

87. **(a)** $\cosh 3x = \cosh(2x + x) = \cosh 2x \cosh x + \sinh 2x \sinh x$

$= (2\cosh^2 x - 1)\cosh x + (2\sinh x \cosh x)\sinh x$

$= 2\cosh^3 x - \cosh x + 2\sinh^2 x \cosh x$

$= 2\cosh^3 x - \cosh x + 2(\cosh^2 x - 1)\cosh x = 4\cosh^3 x - 3\cosh x$

(b) from Theorem 7.8.2 with x replaced by $\dfrac{x}{2}$: $\cosh x = 2\cosh^2 \dfrac{x}{2} - 1$,

$2\cosh^2 \dfrac{x}{2} = \cosh x + 1, \cosh^2 \dfrac{x}{2} = \dfrac{1}{2}(\cosh x + 1)$,

$\cosh \dfrac{x}{2} = \sqrt{\dfrac{1}{2}(\cosh x + 1)}$ (because $\cosh \dfrac{x}{2} > 0$)

(c) from Theorem 7.8.2 with x replaced by $\dfrac{x}{2}$: $\cosh x = 2\sinh^2 \dfrac{x}{2} + 1$,

$2\sinh^2 \dfrac{x}{2} = \cosh x - 1, \sinh^2 \dfrac{x}{2} = \dfrac{1}{2}(\cosh x - 1), \sinh \dfrac{x}{2} = \pm\sqrt{\dfrac{1}{2}(\cosh x - 1)}$

EXERCISE SET 8.1

1. $u = 4 - 2x, du = -2dx, -\dfrac{1}{2}\displaystyle\int u^3\,du = -\dfrac{1}{8}u^4 + C = -\dfrac{1}{8}(4-2x)^4 + C$

3. $u = x^2, du = 2x\,dx, \quad \dfrac{1}{2}\displaystyle\int \sec^2 u\,du = \dfrac{1}{2}\tan u + C = \dfrac{1}{2}\tan(x^2) + C$

5. $u = 2 + \cos 3x, du = -3\sin 3x\,dx, \quad -\dfrac{1}{3}\displaystyle\int \dfrac{du}{u} = -\dfrac{1}{3}\ln|u| + C = -\dfrac{1}{3}\ln(2 + \cos 3x) + C$

7. $u = e^x, du = e^x dx, \quad \displaystyle\int \sinh u\,du = \cosh u + C = \cosh e^x + C$

9. $u = \tan x, du = \sec^2 x\,dx, \quad \displaystyle\int e^u\,du = e^u + C = e^{\tan x} + C$

11. $u = \cos 5x, du = -5\sin 5x\,dx, \quad -\dfrac{1}{5}\displaystyle\int u^5\,du = -\dfrac{1}{30}u^6 + C = -\dfrac{1}{30}\cos^6 5x + C$

13. $u = e^x, du = e^x dx, \quad \displaystyle\int \dfrac{du}{\sqrt{4 + u^2}} = \ln\left(u + \sqrt{u^2 + 4}\right) + C = \ln\left(e^x + \sqrt{e^{2x} + 4}\right) + C$

15. $u = \sqrt{x - 1}, du = \dfrac{1}{2\sqrt{x-1}}\,dx, \quad 2\displaystyle\int e^u\,du = 2e^u + C = 2e^{\sqrt{x-1}} + C$

17. $u = \sqrt{x}, du = \dfrac{1}{2\sqrt{x}}\,dx, \quad \displaystyle\int 2\cosh u\,du = 2\sinh u + C = 2\sinh\sqrt{x} + C$

19. $u = \sqrt{x}, du = \dfrac{1}{2\sqrt{x}}\,dx, \quad \displaystyle\int \dfrac{2\,du}{3^u} = 2\displaystyle\int e^{-u\ln 3}\,du = -\dfrac{2}{\ln 3}e^{-u\ln 3} + C = -\dfrac{2}{\ln 3}3^{-\sqrt{x}} + C$

21. $u = \dfrac{2}{x}, du = -\dfrac{2}{x^2}\,dx, \quad -\dfrac{1}{2}\displaystyle\int \operatorname{csch}^2 u\,du = \dfrac{1}{2}\coth u + C = \dfrac{1}{2}\coth\dfrac{2}{x} + C$

23. $u = e^{-x}, du = -e^{-x}dx, \quad -\displaystyle\int \dfrac{du}{4 - u^2} = -\dfrac{1}{4}\ln\left|\dfrac{2 + u}{2 - u}\right| + C = -\dfrac{1}{4}\ln\left|\dfrac{2 + e^{-x}}{2 - e^{-x}}\right| + C$

25. $u = e^x, du = e^x dx, \quad \displaystyle\int \dfrac{e^x\,dx}{\sqrt{1 - e^{2x}}} = \displaystyle\int \dfrac{du}{\sqrt{1 - u^2}} = \sin^{-1} u + C = \sin^{-1} e^x + C$

27. $u = x^2, du = 2x\,dx, \quad \dfrac{1}{2}\displaystyle\int \dfrac{du}{\csc u} = \dfrac{1}{2}\displaystyle\int \sin u\,du = -\dfrac{1}{2}\cos u + C = -\dfrac{1}{2}\cos(x^2) + C$

29. $4^{-x^2} = e^{-x^2\ln 4}, u = -x^2\ln 4, du = -2x\ln 4\,dx = -x\ln 16\,dx,$

$-\dfrac{1}{\ln 16}\displaystyle\int e^u\,du = -\dfrac{1}{\ln 16}e^u + C = -\dfrac{1}{\ln 16}e^{-x^2\ln 4} + C = -\dfrac{1}{\ln 16}4^{-x^2} + C$

31. (a) $u = \sin x, du = \cos x\,dx, \quad \displaystyle\int u\,du = \dfrac{1}{2}u^2 + C = \dfrac{1}{2}\sin^2 x + C$

(b) $\displaystyle\int \sin x \cos x\,dx = \dfrac{1}{2}\displaystyle\int \sin 2x\,dx = -\dfrac{1}{4}\cos 2x + C = -\dfrac{1}{4}(\cos^2 x - \sin^2 x) + C$

(c) $-\frac{1}{4}(\cos^2 x - \sin^2 x) + C = -\frac{1}{4}(1 - \sin^2 x - \sin^2 x) + C = -\frac{1}{4} + \frac{1}{2}\sin^2 x + C,$

and this is the same as the answer in part (a) except for the constants.

33. (a) $\dfrac{\sec^2 x}{\tan x} = \dfrac{1}{\cos^2 x \tan x} = \dfrac{1}{\cos x \sin x}$

(b) $\csc 2x = \dfrac{1}{\sin 2x} = \dfrac{1}{2\sin x \cos x} = \dfrac{1}{2}\dfrac{\sec^2 x}{\tan x},$ so $\displaystyle\int \csc 2x\, dx = \dfrac{1}{2}\ln\tan x + C$

(c) $\sec x = \dfrac{1}{\cos x} = \dfrac{1}{\sin(\pi/2 - x)} = \csc(\pi/2 - x),$ so

$\displaystyle\int \sec x\, dx = \int \csc(\pi/2 - x)\, dx = -\frac{1}{2}\ln\tan(\pi/2 - x) + C$

EXERCISE SET 8.2

1. $u = x,\ dv = e^{-2x}dx,\ du = dx,\ v = -\frac{1}{2}e^{-2x};$

$\displaystyle\int xe^{-2x}dx = -\frac{1}{2}xe^{-2x} + \int \frac{1}{2}e^{-2x}dx = -\frac{1}{2}xe^{-2x} - \frac{1}{4}e^{-2x} + C$

3. $u = x^2,\ dv = e^x dx,\ du = 2x\, dx,\ v = e^x;\ \displaystyle\int x^2 e^x dx = x^2 e^x - 2\int xe^x dx.$

For $\displaystyle\int xe^x dx$ use $u = x,\ dv = e^x dx,\ du = dx,\ v = e^x$ to get

$\displaystyle\int xe^x dx = xe^x - e^x + C_1$ so $\displaystyle\int x^2 e^x dx = x^2 e^x - 2xe^x + 2e^x + C$

5. $u = x,\ dv = \sin 3x\, dx,\ du = dx,\ v = -\dfrac{1}{3}\cos 3x;$

$\displaystyle\int x\sin 3x\, dx = -\frac{1}{3}x\cos 3x + \frac{1}{3}\int \cos 3x\, dx = -\frac{1}{3}x\cos 3x + \frac{1}{9}\sin 3x + C$

7. $u = x^2,\ dv = \cos x\, dx,\ du = 2x\, dx,\ v = \sin x;\ \displaystyle\int x^2\cos x\, dx = x^2\sin x - 2\int x\sin x\, dx$

For $\displaystyle\int x\sin x\, dx$ use $u = x,\ dv = \sin x\, dx$ to get

$\displaystyle\int x\sin x\, dx = -x\cos x + \sin x + C_1$ so $\displaystyle\int x^2\cos x\, dx = x^2\sin x + 2x\cos x - 2\sin x + C$

9. $u = \ln x,\ dv = x\, dx,\ du = \dfrac{1}{x}dx,\ v = \dfrac{1}{2}x^2;\ \displaystyle\int x\ln x\, dx = \frac{1}{2}x^2\ln x - \frac{1}{2}\int x\, dx = \frac{1}{2}x^2\ln x - \frac{1}{4}x^2 + C$

11. $u = (\ln x)^2,\ dv = dx,\ du = 2\dfrac{\ln x}{x}dx,\ v = x;\ \displaystyle\int (\ln x)^2 dx = x(\ln x)^2 - 2\int \ln x\, dx.$

Use $u = \ln x,\ dv = dx$ to get $\displaystyle\int \ln x\, dx = x\ln x - \int dx = x\ln x - x + C_1$ so

$\displaystyle\int (\ln x)^2 dx = x(\ln x)^2 - 2x\ln x + 2x + C$

13. $u = \ln(3x - 2)$, $dv = dx$, $du = \dfrac{3}{3x - 2}dx$, $v = x$; $\displaystyle\int \ln(3x - 2)dx = x\ln(3x - 2) - \int \dfrac{3x}{3x - 2}dx$

but $\displaystyle\int \dfrac{3x}{3x - 2}dx = \int \left(1 + \dfrac{2}{3x - 2}\right)dx = x + \dfrac{2}{3}\ln(3x - 2) + C_1$ so

$\displaystyle\int \ln(3x - 2)dx = x\ln(3x - 2) - x - \dfrac{2}{3}\ln(3x - 2) + C$

15. $u = \sin^{-1} x$, $dv = dx$, $du = 1/\sqrt{1 - x^2}dx$, $v = x$;

$\displaystyle\int \sin^{-1} x\, dx = x\sin^{-1} x - \int x/\sqrt{1 - x^2}dx = x\sin^{-1} x + \sqrt{1 - x^2} + C$

17. $u = \tan^{-1}(3x)$, $dv = dx$, $du = \dfrac{3}{1 + 9x^2}dx$, $v = x$;

$\displaystyle\int \tan^{-1}(3x)dx = x\tan^{-1}(3x) - \int \dfrac{3x}{1 + 9x^2}dx = x\tan^{-1}(3x) - \dfrac{1}{6}\ln(1 + 9x^2) + C$

19. $u = e^x$, $dv = \sin x\, dx$, $du = e^x dx$, $v = -\cos x$; $\displaystyle\int e^x \sin x\, dx = -e^x \cos x + \int e^x \cos x\, dx$.

For $\displaystyle\int e^x \cos x\, dx$ use $u = e^x$, $dv = \cos x\, dx$ to get $\displaystyle\int e^x \cos x = e^x \sin x - \int e^x \sin x\, dx$ so

$\displaystyle\int e^x \sin x\, dx = -e^x \cos x + e^x \sin x - \int e^x \sin x\, dx$,

$2\displaystyle\int e^x \sin x\, dx = e^x(\sin x - \cos x) + C_1$, $\displaystyle\int e^x \sin x\, dx = \dfrac{1}{2}e^x(\sin x - \cos x) + C$

21. $u = e^{ax}$, $dv = \sin bx\, dx$, $du = ae^{ax}dx$, $v = -\dfrac{1}{b}\cos bx$ $\quad(b \neq 0)$;

$\displaystyle\int e^{ax} \sin bx\, dx = -\dfrac{1}{b}e^{ax}\cos bx + \dfrac{a}{b}\int e^{ax}\cos bx\, dx$. Use $u = e^{ax}$, $dv = \cos bx\, dx$ to get

$\displaystyle\int e^{ax}\cos bx\, dx = \dfrac{1}{b}e^{ax}\sin bx - \dfrac{a}{b}\int e^{ax}\sin bx\, dx$ so

$\displaystyle\int e^{ax}\sin bx\, dx = -\dfrac{1}{b}e^{ax}\cos bx + \dfrac{a}{b^2}e^{ax}\sin bx - \dfrac{a^2}{b^2}\int e^{ax}\sin bx\, dx$,

$\displaystyle\int e^{ax}\sin bx\, dx = \dfrac{e^{ax}}{a^2 + b^2}(a\sin bx - b\cos bx) + C$

23. $u = \sin(\ln x)$, $dv = dx$, $du = \dfrac{\cos(\ln x)}{x}dx$, $v = x$;

$\displaystyle\int \sin(\ln x)dx = x\sin(\ln x) - \int \cos(\ln x)dx$. Use $u = \cos(\ln x)$, $dv = dx$ to get

$\displaystyle\int \cos(\ln x)dx = x\cos(\ln x) + \int \sin(\ln x)dx$ so

$\displaystyle\int \sin(\ln x)dx = x\sin(\ln x) - x\cos(\ln x) - \int \sin(\ln x)dx$,

$\displaystyle\int \sin(\ln x)dx = \dfrac{1}{2}x[\sin(\ln x) - \cos(\ln x)] + C$

25. $u = x$, $dv = \sec^2 x\, dx$, $du = dx$, $v = \tan x$;

$$\int x \sec^2 x\, dx = x \tan x - \int \tan x\, dx = x \tan x - \int \frac{\sin x}{\cos x} dx = x \tan x + \ln|\cos x| + C$$

27. $u = x^2$, $dv = xe^{x^2} dx$, $du = 2x\, dx$, $v = \frac{1}{2}e^{x^2}$;

$$\int x^3 e^{x^2} dx = \frac{1}{2}x^2 e^{x^2} - \int xe^{x^2} dx = \frac{1}{2}x^2 e^{x^2} - \frac{1}{2}e^{x^2} + C$$

29. $u = x$, $dv = e^{2x} dx$, $du = dx$, $v = \frac{1}{2}e^{2x}$;

$$\int_0^2 xe^{2x} dx = \frac{1}{2}xe^{2x}\Big]_0^2 - \frac{1}{2}\int_0^2 e^{2x} dx = e^4 - \frac{1}{4}e^{2x}\Big]_0^2 = e^4 - \frac{1}{4}(e^4 - 1) = (3e^4 + 1)/4$$

31. $u = \ln x$, $dv = x^2 dx$, $du = \frac{1}{x}dx$, $v = \frac{1}{3}x^3$;

$$\int_1^e x^2 \ln x\, dx = \frac{1}{3}x^3 \ln x\Big]_1^e - \frac{1}{3}\int_1^e x^2 dx = \frac{1}{3}e^3 - \frac{1}{9}x^3\Big]_1^e = \frac{1}{3}e^3 - \frac{1}{9}(e^3 - 1) = (2e^3 + 1)/9$$

33. $u = \ln(x + 2)$, $dv = dx$, $du = \frac{1}{x + 2}dx$, $v = x$;

$$\int_{-1}^1 \ln(x + 2)dx = x \ln(x + 2)\Big]_{-1}^1 - \int_{-1}^1 \frac{x}{x + 2}dx = \ln 3 + \ln 1 - \int_{-1}^1 \left[1 - \frac{2}{x + 2}\right] dx$$

$$= \ln 3 - [x - 2\ln(x + 2)]\Big]_{-1}^1 = \ln 3 - (1 - 2\ln 3) + (-1 - 2\ln 1) = 3\ln 3 - 2$$

35. $u = \sec^{-1}\sqrt{\theta}$, $dv = d\theta$, $du = \frac{1}{2\theta\sqrt{\theta - 1}}d\theta$, $v = \theta$;

$$\int_2^4 \sec^{-1}\sqrt{\theta}d\theta = \theta \sec^{-1}\sqrt{\theta}\Big]_2^4 - \frac{1}{2}\int_2^4 \frac{1}{\sqrt{\theta - 1}}d\theta = 4\sec^{-1}2 - 2\sec^{-1}\sqrt{2} - \sqrt{\theta - 1}\Big]_2^4$$

$$= 4\left(\frac{\pi}{3}\right) - 2\left(\frac{\pi}{4}\right) - \sqrt{3} + 1 = \frac{5\pi}{6} - \sqrt{3} + 1$$

37. $u = x$, $dv = \sin 2x\, dx$, $du = dx$, $v = -\frac{1}{2}\cos 2x$;

$$\int_0^\pi x \sin 2x\, dx = -\frac{1}{2}x \cos 2x\Big]_0^\pi + \frac{1}{2}\int_0^\pi \cos 2x\, dx = -\pi/2 + \frac{1}{4}\sin 2x\Big]_0^\pi = -\pi/2$$

39. $u = \tan^{-1}\sqrt{x}$, $dv = \sqrt{x}dx$, $du = \frac{1}{2\sqrt{x}(1 + x)}dx$, $v = \frac{2}{3}x^{3/2}$;

$$\int_1^3 \sqrt{x}\tan^{-1}\sqrt{x}dx = \frac{2}{3}x^{3/2}\tan^{-1}\sqrt{x}\Big]_1^3 - \frac{1}{3}\int_1^3 \frac{x}{1 + x}dx$$

$$= \frac{2}{3}x^{3/2}\tan^{-1}\sqrt{x}\Big]_1^3 - \frac{1}{3}\int_1^3 \left[1 - \frac{1}{1 + x}\right] dx$$

$$= \left[\frac{2}{3}x^{3/2}\tan^{-1}\sqrt{x} - \frac{1}{3}x + \frac{1}{3}\ln|1 + x|\right]_1^3 = (2\sqrt{3}\pi - \pi/2 - 2 + \ln 2)/3$$

41. $t = \sqrt{x}$, $t^2 = x$, $dx = 2t\, dt$

(a) $\int e^{\sqrt{x}} dx = 2 \int t e^t \, dt; \; u = t, dv = e^t dt, du = dt, v = e^t,$

$\int e^{\sqrt{x}} dx = 2te^t - 2 \int e^t \, dt = 2(t-1)e^t + C = 2(\sqrt{x}-1)e^{\sqrt{x}} + C$

(b) $\int \cos \sqrt{x} \, dx = 2 \int t \cos t \, dt; \; u = t, dv = \cos t \, dt, du = dt, v = \sin t,$

$\int \cos \sqrt{x} \, dx = 2t \sin t - 2 \int \sin t \, dt = 2t \sin t + 2 \cos t + C = 2\sqrt{x} \sin \sqrt{x} + 2 \cos \sqrt{x} + C$

43.

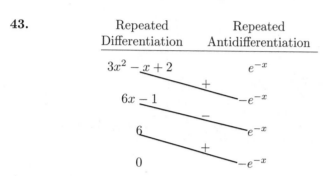

Repeated Differentiation	Repeated Antidifferentiation
$3x^2 - x + 2$	e^{-x}
$6x - 1$	$-e^{-x}$
6	e^{-x}
0	$-e^{-x}$

$\int (3x^2 - x + 2)e^{-x} = -(3x^2 - x + 2)e^{-x} - (6x-1)e^{-x} - 6e^{-x} + C = -e^{-x}[3x^2 + 5x + 7] + C$

45.

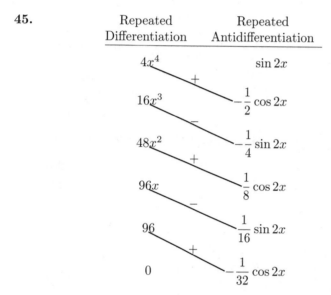

Repeated Differentiation	Repeated Antidifferentiation
$4x^4$	$\sin 2x$
$16x^3$	$-\dfrac{1}{2} \cos 2x$
$48x^2$	$-\dfrac{1}{4} \sin 2x$
$96x$	$\dfrac{1}{8} \cos 2x$
96	$\dfrac{1}{16} \sin 2x$
0	$-\dfrac{1}{32} \cos 2x$

$\int 4x^4 \sin 2x \, dx = (-2x^4 + 6x^2 - 3) \cos 2x + -(4x^3 + 6x) \sin 2x + C$

47. (a) We perform a single integration by parts:

$u = \cos x, \; dv = \sin x \, dx, \; du = -\sin x \, dx, \; v = -\cos x,$

$\int \sin x \cos x \, dx = -\cos^2 x - \int \sin x \cos x \, dx.$ Thus

$2 \int \sin x \cos x \, dx = -\cos^2 x + C, \int \sin x \cos x \, dx = -\dfrac{1}{2} \cos^2 x + C$

(b) $u = \sin x, du = \cos x \, dx, \int \sin x \cos x \, dx = \int u \, du = \dfrac{1}{2} u^2 + C = \dfrac{1}{2} \sin^2 x + C$

49. (a) $A = \int_1^e \ln x\, dx = (x\ln x - x)\Big]_1^e = 1$

(b) $V = \pi \int_1^e (\ln x)^2 dx = \pi\Big[(x(\ln x)^2 - 2x\ln x + 2x)\Big]_1^e = \pi(e-2)$

51. $V = 2\pi \int_0^\pi x\sin x\, dx = 2\pi(-x\cos x + \sin x)\Big]_0^\pi = 2\pi^2$

53. distance $= \int_0^\pi t^3 \sin t\, dt;$

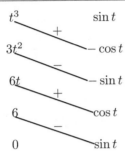

Repeated Differentiation	Repeated Antidifferentiation
t^3	$\sin t$
$3t^2$	$-\cos t$
$6t$	$-\sin t$
6	$\cos t$
0	$\sin t$

$$\int_0^\pi t^3 \sin t\, dx = \Big[(-t^3\cos t + 3t^2\sin t + 6t\cos t - 6\sin t)\Big]_0^\pi = \pi^3 - 6\pi$$

55. (a) $\int \sin^4 x\, dx = -\frac{1}{4}\sin^3 x\cos x + \frac{3}{4}\int \sin^2 x\, dx$

$$= -\frac{1}{4}\sin^3 x\cos x + \frac{3}{4}\Big[-\frac{1}{2}\sin x\cos x + \frac{1}{2}x\Big] + C$$

$$= -\frac{1}{4}\sin^3 x\cos x - \frac{3}{8}\sin x\cos x + \frac{3}{8}x + C$$

(b) $\int_0^{\pi/2} \sin^5 x\, dx = -\frac{1}{5}\sin^4 x\cos x\Big]_0^{\pi/2} + \frac{4}{5}\int_0^{\pi/2}\sin^3 x\, dx$

$$= \frac{4}{5}\Big[-\frac{1}{3}\sin^2 x\cos x\Big]_0^{\pi/2} + \frac{2}{3}\int_0^{\pi/2}\sin x\, dx\Big]$$

$$= -\frac{8}{15}\cos x\Big]_0^{\pi/2} = \frac{8}{15}$$

57. $u = \sin^{n-1} x,\ dv = \sin x\, dx,\ du = (n-1)\sin^{n-2} x\cos x\, dx,\ v = -\cos x;$

$$\int \sin^n x\, dx = -\sin^{n-1} x\cos x + (n-1)\int \sin^{n-2} x\cos^2 x\, dx$$

$$= -\sin^{n-1} x\cos x + (n-1)\int \sin^{n-2} x\,(1 - \sin^2 x)dx$$

$$= -\sin^{n-1} x\cos x + (n-1)\int \sin^{n-2} x\, dx - (n-1)\int \sin^n x\, dx,$$

$$n \int \sin^n x \, dx = -\sin^{n-1} x \cos x + (n-1) \int \sin^{n-2} x \, dx,$$

$$\int \sin^n x \, dx = -\frac{1}{n} \sin^{n-1} x \cos x + \frac{n-1}{n} \int \sin^{n-2} x \, dx$$

59. (a) $\displaystyle \int \tan^4 x \, dx = \frac{1}{3} \tan^3 x - \int \tan^2 x \, dx = \frac{1}{3} \tan^3 x - \tan x + \int dx = \frac{1}{3} \tan^3 x - \tan x + x + C$

(b) $\displaystyle \int \sec^4 x \, dx = \frac{1}{3} \sec^2 x \tan x + \frac{2}{3} \int \sec^2 x \, dx = \frac{1}{3} \sec^2 x \tan x + \frac{2}{3} \tan x + C$

(c) $\displaystyle \int x^3 e^x \, dx = x^3 e^x - 3 \int x^2 e^x \, dx = x^3 e^x - 3 \left[x^2 e^x - 2 \int x e^x \, dx \right]$

$$= x^3 e^x - 3x^2 e^x + 6 \left[x e^x - \int e^x \, dx \right] = x^3 e^x - 3x^2 e^x + 6x e^x - 6e^x + C$$

61. $u = x, \, dv = f''(x)dx, \, du = dx, \, v = f'(x);$

$$\int_{-1}^{1} x f''(x) dx = x f'(x) \Big]_{-1}^{1} - \int_{-1}^{1} f'(x) dx$$

$$= f'(1) + f'(-1) - f(x) \Big]_{-1}^{1} = f'(1) + f'(-1) - f(1) + f(-1)$$

63. $u = \ln(x+1), \, dv = dx, \, du = \dfrac{dx}{x+1}, \, v = x+1;$

$$\int \ln(x+1) \, dx = \int u \, dv = uv - \int v \, du = (x+1) \ln(x+1) - \int dx = (x+1) \ln(x+1) - x + C$$

65. $u = \tan^{-1} x, \, dv = x \, dx, \, du = \dfrac{1}{1+x^2} \, dx, \, v = \dfrac{1}{2}(x^2+1)$

$$\int x \tan^{-1} x \, dx = \int u \, dv = uv - \int v \, du = \frac{1}{2}(x^2+1) \tan^{-1} x - \frac{1}{2} \int dx$$

$$= \frac{1}{2}(x^2+1) \tan^{-1} x - \frac{1}{2}x + C$$

67. (a) $u = f(x), \, dv = dx, \, du = f'(x), \, v = x;$

$$\int_a^b f(x) \, dx = x f(x) \Big]_a^b - \int_a^b x f'(x) \, dx = b f(b) - a f(a) - \int_a^b x f'(x) \, dx$$

(b) Substitute $y = f(x), \, dy = f'(x) \, dx, \, x = a$ when $y = f(a), \, x = b$ when $y = f(b)$,

$$\int_a^b x f'(x) \, dx = \int_{f(a)}^{f(b)} x \, dy = \int_{f(a)}^{f(b)} f^{-1}(y) \, dy$$

(c) From $a = f^{-1}(\alpha)$ and $b = f^{-1}(\beta)$ we get
$b f(b) - a f(a) = \beta f^{-1}(\beta) - \alpha f^{-1}(\alpha);$ then

$$\int_\alpha^\beta f^{-1}(x) \, dx = \int_\alpha^\beta f^{-1}(y) \, dy = \int_{f(a)}^{f(b)} f^{-1}(y) \, dy,$$

which, by Part (b), yields

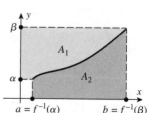

$$\int_\alpha^\beta f^{-1}(x)\,dx = bf(b) - af(a) - \int_a^b f(x)\,dx$$

$$= \beta f^{-1}(\beta) - \alpha f^{-1}(\alpha) - \int_{f^{-1}(\alpha)}^{f^{-1}(\beta)} f(x)\,dx$$

Note from the figure that $A_1 = \int_\alpha^\beta f^{-1}(x)\,dx$, $A_2 = \int_{f^{-1}(\alpha)}^{f^{-1}(\beta)} f(x)\,dx$, and

$A_1 + A_2 = \beta f^{-1}(\beta) - \alpha f^{-1}(\alpha)$, a "picture proof".

EXERCISE SET 8.3

1. $u = \cos x$, $-\int u^3\,du = -\dfrac{1}{4}\cos^4 x + C$

3. $\displaystyle\int \sin^2 5\theta = \frac{1}{2}\int (1 - \cos 10\theta)\,d\theta = \frac{1}{2}\theta - \frac{1}{20}\sin 10\theta + C$

5. $\displaystyle\int \sin^3 a\theta\,d\theta = \int \sin a\theta(1 - \cos^2 a\theta)\,d\theta = -\frac{1}{a}\cos a\theta - \frac{1}{3a}\cos^3 a\theta + C \quad (a \neq 0)$

7. $u = \sin ax$, $\dfrac{1}{a}\displaystyle\int u\,du = \dfrac{1}{2a}\sin^2 ax + C, a \neq 0$

9. $\displaystyle\int \sin^2 t\cos^3 t\,dt = \int \sin^2 t(1 - \sin^2 t)\cos t\,dt = \int (\sin^2 t - \sin^4 t)\cos t\,dt$

$$= \frac{1}{3}\sin^3 t - \frac{1}{5}\sin^5 t + C$$

11. $\displaystyle\int \sin^2 x\cos^2 x\,dx = \frac{1}{4}\int \sin^2 2x\,dx = \frac{1}{8}\int (1 - \cos 4x)\,dx = \frac{1}{8}x - \frac{1}{32}\sin 4x + C$

13. $\displaystyle\int \sin 2x\cos 3x\,dx = \frac{1}{2}\int (\sin 5x - \sin x)\,dx = -\frac{1}{10}\cos 5x + \frac{1}{2}\cos x + C$

15. $\displaystyle\int \sin x\cos(x/2)\,dx = \frac{1}{2}\int [\sin(3x/2) + \sin(x/2)]\,dx = -\frac{1}{3}\cos(3x/2) - \cos(x/2) + C$

17. $\displaystyle\int_0^{\pi/2} \cos^3 x\,dx = \int_0^{\pi/2}(1 - \sin^2 x)\cos x\,dx$

$$= \left[\sin x - \frac{1}{3}\sin^3 x\right]_0^{\pi/2} = \frac{2}{3}$$

19. $\displaystyle\int_0^{\pi/3} \sin^4 3x\cos^3 3x\,dx = \int_0^{\pi/3} \sin^4 3x(1 - \sin^2 3x)\cos 3x\,dx = \left[\frac{1}{15}\sin^5 3x - \frac{1}{21}\sin^7 3x\right]_0^{\pi/3} = 0$

21. $\displaystyle\int_0^{\pi/6} \sin 4x\cos 2x\,dx = \frac{1}{2}\int_0^{\pi/6}(\sin 2x + \sin 6x)\,dx = \left[-\frac{1}{4}\cos 2x - \frac{1}{12}\cos 6x\right]_0^{\pi/6}$

$$= [(-1/4)(1/2) - (1/12)(-1)] - [-1/4 - 1/12] = 7/24$$

23. $\dfrac{1}{2}\tan(2x - 1) + C$

25. $u = e^{-x}, du = -e^{-x} dx; \; -\int \tan u \, du = \ln|\cos u| + C = \ln|\cos(e^{-x})| + C$

27. $\dfrac{1}{4}\ln|\sec 4x + \tan 4x| + C$

29. $u = \tan x, \; \int u^2 du = \dfrac{1}{3}\tan^3 x + C$

31. $\displaystyle\int \tan 4x(1 + \tan^2 4x)\sec^2 4x \, dx = \int (\tan 4x + \tan^3 4x)\sec^2 4x \, dx = \dfrac{1}{8}\tan^2 4x + \dfrac{1}{16}\tan^4 4x + C$

33. $\displaystyle\int \sec^4 x(\sec^2 x - 1)\sec x \tan x \, dx = \int (\sec^6 x - \sec^4 x)\sec x \tan x \, dx = \dfrac{1}{7}\sec^7 x - \dfrac{1}{5}\sec^5 x + C$

35. $\displaystyle\int (\sec^2 x - 1)^2 \sec x \, dx = \int (\sec^5 x - 2\sec^3 x + \sec x)dx = \int \sec^5 x \, dx - 2\int \sec^3 x \, dx + \int \sec x \, dx$

$$= \dfrac{1}{4}\sec^3 x \tan x + \dfrac{3}{4}\int \sec^3 x \, dx - 2\int \sec^3 x \, dx + \ln|\sec x + \tan x|$$

$$= \dfrac{1}{4}\sec^3 x \tan x - \dfrac{5}{4}\left[\dfrac{1}{2}\sec x \tan x + \dfrac{1}{2}\ln|\sec x + \tan x|\right] + \ln|\sec x + \tan x| + C$$

$$= \dfrac{1}{4}\sec^3 x \tan x - \dfrac{5}{8}\sec x \tan x + \dfrac{3}{8}\ln|\sec x + \tan x| + C$$

37. $\displaystyle\int \sec^2 t(\sec t \tan t)dt = \dfrac{1}{3}\sec^3 t + C$

39. $\displaystyle\int \sec^4 x \, dx = \int (1 + \tan^2 x)\sec^2 x \, dx = \int (\sec^2 x + \tan^2 x \sec^2 x)dx = \tan x + \dfrac{1}{3}\tan^3 x + C$

41. $u = 4x$, use equation (19) to get

$$\dfrac{1}{4}\int \tan^3 u \, du = \dfrac{1}{4}\left[\dfrac{1}{2}\tan^2 u + \ln|\cos u|\right] + C = \dfrac{1}{8}\tan^2 4x + \dfrac{1}{4}\ln|\cos 4x| + C$$

43. $\displaystyle\int \sqrt{\tan x}(1 + \tan^2 x)\sec^2 x \, dx = \dfrac{2}{3}\tan^{3/2} x + \dfrac{2}{7}\tan^{7/2} x + C$

45. $\displaystyle\int_0^{\pi/8} (\sec^2 2x - 1)dx = \left[\dfrac{1}{2}\tan 2x - x\right]_0^{\pi/8} = 1/2 - \pi/8$

47. $u = x/2$,

$$2\int_0^{\pi/4} \tan^5 u \, du = \left[\dfrac{1}{2}\tan^4 u - \tan^2 u - 2\ln|\cos u|\right]_0^{\pi/4} = 1/2 - 1 - 2\ln(1/\sqrt{2}) = -1/2 + \ln 2$$

49. $\displaystyle\int (\csc^2 x - 1)\csc^2 x(\csc x \cot x)dx = \int (\csc^4 x - \csc^2 x)(\csc x \cot x)dx = -\dfrac{1}{5}\csc^5 x + \dfrac{1}{3}\csc^3 x + C$

51. $\displaystyle\int (\csc^2 x - 1)\cot x \, dx = \int \csc x(\csc x \cot x)dx - \int \dfrac{\cos x}{\sin x}dx = -\dfrac{1}{2}\csc^2 x - \ln|\sin x| + C$

53. **(a)** $\displaystyle\int_0^{2\pi} \sin mx \cos nx \, dx = \dfrac{1}{2}\int_0^{2\pi} [\sin(m+n)x + \sin(m-n)x]dx$

$$= \left[-\dfrac{\cos(m+n)x}{2(m+n)} - \dfrac{\cos(m-n)x}{2(m-n)}\right]_0^{2\pi}$$

$$\text{but } \cos(m+n)x\Big]_0^{2\pi} = 0, \cos(m-n)x\Big]_0^{2\pi} = 0.$$

(b) $\displaystyle\int_0^{2\pi} \cos mx \cos nx\,dx = \frac{1}{2}\int_0^{2\pi} [\cos(m+n)x + \cos(m-n)x]dx;$

since $m \neq n$, evaluate sin at integer multiples of 2π to get 0.

(c) $\displaystyle\int_0^{2\pi} \sin mx \sin nx\,dx = \frac{1}{2}\int_0^{2\pi} [\cos(m-n)x - \cos(m+n)x]\,dx;$

since $m \neq n$, evaluate sin at integer multiples of 2π to get 0.

55. $y' = \tan x,\ 1 + (y')^2 = 1 + \tan^2 x = \sec^2 x,$

$$L = \int_0^{\pi/4} \sqrt{\sec^2 x}\,dx = \int_0^{\pi/4} \sec x\,dx = \ln|\sec x + \tan x|\Big]_0^{\pi/4} = \ln(\sqrt{2}+1)$$

57. $\displaystyle V = \pi\int_0^{\pi/4} (\cos^2 x - \sin^2 x)dx = \pi\int_0^{\pi/4} \cos 2x\,dx = \frac{1}{2}\pi\sin 2x\Big]_0^{\pi/4} = \pi/2$

59. With $0 < \alpha < \beta, D = D_\beta - D_\alpha = \dfrac{L}{2\pi}\displaystyle\int_\alpha^\beta \sec x\,dx = \dfrac{L}{2\pi}\ln|\sec x + \tan x|\Big]_\alpha^\beta = \dfrac{L}{2\pi}\ln\left|\dfrac{\sec\beta + \tan\beta}{\sec\alpha + \tan\alpha}\right|$

61. (a) $\displaystyle\int \csc x\,dx = \int \sec(\pi/2 - x)dx = -\ln|\sec(\pi/2 - x) + \tan(\pi/2 - x)| + C$

$$= -\ln|\csc x + \cot x| + C$$

(b) $-\ln|\csc x + \cot x| = \ln\dfrac{1}{|\csc x + \cot x|} = \ln\dfrac{|\csc x - \cot x|}{|\csc^2 x - \cot^2 x|} = \ln|\csc x - \cot x|,$

$$-\ln|\csc x + \cot x| = -\ln\left|\dfrac{1}{\sin x} + \dfrac{\cos x}{\sin x}\right| = \ln\left|\dfrac{\sin x}{1 + \cos x}\right|$$

$$= \ln\left|\dfrac{2\sin(x/2)\cos(x/2)}{2\cos^2(x/2)}\right| = \ln|\tan(x/2)|$$

63. $a\sin x + b\cos x = \sqrt{a^2 + b^2}\left[\dfrac{a}{\sqrt{a^2 + b^2}}\sin x + \dfrac{b}{\sqrt{a^2 + b^2}}\cos x\right] = \sqrt{a^2 + b^2}(\sin x\cos\theta + \cos x\sin\theta)$

where $\cos\theta = a/\sqrt{a^2 + b^2}$ and $\sin\theta = b/\sqrt{a^2 + b^2}$ so $a\sin x + b\cos x = \sqrt{a^2 + b^2}\sin(x + \theta)$

and $\displaystyle\int \dfrac{dx}{a\sin x + b\cos x} = \dfrac{1}{\sqrt{a^2 + b^2}}\int \csc(x + \theta)dx = -\dfrac{1}{\sqrt{a^2 + b^2}}\ln|\csc(x + \theta) + \cot(x + \theta)| + C$

$$= -\dfrac{1}{\sqrt{a^2 + b^2}}\ln\left|\dfrac{\sqrt{a^2 + b^2} + a\cos x - b\sin x}{a\sin x + b\cos x}\right| + C$$

65. (a) $\displaystyle\int_0^{\pi/2} \sin^3 x\,dx = \dfrac{2}{3}$　　　　　　　　**(b)** $\displaystyle\int_0^{\pi/2} \sin^4 x\,dx = \dfrac{1\cdot 3}{2\cdot 4}\cdot\dfrac{\pi}{2} = 3\pi/16$

(c) $\displaystyle\int_0^{\pi/2} \sin^5 x\,dx = \dfrac{2\cdot 4}{3\cdot 5} = 8/15$　　　　**(d)** $\displaystyle\int_0^{\pi/2} \sin^6 x\,dx = \dfrac{1\cdot 3\cdot 5}{2\cdot 4\cdot 6}\cdot\dfrac{\pi}{2} = 5\pi/32$

EXERCISE SET 8.4

1. $x = 2\sin\theta$, $dx = 2\cos\theta\,d\theta$,

$$4\int \cos^2\theta\,d\theta = 2\int(1+\cos 2\theta)d\theta = 2\theta + \sin 2\theta + C$$

$$= 2\theta + 2\sin\theta\cos\theta + C = 2\sin^{-1}(x/2) + \frac{1}{2}x\sqrt{4-x^2} + C$$

3. $x = 4\sin\theta$, $dx = 4\cos\theta\,d\theta$,

$$16\int \sin^2\theta\,d\theta = 8\int(1-\cos 2\theta)d\theta = 8\theta - 4\sin 2\theta + C = 8\theta - 8\sin\theta\cos\theta + C$$

$$= 8\sin^{-1}(x/4) - \frac{1}{2}x\sqrt{16-x^2} + C$$

5. $x = 2\tan\theta$, $dx = 2\sec^2\theta\,d\theta$,

$$\frac{1}{8}\int \frac{1}{\sec^2\theta}d\theta = \frac{1}{8}\int \cos^2\theta\,d\theta = \frac{1}{16}\int(1+\cos 2\theta)d\theta = \frac{1}{16}\theta + \frac{1}{32}\sin 2\theta + C$$

$$= \frac{1}{16}\theta + \frac{1}{16}\sin\theta\cos\theta + C = \frac{1}{16}\tan^{-1}\frac{x}{2} + \frac{x}{8(4+x^2)} + C$$

7. $x = 3\sec\theta$, $dx = 3\sec\theta\tan\theta\,d\theta$,

$$3\int \tan^2\theta\,d\theta = 3\int(\sec^2\theta - 1)d\theta = 3\tan\theta - 3\theta + C = \sqrt{x^2-9} - 3\sec^{-1}\frac{x}{3} + C$$

9. $x = \sin\theta$, $dx = \cos\theta\,d\theta$,

$$3\int \sin^3\theta\,d\theta = 3\int\left[1-\cos^2\theta\right]\sin\theta\,d\theta$$

$$= 3\left(-\cos\theta + \cos^3\theta\right) + C = -3\sqrt{1-x^2} + (1-x^2)^{3/2} + C$$

11. $x = \frac{2}{3}\sec\theta$, $dx = \frac{2}{3}\sec\theta\tan\theta\,d\theta$, $\frac{3}{4}\int \frac{1}{\sec\theta}d\theta = \frac{3}{4}\int \cos\theta\,d\theta = \frac{3}{4}\sin\theta + C = \frac{1}{4x}\sqrt{9x^2-4} + C$

13. $x = \sin\theta$, $dx = \cos\theta\,d\theta$, $\int \frac{1}{\cos^2\theta}d\theta = \int \sec^2\theta\,d\theta = \tan\theta + C = x/\sqrt{1-x^2} + C$

15. $x = 3\sec\theta$, $dx = 3\sec\theta\tan\theta\,d\theta$, $\int \sec\theta\,d\theta = \ln|\sec\theta + \tan\theta| + C = \ln\left|\frac{1}{3}x + \frac{1}{3}\sqrt{x^2-9}\right| + C$

17. $x = \frac{3}{2}\sec\theta$, $dx = \frac{3}{2}\sec\theta\tan\theta\,d\theta$,

$$\frac{3}{2}\int \frac{\sec\theta\tan\theta\,d\theta}{27\tan^3\theta} = \frac{1}{18}\int \frac{\cos\theta}{\sin^2\theta}d\theta = -\frac{1}{18}\frac{1}{\sin\theta} + C = -\frac{1}{18}\csc\theta + C = -\frac{x}{9\sqrt{4x^2-9}} + C$$

19. $e^x = \sin\theta$, $e^x dx = \cos\theta\,d\theta$,

$$\int \cos^2\theta\,d\theta = \frac{1}{2}\int(1+\cos 2\theta)d\theta = \frac{1}{2}\theta + \frac{1}{4}\sin 2\theta + C = \frac{1}{2}\sin^{-1}(e^x) + \frac{1}{2}e^x\sqrt{1-e^{2x}} + C$$

21. $x = \sin\theta$, $dx = \cos\theta\,d\theta$,

$$5\int_0^1 \sin^3\theta\cos^2\theta\,d\theta = 5\left[-\frac{1}{3}\cos^3\theta + \frac{1}{5}\cos^5\theta\right]_0^{\pi/2} = 5(1/3 - 1/5) = 2/3$$

23. $x = \sec\theta,\ dx = \sec\theta\tan\theta\,d\theta,\ \displaystyle\int_{\pi/4}^{\pi/3}\frac{1}{\sec\theta}d\theta = \int_{\pi/4}^{\pi/3}\cos\theta\,d\theta = \sin\theta\Big]_{\pi/4}^{\pi/3} = (\sqrt{3}-\sqrt{2})/2$

25. $x = \sqrt{3}\tan\theta,\ dx = \sqrt{3}\sec^2\theta\,d\theta,$

$$\frac{1}{9}\int_{\pi/6}^{\pi/3}\frac{\sec\theta}{\tan^4\theta}d\theta = \frac{1}{9}\int_{\pi/6}^{\pi/3}\frac{\cos^3\theta}{\sin^4\theta}d\theta = \frac{1}{9}\int_{\pi/6}^{\pi/3}\frac{1-\sin^2\theta}{\sin^4\theta}\cos\theta\,d\theta = \frac{1}{9}\int_{1/2}^{\sqrt{3}/2}\frac{1-u^2}{u^4}du\ (u=\sin\theta)$$

$$= \frac{1}{9}\int_{1/2}^{\sqrt{3}/2}(u^{-4}-u^{-2})du = \frac{1}{9}\left[-\frac{1}{3u^3}+\frac{1}{u}\right]_{1/2}^{\sqrt{3}/2} = \frac{10\sqrt{3}+18}{243}$$

27. $u = x^2+4,\ du = 2x\,dx,$

$$\frac{1}{2}\int\frac{1}{u}du = \frac{1}{2}\ln|u|+C = \frac{1}{2}\ln(x^2+4)+C;\ \text{or } x = 2\tan\theta,\ dx = 2\sec^2\theta\,d\theta,$$

$$\int\tan\theta\,d\theta = \ln|\sec\theta|+C_1 = \ln\frac{\sqrt{x^2+4}}{2}+C_1 = \ln(x^2+4)^{1/2}-\ln 2+C_1$$

$$= \frac{1}{2}\ln(x^2+4)+C \text{ with } C = C_1-\ln 2$$

29. $y' = \dfrac{1}{x},\ 1+(y')^2 = 1+\dfrac{1}{x^2} = \dfrac{x^2+1}{x^2},$

$$L = \int_1^2\sqrt{\frac{x^2+1}{x^2}}dx;\qquad x = \tan\theta,\ dx = \sec^2\theta\,d\theta,$$

$$L = \int_{\pi/4}^{\tan^{-1}(2)}\frac{\sec^3\theta}{\tan\theta}d\theta = \int_{\pi/4}^{\tan^{-1}(2)}\frac{\tan^2\theta+1}{\tan\theta}\sec\theta\,d\theta = \int_{\pi/4}^{\tan^{-1}(2)}(\sec\theta\tan\theta+\csc\theta)d\theta$$

$$= \Big[\sec\theta+\ln|\csc\theta-\cot\theta|\Big]_{\pi/4}^{\tan^{-1}(2)} = \sqrt{5}+\ln\left(\frac{\sqrt{5}}{2}-\frac{1}{2}\right)-\Big[\sqrt{2}+\ln|\sqrt{2}-1|\Big]$$

$$= \sqrt{5}-\sqrt{2}+\ln\frac{2+2\sqrt{2}}{1+\sqrt{5}}$$

31. $y' = 2x,\ 1+(y')^2 = 1+4x^2,$

$$S = 2\pi\int_0^1 x^2\sqrt{1+4x^2}dx;\ x = \frac{1}{2}\tan\theta,\ dx = \frac{1}{2}\sec^2\theta\,d\theta,$$

$$S = \frac{\pi}{4}\int_0^{\tan^{-1}2}\tan^2\theta\sec^3\theta\,d\theta = \frac{\pi}{4}\int_0^{\tan^{-1}2}(\sec^2\theta-1)\sec^3\theta\,d\theta = \frac{\pi}{4}\int_0^{\tan^{-1}2}(\sec^5\theta-\sec^3\theta)d\theta$$

$$= \frac{\pi}{4}\left[\frac{1}{4}\sec^3\theta\tan\theta-\frac{1}{8}\sec\theta\tan\theta-\frac{1}{8}\ln|\sec\theta+\tan\theta|\right]_0^{\tan^{-1}2} = \frac{\pi}{32}[18\sqrt{5}-\ln(2+\sqrt{5})]$$

33. $\displaystyle\int\frac{1}{(x-2)^2+1}dx = \tan^{-1}(x-2)+C$

35. $\displaystyle\int\frac{1}{\sqrt{4-(x-1)^2}}dx = \sin^{-1}\left(\frac{x-1}{2}\right)+C$

37. $\displaystyle\int\frac{1}{\sqrt{(x-3)^2+1}}dx = \ln\left(x-3+\sqrt{(x-3)^2+1}\right)+C$

39. $\int \sqrt{4 - (x+1)^2}\, dx$, let $x + 1 = 2\sin\theta$,

$$= \int 4\cos^2\theta\, d\theta = \int 2(1 + \cos 2\theta)\, d\theta$$

$$= 2\theta + \sin 2\theta + C = 2\sin^{-1}\left(\frac{x+1}{2}\right) + \frac{1}{2}(x+1)\sqrt{3 - 2x - x^2} + C$$

41. $\int \dfrac{1}{2(x+1)^2 + 5}\, dx = \dfrac{1}{2}\int \dfrac{1}{(x+1)^2 + 5/2}\, dx = \dfrac{1}{\sqrt{10}}\tan^{-1}\sqrt{2/5}(x+1) + C$

43. $\displaystyle\int_1^2 \dfrac{1}{\sqrt{4x - x^2}}\, dx = \int_1^2 \dfrac{1}{\sqrt{4 - (x-2)^2}}\, dx = \sin^{-1}\dfrac{x-2}{2}\bigg]_1^2 = \pi/6$

45. $u = \sin^2 x, du = 2\sin x\cos x\, dx$;

$$\frac{1}{2}\int \sqrt{1 - u^2}\, du = \frac{1}{4}\left[u\sqrt{1 - u^2} + \sin^{-1} u\right] + C = \frac{1}{4}\left[\sin^2 x\sqrt{1 - \sin^4 x} + \sin^{-1}(\sin^2 x)\right] + C$$

47. **(a)** $x = 3\sinh u$, $dx = 3\cosh u\, du$, $\displaystyle\int du = u + C = \sinh^{-1}(x/3) + C$

(b) $x = 3\tan\theta$, $dx = 3\sec^2\theta\, d\theta$,

$$\int \sec\theta\, d\theta = \ln|\sec\theta + \tan\theta| + C = \ln\left(\sqrt{x^2 + 9}/3 + x/3\right) + C$$

but $\sinh^{-1}(x/3) = \ln\left(x/3 + \sqrt{x^2/9 + 1}\right) = \ln\left(x/3 + \sqrt{x^2 + 9}/3\right)$ so the results agree.

EXERCISE SET 8.5

1. $\dfrac{A}{(x-3)} + \dfrac{B}{(x+4)}$

3. $\dfrac{2x - 3}{x^2(x-1)} = \dfrac{A}{x} + \dfrac{B}{x^2} + \dfrac{C}{x-1}$

5. $\dfrac{A}{x} + \dfrac{B}{x^2} + \dfrac{C}{x^3} + \dfrac{Dx + E}{x^2 + 2}$

7. $\dfrac{Ax + B}{x^2 + 5} + \dfrac{Cx + D}{(x^2 + 5)^2}$

9. $\dfrac{1}{(x-4)(x+1)} = \dfrac{A}{x-4} + \dfrac{B}{x+1}$; $A = \dfrac{1}{5}$, $B = -\dfrac{1}{5}$ so

$$\frac{1}{5}\int \frac{1}{x-4}\, dx - \frac{1}{5}\int \frac{1}{x+1}\, dx = \frac{1}{5}\ln|x-4| - \frac{1}{5}\ln|x+1| + C = \frac{1}{5}\ln\left|\frac{x-4}{x+1}\right| + C$$

11. $\dfrac{11x + 17}{(2x-1)(x+4)} = \dfrac{A}{2x-1} + \dfrac{B}{x+4}$; $A = 5$, $B = 3$ so

$$5\int \frac{1}{2x-1}\, dx + 3\int \frac{1}{x+4}\, dx = \frac{5}{2}\ln|2x-1| + 3\ln|x+4| + C$$

13. $\dfrac{2x^2 - 9x - 9}{x(x+3)(x-3)} = \dfrac{A}{x} + \dfrac{B}{x+3} + \dfrac{C}{x-3}$; $A = 1$, $B = 2$, $C = -1$ so

$$\int \frac{1}{x}\, dx + 2\int \frac{1}{x+3}\, dx - \int \frac{1}{x-3}\, dx = \ln|x| + 2\ln|x+3| - \ln|x-3| + C = \ln\left|\frac{x(x+3)^2}{x-3}\right| + C$$

Note that the symbol C has been recycled; to save space this recycling is usually not mentioned.

15. $\dfrac{x^2-8}{x+3} = x - 3 + \dfrac{1}{x+3}$, $\displaystyle\int\left(x - 3 + \dfrac{1}{x+3}\right)dx = \dfrac{1}{2}x^2 - 3x + \ln|x+3| + C$

17. $\dfrac{3x^2-10}{x^2-4x+4} = 3 + \dfrac{12x-22}{x^2-4x+4}$, $\dfrac{12x-22}{(x-2)^2} = \dfrac{A}{x-2} + \dfrac{B}{(x-2)^2}$; $A=12$, $B=2$ so

$\displaystyle\int 3\,dx + 12\int\dfrac{1}{x-2}\,dx + 2\int\dfrac{1}{(x-2)^2}\,dx = 3x + 12\ln|x-2| - 2/(x-2) + C$

19. $\dfrac{x^5+x^2+2}{x^3-x} = x^2 + 1 + \dfrac{x^2+x+2}{x^3-x}$,

$\dfrac{x^2+x+2}{x(x+1)(x-1)} = \dfrac{A}{x} + \dfrac{B}{x+1} + \dfrac{C}{x-1}$; $A=-2$, $B=1$, $C=2$ so

$\displaystyle\int(x^2+1)\,dx - \int\dfrac{2}{x}\,dx + \int\dfrac{1}{x+1}\,dx + \int\dfrac{2}{x-1}\,dx$

$\quad = \dfrac{1}{3}x^3 + x - 2\ln|x| + \ln|x+1| + 2\ln|x-1| + C = \dfrac{1}{3}x^3 + x + \ln\left|\dfrac{(x+1)(x-1)^2}{x^2}\right| + C$

21. $\dfrac{2x^2+3}{x(x-1)^2} = \dfrac{A}{x} + \dfrac{B}{x-1} + \dfrac{C}{(x-1)^2}$; $A=3$, $B=-1$, $C=5$ so

$3\displaystyle\int\dfrac{1}{x}\,dx - \int\dfrac{1}{x-1}\,dx + 5\int\dfrac{1}{(x-1)^2}\,dx = 3\ln|x| - \ln|x-1| - 5/(x-1) + C$

23. $\dfrac{2x^2-10x+4}{(x+1)(x-3)^2} = \dfrac{A}{x+1} + \dfrac{B}{x-3} + \dfrac{C}{(x-3)^2}$; $A=1$, $B=1$, $C=-2$ so

$\displaystyle\int\dfrac{1}{x+1}\,dx + \int\dfrac{1}{x-3}\,dx - \int\dfrac{2}{(x-3)^2}\,dx = \ln|x+1| + \ln|x-3| + \dfrac{2}{x-3} + C_1$

25. $\dfrac{x^2}{(x+1)^3} = \dfrac{A}{x+1} + \dfrac{B}{(x+1)^2} + \dfrac{C}{(x+1)^3}$; $A=1$, $B=-2$, $C=1$ so

$\displaystyle\int\dfrac{1}{x+1}\,dx - \int\dfrac{2}{(x+1)^2}\,dx + \int\dfrac{1}{(x+1)^3}\,dx = \ln|x+1| + \dfrac{2}{x+1} - \dfrac{1}{2(x+1)^2} + C$

27. $\dfrac{2x^2-1}{(4x-1)(x^2+1)} = \dfrac{A}{4x-1} + \dfrac{Bx+C}{x^2+1}$; $A=-14/17$, $B=12/17$, $C=3/17$ so

$\displaystyle\int\dfrac{2x^2-1}{(4x-1)(x^2+1)}\,dx = -\dfrac{7}{34}\ln|4x-1| + \dfrac{6}{17}\ln(x^2+1) + \dfrac{3}{17}\tan^{-1}x + C$

29. $\dfrac{x^3+3x^2+x+9}{(x^2+1)(x^2+3)} = \dfrac{Ax+B}{x^2+1} + \dfrac{Cx+D}{x^2+3}$; $A=0$, $B=3$, $C=1$, $D=0$ so

$\displaystyle\int\dfrac{x^3+3x^2+x+9}{(x^2+1)(x^2+3)}\,dx = 3\tan^{-1}x + \dfrac{1}{2}\ln(x^2+3) + C$

31. $\dfrac{x^3-2x^2+2x-2}{x^2+1} = x - 2 + \dfrac{x}{x^2+1}$,

$\displaystyle\int\dfrac{x^3-3x^2+2x-3}{x^2+1}\,dx = \dfrac{1}{2}x^2 - 2x + \dfrac{1}{2}\ln(x^2+1) + C$

33. Let $x = \sin\theta$ to get $\displaystyle\int \frac{1}{x^2 + 4x - 5}\, dx$, and $\displaystyle\frac{1}{(x+5)(x-1)} = \frac{A}{x+5} + \frac{B}{x-1}$; $A = -1/6$,

$B = 1/6$ so we get $\displaystyle -\frac{1}{6}\int \frac{1}{x+5}\, dx + \frac{1}{6}\int \frac{1}{x-1}\, dx = \frac{1}{6}\ln\left|\frac{x-1}{x+5}\right| + C = \frac{1}{6}\ln\left(\frac{1-\sin\theta}{5+\sin\theta}\right) + C.$

35. $\displaystyle V = \pi\int_0^2 \frac{x^4}{(9 - x^2)^2}\, dx$, $\displaystyle\frac{x^4}{x^4 - 18x^2 + 81} = 1 + \frac{18x^2 - 81}{x^4 - 18x^2 + 81}$,

$\displaystyle\frac{18x^2 - 81}{(9 - x^2)^2} = \frac{18x^2 - 81}{(x+3)^2(x-3)^2} = \frac{A}{x+3} + \frac{B}{(x+3)^2} + \frac{C}{x-3} + \frac{D}{(x-3)^2}$;

$\displaystyle A = -\frac{9}{4}, B = \frac{9}{4}, C = \frac{9}{4}, D = \frac{9}{4}$ so

$\displaystyle V = \pi\left[x - \frac{9}{4}\ln|x+3| - \frac{9/4}{x+3} + \frac{9}{4}\ln|x-3| - \frac{9/4}{x-3}\right]_0^2 = \pi\left(\frac{19}{5} - \frac{9}{4}\ln 5\right)$

37. $\displaystyle\frac{x^2 + 1}{(x^2 + 2x + 3)^2} = \frac{Ax + B}{x^2 + 2x + 3} + \frac{Cx + D}{(x^2 + 2x + 3)^2}$; $A = 0, B = 1, C = D = -2$ so

$\displaystyle\int \frac{x^2 + 1}{(x^2 + 2x + 3)^2}\, dx = \int \frac{1}{(x+1)^2 + 2}\, dx - \int \frac{2x + 2}{(x^2 + 2x + 3)^2}\, dx$

$\displaystyle = \frac{1}{\sqrt{2}}\tan^{-1}\frac{x+1}{\sqrt{2}} + 1/(x^2 + 2x + 3) + C$

39. $x^4 - 3x^3 - 7x^2 + 27x - 18 = (x-1)(x-2)(x-3)(x+3)$,

$\displaystyle\frac{1}{(x-1)(x-2)(x-3)(x+3)} = \frac{A}{x-1} + \frac{B}{x-2} + \frac{C}{x-3} + \frac{D}{x+3}$;

$A = 1/8, B = -1/5, C = 1/12, D = -1/120$ so

$\displaystyle\int \frac{dx}{x^4 - 3x^3 - 7x^2 + 27x - 18} = \frac{1}{8}\ln|x-1| - \frac{1}{5}\ln|x-2| + \frac{1}{12}\ln|x-3| - \frac{1}{120}\ln|x+3| + C$

41. Let $u = x^2, du = 2x\, dx$, $\displaystyle\int_0^1 \frac{x}{x^4 + 1}\, dx = \frac{1}{2}\int_0^1 \frac{1}{1 + u^2}\, du = \frac{1}{2}\tan^{-1}u\Big]_0^1 = \frac{1}{2}\frac{\pi}{4} = \frac{\pi}{8}.$

43. If the polynomial has distinct roots $r_1, r_2, r_1 \neq r_2$, then the partial fraction decomposition will contain terms of the form $\displaystyle\frac{A}{x - r_1}, \frac{B}{x - r_2}$, and they will give logarithms and no inverse tangents. If there are two roots not distinct, say $x = r$, then the terms $\displaystyle\frac{A}{x - r}, \frac{B}{(x - r)^2}$ will appear, and neither will give an inverse tangent term. The only other possibility is no real roots, and the integrand can be written in the form $\displaystyle\frac{1}{a\left(x + \frac{b}{2a}\right)^2 + c - \frac{b^2}{4a}}$, which will yield an inverse tangent, specifically of the form $\tan^{-1}\left[A\left(x + \frac{b}{2a}\right)\right]$ for some constant A.

45. Yes, for instance the integrand $\displaystyle\frac{1}{x^2 + 1}$, whose integral is precisely $\tan^{-1}x + C$.

EXERCISE SET 8.6

1. Formula (60): $\dfrac{4}{9}\Big[3x + \ln|-1 + 3x|\Big] + C$ **3.** Formula (65): $\dfrac{1}{5}\ln\left|\dfrac{x}{5 + 2x}\right| + C$

5. Formula (102): $\dfrac{1}{5}(x - 1)(2x + 3)^{3/2} + C$ **7.** Formula (108): $\dfrac{1}{2}\ln\left|\dfrac{\sqrt{4 - 3x} - 2}{\sqrt{4 - 3x} + 2}\right| + C$

9. Formula (69): $\dfrac{1}{8}\ln\left|\dfrac{x + 4}{x - 4}\right| + C$

11. Formula (73): $\dfrac{x}{2}\sqrt{x^2 - 3} - \dfrac{3}{2}\ln\left|x + \sqrt{x^2 - 3}\right| + C$

13. Formula (95): $\dfrac{x}{2}\sqrt{x^2 + 4} - 2\ln(x + \sqrt{x^2 + 4}) + C$

15. Formula (74): $\dfrac{x}{2}\sqrt{9 - x^2} + \dfrac{9}{2}\sin^{-1}\dfrac{x}{3} + C$

17. Formula (79): $\sqrt{4 - x^2} - 2\ln\left|\dfrac{2 + \sqrt{4 - x^2}}{x}\right| + C$

19. Formula (38): $-\dfrac{1}{14}\sin(7x) + \dfrac{1}{2}\sin x + C$ **21.** Formula (50): $\dfrac{x^4}{16}\big[4\ln x - 1\big] + C$

23. Formula (42): $\dfrac{e^{-2x}}{13}(-2\sin(3x) - 3\cos(3x)) + C$

25. $u = e^{2x}, du = 2e^{2x}dx$, Formula (62): $\dfrac{1}{2}\displaystyle\int \dfrac{u\,du}{(4 - 3u)^2} = \dfrac{1}{18}\left[\dfrac{4}{4 - 3e^{2x}} + \ln\left|4 - 3e^{2x}\right|\right] + C$

27. $u = 3\sqrt{x}, du = \dfrac{3}{2\sqrt{x}}dx$, Formula (68): $\dfrac{2}{3}\displaystyle\int \dfrac{du}{u^2 + 4} = \dfrac{1}{3}\tan^{-1}\dfrac{3\sqrt{x}}{2} + C$

29. $u = 2x, du = 2dx$, Formula (76): $\dfrac{1}{2}\displaystyle\int \dfrac{du}{\sqrt{u^2 - 9}} = \dfrac{1}{2}\ln\left|2x + \sqrt{4x^2 - 9}\right| + C$

31. $u = 2x^2, du = 4xdx, u^2 du = 16x^5\,dx$, Formula (81):

$\dfrac{1}{4}\displaystyle\int \dfrac{u^2\,du}{\sqrt{2 - u^2}} = -\dfrac{x^2}{4}\sqrt{2 - 4x^4} + \dfrac{1}{4}\sin^{-1}(\sqrt{2}x^2) + C$

33. $u = \ln x, du = dx/x$, Formula (26): $\displaystyle\int \sin^2 u\,du = \dfrac{1}{2}\ln x + \dfrac{1}{4}\sin(2\ln x) + C$

35. $u = -2x, du = -2dx$, Formula (51): $\dfrac{1}{4}\displaystyle\int ue^u\,du = \dfrac{1}{4}(-2x - 1)e^{-2x} + C$

37. $u = \sin 3x, du = 3\cos 3x\,dx$, Formula (67): $\dfrac{1}{3}\displaystyle\int \dfrac{du}{u(u + 1)^2} = \dfrac{1}{3}\left[\dfrac{1}{1 + \sin 3x} + \ln\left|\dfrac{\sin 3x}{1 + \sin 3x}\right|\right] + C$

39. $u = 4x^2, du = 8xdx$, Formula (70): $\dfrac{1}{8}\displaystyle\int \dfrac{du}{u^2 - 1} = \dfrac{1}{16}\ln\left|\dfrac{4x^2 - 1}{4x^2 + 1}\right| + C$

41. $u = 2e^x, du = 2e^x dx$, Formula (74):

$$\frac{1}{2}\int \sqrt{3-u^2}\, du = \frac{1}{4}u\sqrt{3-u^2} + \frac{3}{4}\sin^{-1}(u/\sqrt{3}) + C = \frac{1}{2}e^x\sqrt{3-4e^{2x}} + \frac{3}{4}\sin^{-1}(2e^x/\sqrt{3}) + C$$

43. $u = 3x, du = 3dx$, Formula (112):

$$\frac{1}{3}\int \sqrt{\frac{5}{3}u - u^2}\, du = \frac{1}{6}\left(u - \frac{5}{6}\right)\sqrt{\frac{5}{3}u - u^2} + \frac{25}{216}\sin^{-1}\left(\frac{u-5}{5}\right) + C$$

$$= \frac{18x-5}{36}\sqrt{5x-9x^2} + \frac{25}{216}\sin^{-1}\left(\frac{18x-5}{5}\right) + C$$

45. $u = 2x, du = 2dx$, Formula (44):

$$\int u\sin u\, du = (\sin u - u\cos u) + C = \sin 2x - 2x\cos 2x + C$$

47. $u = -\sqrt{x}, u^2 = x, 2u\, du = dx$, Formula (51): $2\int ue^u du = -2(\sqrt{x}+1)e^{-\sqrt{x}} + C$

49. $x^2 + 6x - 7 = (x+3)^2 - 16; u = x+3, du = dx$, Formula (70):

$$\int \frac{du}{u^2-16} = \frac{1}{8}\ln\left|\frac{u-4}{u+4}\right| + C = \frac{1}{8}\ln\left|\frac{x-1}{x+7}\right| + C$$

51. $x^2 - 4x - 5 = (x-2)^2 - 9, u = x-2, du = dx$, Formula (77):

$$\int \frac{u+2}{\sqrt{9-u^2}}\, du = \int \frac{u\, du}{\sqrt{9-u^2}} + 2\int \frac{du}{\sqrt{9-u^2}} = -\sqrt{9-u^2} + 2\sin^{-1}\frac{u}{3} + C$$

$$= -\sqrt{5+4x-x^2} + 2\sin^{-1}\left(\frac{x-2}{3}\right) + C$$

53. $u = \sqrt{x-2}, x = u^2 + 2, dx = 2u\, du$;

$$\int 2u^2(u^2+2)du = 2\int (u^4 + 2u^2)du = \frac{2}{5}u^5 + \frac{4}{3}u^3 + C = \frac{2}{5}(x-2)^{5/2} + \frac{4}{3}(x-2)^{3/2} + C$$

55. $u = \sqrt{x^3+1}, x^3 = u^2 - 1, 3x^2 dx = 2u\, du$;

$$\frac{2}{3}\int u^2(u^2-1)du = \frac{2}{3}\int (u^4 - u^2)du = \frac{2}{15}u^5 - \frac{2}{9}u^3 + C = \frac{2}{15}(x^3+1)^{5/2} - \frac{2}{9}(x^3+1)^{3/2} + C$$

57. $u = x^{1/3}, x = u^3, dx = 3u^2\, du$;

$$\int \frac{3u^2}{u^3 - u}du = 3\int \frac{u}{u^2-1}du = 3\int \left[\frac{1}{2(u+1)} + \frac{1}{2(u-1)}\right]du$$

$$= \frac{3}{2}\ln|x^{1/3}+1| + \frac{3}{2}\ln|x^{1/3}-1| + C$$

59. $u = x^{1/4}, x = u^4, dx = 4u^3 du$; $4\int \frac{1}{u(1-u)}du = 4\int \left[\frac{1}{u} + \frac{1}{1-u}\right]du = 4\ln\frac{x^{1/4}}{|1-x^{1/4}|} + C$

61. $u = x^{1/6}, x = u^6, dx = 6u^5 du$;

$$6\int \frac{u^3}{u-1}du = 6\int \left[u^2 + u + 1 + \frac{1}{u-1}\right]du = 2x^{1/2} + 3x^{1/3} + 6x^{1/6} + 6\ln|x^{1/6}-1| + C$$

63. $u = \sqrt{1+x^2}, x^2 = u^2 - 1, 2x\, dx = 2u\, du, x\, dx = u\, du$;

$$\int (u^2-1)du = \frac{1}{3}(1+x^2)^{3/2} - (1+x^2)^{1/2} + C$$

65. $\displaystyle \int \frac{1}{1 + \dfrac{2u}{1+u^2} + \dfrac{1-u^2}{1+u^2}} \, \frac{2}{1+u^2} \, du = \int \frac{1}{u+1} \, du = \ln|\tan(x/2) + 1| + C$

67. $u = \tan(\theta/2)$, $\displaystyle \int \frac{d\theta}{1-\cos\theta} = \int \frac{1}{u^2} \, du = -\frac{1}{u} + C = -\cot(\theta/2) + C$

69. $u = \tan(x/2)$, $\displaystyle \frac{1}{2}\int \frac{1-u^2}{u}\, du = \frac{1}{2}\int (1/u - u)\, du = \frac{1}{2}\ln|\tan(x/2)| - \frac{1}{4}\tan^2(x/2) + C$

71. $\displaystyle \int_2^x \frac{1}{t(4-t)} \, dt = \frac{1}{4}\ln\frac{t}{4-t}\Big]_2^x$ (Formula (65), $a = 4, b = -1$)

$$= \frac{1}{4}\left[\ln\frac{x}{4-x} - \ln 1\right] = \frac{1}{4}\ln\frac{x}{4-x}, \, \frac{1}{4}\ln\frac{x}{4-x} = 0.5, \ln\frac{x}{4-x} = 2,$$

$$\frac{x}{4-x} = e^2, \, x = 4e^2 - e^2 x, \, x(1 + e^2) = 4e^2, \, x = 4e^2/(1+e^2) \approx 3.523188312$$

73. $\displaystyle A = \int_0^4 \sqrt{25 - x^2}\, dx = \left(\frac{1}{2}x\sqrt{25-x^2} + \frac{25}{2}\sin^{-1}\frac{x}{5}\right)\Big]_0^4$ (Formula (74), $a = 5$)

$$= 6 + \frac{25}{2}\sin^{-1}\frac{4}{5} \approx 17.59119023$$

75. $\displaystyle A = \int_0^1 \frac{1}{25-16x^2}\, dx; \, u = 4x,$

$$A = \frac{1}{4}\int_0^4 \frac{1}{25-u^2}\, du = \frac{1}{40}\ln\left|\frac{u+5}{u-5}\right|\,\Big]_0^4 = \frac{1}{40}\ln 9 \approx 0.054930614 \text{ (Formula (69), } a = 5)$$

77. $\displaystyle V = 2\pi\int_0^{\pi/2} x\cos x\, dx = 2\pi(\cos x + x\sin x)\Big]_0^{\pi/2} = \pi(\pi - 2) \approx 3.586419094$ (Formula (45))

79. $\displaystyle V = 2\pi\int_0^3 xe^{-x}\, dx; \, u = -x,$

$$V = 2\pi\int_0^{-3} ue^u\, du = 2\pi e^u(u-1)\Big]_0^{-3} = 2\pi(1 - 4e^{-3}) \approx 5.031899801$$ (Formula (51))

81. $\displaystyle L = \int_0^2 \sqrt{1 + 16x^2}\, dx; \, u = 4x,$

$$L = \frac{1}{4}\int_0^8 \sqrt{1+u^2}\, du = \frac{1}{4}\left(\frac{u}{2}\sqrt{1+u^2} + \frac{1}{2}\ln\left(u + \sqrt{1+u^2}\right)\right)\Big]_0^8$$ (Formula (72), $a^2 = 1$)

$$= \sqrt{65} + \frac{1}{8}\ln(8 + \sqrt{65}) \approx 8.409316783$$

83. $\displaystyle S = 2\pi\int_0^\pi (\sin x)\sqrt{1 + \cos^2 x}\, dx; \, u = \cos x,$

$$S = -2\pi\int_1^{-1} \sqrt{1+u^2}\, du = 4\pi\int_0^1 \sqrt{1+u^2}\, du = 4\pi\left(\frac{u}{2}\sqrt{1+u^2} + \frac{1}{2}\ln\left(u + \sqrt{1+u^2}\right)\right)\Big]_0^1 a^2 = 1$$

$$= 2\pi\left[\sqrt{2} + \ln(1 + \sqrt{2})\right] \approx 14.42359945$$ (Formula (72))

85. (a) $s(t) = 2 + \displaystyle\int_0^t 20\cos^6 u \sin^3 u\, du$

$$= -\frac{20}{9}\sin^2 t \cos^7 t - \frac{40}{63}\cos^7 t + \frac{166}{63}$$

(b)

87. (a) $\displaystyle\int \sec x\, dx = \int \frac{1}{\cos x}\, dx = \int \frac{2}{1-u^2}\, du = \ln\left|\frac{1+u}{1-u}\right| + C = \ln\left|\frac{1+\tan(x/2)}{1-\tan(x/2)}\right| + C$

$$= \ln\left\{\left|\frac{\cos(x/2)+\sin(x/2)}{\cos(x/2)-\sin(x/2)}\right|\,\left|\frac{\cos(x/2)+\sin(x/2)}{\cos(x/2)+\sin(x/2)}\right|\right\} + C = \ln\left|\frac{1+\sin x}{\cos x}\right| + C$$

$$= \ln|\sec x + \tan x| + C$$

(b) $\tan\left(\dfrac{\pi}{4} + \dfrac{x}{2}\right) = \dfrac{\tan\dfrac{\pi}{4} + \tan\dfrac{x}{2}}{1 - \tan\dfrac{\pi}{4}\tan\dfrac{x}{2}} = \dfrac{1+\tan\dfrac{x}{2}}{1-\tan\dfrac{x}{2}}$

89. Let $u = \tanh(x/2)$ then $\cosh(x/2) = 1/\operatorname{sech}(x/2) = 1/\sqrt{1-\tanh^2(x/2)} = 1/\sqrt{1-u^2}$,

$\sinh(x/2) = \tanh(x/2)\cosh(x/2) = u/\sqrt{1-u^2}$, so $\sinh x = 2\sinh(x/2)\cosh(x/2) = 2u/(1-u^2)$,

$\cosh x = \cosh^2(x/2) + \sinh^2(x/2) = (1+u^2)/(1-u^2)$, $x = 2\tanh^{-1} u$, $dx = [2/(1-u^2)]du$;

$$\int \frac{dx}{2\cosh x + \sinh x} = \int \frac{1}{u^2 + u + 1}\, du = \frac{2}{\sqrt{3}}\tan^{-1}\frac{2u+1}{\sqrt{3}} + C = \frac{2}{\sqrt{3}}\tan^{-1}\frac{2\tanh(x/2)+1}{\sqrt{3}} + C.$$

91. $\displaystyle\int (\cos^{32} x \sin^{30} x - \cos^{30} x \sin^{32} x)dx = \int \cos^{30} x \sin^{30} x(\cos^2 x - \sin^2 x)dx$

$$= \frac{1}{2^{30}}\int \sin^{30} 2x \cos 2x\, dx = \frac{\sin^{31} 2x}{31(2^{31})} + C$$

93. $\displaystyle\int \frac{1}{x^{10}(1+x^{-9})}\, dx = -\frac{1}{9}\int \frac{1}{u}\, du = -\frac{1}{9}\ln|u| + C = -\frac{1}{9}\ln|1+x^{-9}| + C$

EXERCISE SET 8.7

1. exact value $= 14/3 \approx 4.666666667$
   (a) 4.667600663, $|E_M| \approx 0.000933996$
   (b) 4.664795679, $|E_T| \approx 0.001870988$
   (c) 4.666651630, $|E_S| \approx 0.000015037$

3. exact value $= 2$
   (a) 2.008248408, $|E_M| \approx 0.008248408$
   (b) 1.983523538, $|E_T| \approx 0.016476462$
   (c) 2.000109517, $|E_S| \approx 0.000109517$

5. exact value $= e^{-1} - e^{-4} \approx 0.3495638023$
   (a) 0.3482563710, $|E_M| \approx 0.0013074313$
   (b) 0.3521816066, $|E_T| \approx 0.0026178043$
   (c) 0.3495793657, $|E_S| \approx 0.0000155634$

7. $f(x) = \sqrt{x-1}$, $f''(x) = -\frac{1}{4}(x-1)^{-3/2}$, $f^{(4)}(x) = -\frac{15}{16}(x-1)^{-7/2}$; $K_2 = 1/4$, $K_4 = 15/16$

(a) $|E_M| \le \dfrac{27}{2400}(1/4) = 0.002812500$ (b) $|E_T| \le \dfrac{27}{1200}(1/4) = 0.005625000$

(c) $|E_S| \le \dfrac{243}{180 \times 10^4}(15/16) \approx 0.000126563$

9. $f(x) = \sin x$, $f''(x) = -\sin x$, $f^{(4)}(x) = \sin x$; $K_2 = K_4 = 1$

(a) $|E_M| \le \dfrac{\pi^3}{2400}(1) \approx 0.012919282$ (b) $|E_T| \le \dfrac{\pi^3}{1200}(1) \approx 0.025838564$

(c) $|E_S| \le \dfrac{\pi^5}{180 \times 10^4}(1) \approx 0.000170011$

11. $f(x) = e^{-x}$, $f''(x) = f^{(4)}(x) = e^{-x}$; $K_2 = K_4 = e^{-1}$

(a) $|E_M| \le \dfrac{9}{800}(e^{-1}) \approx 0.001226265$ (b) $|E_T| \le \dfrac{9}{400}(e^{-1}) \approx 0.002452530$

(c) $|E_S| \le \dfrac{27}{2 \times 10^5}(e^{-1}) \approx 0.000049664$

13. (a) $n > \left[\dfrac{(27)(1/4)}{(24)(5 \times 10^{-4})}\right]^{1/2} \approx 23.7$; $n = 24$ (b) $n > \left[\dfrac{(27)(1/4)}{(12)(5 \times 10^{-4})}\right]^{1/2} \approx 33.5$; $n = 34$

(c) $n > \left[\dfrac{(243)(15/16)}{(180)(5 \times 10^{-4})}\right]^{1/4} \approx 7.1$; $2n = 8$

15. (a) $n > \left[\dfrac{(\pi^3)(1)}{(24)(10^{-3})}\right]^{1/2} \approx 35.9$; $n = 36$ (b) $n > \left[\dfrac{(\pi^3)(1)}{(12)(10^{-3})}\right]^{1/2} \approx 50.8$; $n = 51$

(c) $n > \left[\dfrac{(\pi^5)(1)}{(180)(10^{-3})}\right]^{1/4} \approx 6.4$; $2n = 8$

17. (a) $n > \left[\dfrac{(8)(e^{-1})}{(24)(10^{-6})}\right]^{1/2} \approx 350.2$; $n = 351$ (b) $n > \left[\dfrac{(8)(e^{-1})}{(12)(10^{-6})}\right]^{1/2} \approx 495.2$; $n = 496$

(c) $n > \left[\dfrac{(32)(e^{-1})}{(180)(10^{-6})}\right]^{1/4} \approx 15.99$; $n = 16$

19. $g(X_0) = aX_0^2 + bX_0 + c = 4a + 2b + c = f(X_0) = 1/X_0 = 1/2$; similarly
$9a + 3b + c = 1/3$, $16a + 4b + c = 1/4$. Three equations in three unknowns, with solution
$a = 1/24$, $b = -3/8$, $c = 13/12$, $g(x) = x^2/24 - 3x/8 + 13/12$.

$$\int_2^4 g(x)\,dx = \int_2^4 \left(\frac{x^2}{24} - \frac{3x}{8} + \frac{13}{12}\right) dx = \frac{25}{36}$$

$$\frac{\Delta x}{3}[f(X_0) + 4f(X_1) + f(X_2)] = \frac{1}{3}\left[\frac{1}{2} + \frac{4}{3} + \frac{1}{4}\right] = \frac{25}{36}$$

21. 0.746824948,
0.746824133

23. 1.511518747,
1.515927142

25. 0.805376152,
0.804776489

27. (a) 3.142425985, $|E_M| \approx 0.000833331$
(b) 3.139925989, $|E_T| \approx 0.001666665$
(c) 3.141592614, $|E_S| \approx 0.000000040$

29. $S_{14} = 0.693147984$, $|E_S| \approx 0.000000803 = 8.03 \times 10^{-7}$; the method used in Example 6 results in a value of n which ensures that the magnitude of the error will be less than 10^{-6}, this is not necessarily the *smallest* value of n.

31. $f(x) = x \sin x$, $f''(x) = 2\cos x - x \sin x$, $|f''(x)| \le 2|\cos x| + |x| |\sin x| \le 2 + 2 = 4$ so $K_2 \le 4$,

$$n > \left[\frac{(8)(4)}{(24)(10^{-4})} \right]^{1/2} \approx 115.5; \ n = 116 \text{ (a smaller } n \text{ might suffice)}$$

33. $f(x) = x\sqrt{x}$, $f''(x) = \dfrac{3}{4\sqrt{x}}$, $\displaystyle\lim_{x \to 0^+} |f''(x)| = +\infty$

35. $L = \displaystyle\int_{-\pi/2}^{\pi/2} \sqrt{1 + \sin^2 x}\, dx \approx 3.820187623$

37. $\displaystyle\int_0^{20} v\, dt \approx \dfrac{15}{(3)(6)}[0 + 4(35.2) + 2(60.1) + 4(79.2) + 2(90.9) + 4(104.1) + 114.4] \approx 1078$ ft

39. $\displaystyle\int_0^{180} v\, dt \approx \dfrac{180}{(3)(6)}[0.00 + 4(0.03) + 2(0.08) + 4(0.16) + 2(0.27) + 4(0.42) + 0.65] = 37.9$ mi

41. $V = \displaystyle\int_0^{16} \pi r^2 dy = \pi \int_0^{16} r^2 dy \approx \pi \dfrac{16}{(3)(4)}[(8.5)^2 + 4(11.5)^2 + 2(13.8)^2 + 4(15.4)^2 + (16.8)^2]$

$$\approx 9270 \text{ cm}^3 \approx 9.3 \text{ L}$$

43. **(a)** The maximum value of $|f''(x)|$ is approximately 3.8442

    (b) $n = 18$

    (c) 0.9047406684

45. **(a)** The maximum value of $|f^{(4)}(x)|$ is approximately 42.5518.

    (b) $2n = 8$

    (c) 0.9045241594

47. **(a)** Left endpoint approximation $\approx \dfrac{b-a}{n}[y_0 + y_1 + \ldots + y_{n-2} + y_{n-1}]$

    Right endpoint approximation $\approx \dfrac{b-a}{n}[y_1 + y_2 + \ldots + y_{n-1} + y_n]$

    Average of the two $= \dfrac{b-a}{n}\dfrac{1}{2}[y_0 + 2y_1 + 2y_2 + \ldots + 2y_{n-2} + 2y_{n-1} + y_n]$

    (b) Area of trapezoid $= (x_{k+1} - x_k)\dfrac{y_k + y_{k+1}}{2}$. If we sum from $k = 0$ to $k = n - 1$ then we get the right hand side of (2).

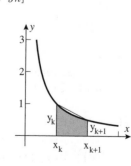

49. Given $g(x) = Ax^2 + Bx + C$, suppose $\Delta x = 1$ and $m = 0$. Then set $Y_0 = g(-1), Y_1 = g(0)$,
$Y_2 = g(1)$. Also $Y_0 = g(-1) = A - B + C, Y_1 = g(0) = C, Y_2 = g(1) = A + B + C$, with solution
$C = Y_1, B = \frac{1}{2}(Y_2 - Y_0)$, and $A = \frac{1}{2}(Y_0 + Y_2) - Y_1$.

Then $\displaystyle\int_{-1}^{1} g(x)\,dx = 2\int_{0}^{1}(Ax^2 + C)\,dx = \frac{2}{3}A + 2C = \frac{1}{3}(Y_0 + Y_2) - \frac{2}{3}Y_1 + 2Y_1 = \frac{1}{3}(Y_0 + 4Y_1 + Y_2)$,

which is exactly what one gets applying the Simpson's Rule.

The general case with the interval $(m - \Delta x, m + \Delta x)$ and values Y_0, Y_1, Y_2, can be converted by
the change of variables $z = \dfrac{x - m}{\Delta x}$. Set $g(x) = h(z) = h((x - m)/\Delta x)$ to get $dx = \Delta x\,dz$ and

$\displaystyle\Delta x \int_{m-\Delta x}^{m+\Delta x} h(z)\,dz = \int_{-1}^{1} g(x)\,dx$. Finally, $Y_0 = g(m - \Delta x) = h(-1)$,

$Y_1 = g(m) = h(0), Y_2 = g(m + \Delta x) = h(1)$.

EXERCISE SET 8.8

1. (a) improper; infinite discontinuity at $x = 3$
  (b) continuous integrand, not improper
  (c) improper; infinite discontinuity at $x = 0$
  (d) improper; infinite interval of integration
  (e) improper; infinite interval of integration and infinite discontinuity at $x = 1$
  (f) continuous integrand, not improper

3. $\displaystyle\lim_{\ell \to +\infty} \left(-\frac{1}{2}e^{-2x}\right)\Big]_{0}^{\ell} = \frac{1}{2}\lim_{\ell \to +\infty}(-e^{-2\ell} + 1) = \frac{1}{2}$

5. $\displaystyle\lim_{\ell \to +\infty} -2\coth^{-1}x\Big]_{3}^{\ell} = \lim_{\ell \to +\infty}\left(2\coth^{-1}3 - 2\coth^{-1}\ell\right) = 2\coth^{-1}3$

7. $\displaystyle\lim_{\ell \to +\infty} -\frac{1}{2\ln^2 x}\Big]_{e}^{\ell} = \lim_{\ell \to +\infty}\left[-\frac{1}{2\ln^2 \ell} + \frac{1}{2}\right] = \frac{1}{2}$

9. $\displaystyle\lim_{\ell \to -\infty} -\frac{1}{4(2x - 1)^2}\Big]_{\ell}^{0} = \lim_{\ell \to -\infty} \frac{1}{4}[-1 + 1/(2\ell - 1)^2] = -1/4$

11. $\displaystyle\lim_{\ell \to -\infty} \frac{1}{3}e^{3x}\Big]_{\ell}^{0} = \lim_{\ell \to -\infty}\left[\frac{1}{3} - \frac{1}{3}e^{3\ell}\right] = \frac{1}{3}$

13. $\displaystyle\int_{-\infty}^{+\infty} x\,dx$ converges if $\displaystyle\int_{-\infty}^{0} x\,dx$ and $\displaystyle\int_{0}^{+\infty} x\,dx$ both converge; it diverges if either (or both)

diverges. $\displaystyle\int_{0}^{+\infty} x\,dx = \lim_{\ell \to +\infty}\frac{1}{2}x^2\Big]_{0}^{\ell} = \lim_{\ell \to +\infty}\frac{1}{2}\ell^2 = +\infty$ so $\displaystyle\int_{-\infty}^{+\infty} x\,dx$ is divergent.

15. $\displaystyle\int_{0}^{+\infty} \frac{x}{(x^2 + 3)^2}\,dx = \lim_{\ell \to +\infty} -\frac{1}{2(x^2 + 3)}\Big]_{0}^{\ell} = \lim_{\ell \to +\infty}\frac{1}{2}[-1/(\ell^2 + 3) + 1/3] = \frac{1}{6}$,

similarly $\displaystyle\int_{-\infty}^{0} \frac{x}{(x^2 + 3)^2}\,dx = -1/6$ so $\displaystyle\int_{-\infty}^{\infty} \frac{x}{(x^2 + 3)^2}\,dx = 1/6 + (-1/6) = 0$

17. $\displaystyle\lim_{\ell \to 4^-} -\frac{1}{x - 4}\Big]_{0}^{\ell} = \lim_{\ell \to 4^-}\left[-\frac{1}{\ell - 4} - \frac{1}{4}\right] = +\infty$, divergent

19. $\displaystyle \lim_{\ell \to \pi/2^-} -\ln(\cos x)\Big]_0^\ell = \lim_{\ell \to \pi/2^-} -\ln(\cos \ell) = +\infty$, divergent

21. $\displaystyle \lim_{\ell \to 1^-} \sin^{-1} x\Big]_0^\ell = \lim_{\ell \to 1^-} \sin^{-1} \ell = \pi/2$

23. $\displaystyle \lim_{\ell \to \pi/3^+} \sqrt{1-2\cos x}\Big]_\ell^{\pi/2} = \lim_{\ell \to \pi/3^+} (1-\sqrt{1-2\cos \ell}) = 1$

25. $\displaystyle \int_0^2 \frac{dx}{x-2} = \lim_{\ell \to 2^-} \ln|x-2|\Big]_0^\ell = \lim_{\ell \to 2^-} (\ln|\ell-2| - \ln 2) = -\infty$, divergent

27. $\displaystyle \int_0^8 x^{-1/3} dx = \lim_{\ell \to 0^+} \frac{3}{2} x^{2/3}\Big]_\ell^8 = \lim_{\ell \to 0^+} \frac{3}{2}(4 - \ell^{2/3}) = 6$,

$\displaystyle \int_{-1}^0 x^{-1/3} dx = \lim_{\ell \to 0^-} \frac{3}{2} x^{2/3}\Big]_{-1}^\ell = \lim_{\ell \to 0^-} \frac{3}{2}(\ell^{2/3} - 1) = -3/2$

so $\displaystyle \int_{-1}^8 x^{-1/3} dx = 6 + (-3/2) = 9/2$

29. Define $\displaystyle \int_0^{+\infty} \frac{1}{x^2} dx = \int_0^a \frac{1}{x^2} dx + \int_a^{+\infty} \frac{1}{x^2} dx$ where $a > 0$; take $a = 1$ for convenience,

$\displaystyle \int_0^1 \frac{1}{x^2} dx = \lim_{\ell \to 0^+} (-1/x)\Big]_\ell^1 = \lim_{\ell \to 0^+} (1/\ell - 1) = +\infty$ so $\displaystyle \int_0^{+\infty} \frac{1}{x^2} dx$ is divergent.

31. $\displaystyle \int_0^{+\infty} \frac{e^{-\sqrt{x}}}{\sqrt{x}} dx = 2\int_0^{+\infty} e^{-u} du = 2 \lim_{\ell \to +\infty} (-e^{-u})\Big]_0^\ell = 2 \lim_{\ell \to +\infty} (1 - e^{-\ell}) = 2$

33. $\displaystyle \int_0^{+\infty} \frac{e^{-x}}{\sqrt{1-e^{-x}}} dx = \int_0^1 \frac{du}{\sqrt{u}} = \lim_{\ell \to 0^+} 2\sqrt{u}\Big]_\ell^1 = \lim_{\ell \to 0^+} 2(1 - \sqrt{\ell}) = 2$

35. $\displaystyle \lim_{\ell \to +\infty} \int_0^\ell e^{-x} \cos x \, dx = \lim_{\ell \to +\infty} \frac{1}{2} e^{-x}(\sin x - \cos x)\Big]_0^\ell = 1/2$

37. (a) 2.726585 **(b)** 2.804364 **(c)** 0.219384 **(d)** 0.504067

39. $\displaystyle 1 + \left(\frac{dy}{dx}\right)^2 = 1 + \frac{4 - x^{2/3}}{x^{2/3}} = \frac{4}{x^{2/3}}$; the arc length is $\displaystyle \int_0^8 \frac{2}{x^{1/3}} dx = 3x^{2/3}\Big|_0^8 = 12$

41. $\displaystyle \int \ln x \, dx = x \ln x - x + C$,

$\displaystyle \int_0^1 \ln x \, dx = \lim_{\ell \to 0^+} \int_\ell^1 \ln x \, dx = \lim_{\ell \to 0^+} (x \ln x - x)\Big]_\ell^1 = \lim_{\ell \to 0^+} (-1 - \ell \ln \ell + \ell)$,

but $\displaystyle \lim_{\ell \to 0^+} \ell \ln \ell = \lim_{\ell \to 0^+} \frac{\ln \ell}{1/\ell} = \lim_{\ell \to 0^+} (-\ell) = 0$ so $\displaystyle \int_0^1 \ln x \, dx = -1$

43. $\displaystyle \int_0^{+\infty} e^{-3x} dx = \lim_{\ell \to +\infty} \int_0^\ell e^{-3x} dx = \lim_{\ell \to +\infty} \left(-\frac{1}{3} e^{-3x}\right)\Big]_0^\ell = \lim_{\ell \to +\infty} \left(-\frac{1}{3} e^{-3\ell} + \frac{1}{9}\right) = \frac{1}{3}$

45. (a) $V = \pi \int_0^{+\infty} e^{-2x}\,dx = -\dfrac{\pi}{2} \lim_{\ell\to+\infty} e^{-2x}\Big]_0^\ell = \pi/2$

(b) $S = 2\pi \int_0^{+\infty} e^{-x}\sqrt{1+e^{-2x}}\,dx$, let $u = e^{-x}$ to get

$S = -2\pi \int_1^0 \sqrt{1+u^2}\,du = 2\pi \left[\dfrac{u}{2}\sqrt{1+u^2} + \dfrac{1}{2}\ln\left|u+\sqrt{1+u^2}\right|\right]_0^1 = \pi\left[\sqrt{2}+\ln(1+\sqrt{2})\right]$

47. (a) For $x \geq 1, x^2 \geq x, e^{-x^2} \leq e^{-x}$

(b) $\displaystyle\int_1^{+\infty} e^{-x}\,dx = \lim_{\ell\to+\infty}\int_1^\ell e^{-x}\,dx = \lim_{\ell\to+\infty} -e^{-x}\Big]_1^\ell = \lim_{\ell\to+\infty}(e^{-1}-e^{-\ell}) = 1/e$

(c) By Parts (a) and (b) and Exercise 46(b), $\displaystyle\int_1^{+\infty} e^{-x^2}\,dx$ is convergent and is $\leq 1/e$.

49. $V = \displaystyle\lim_{\ell\to+\infty}\int_1^\ell (\pi/x^2)\,dx = \lim_{\ell\to+\infty} -(\pi/x)\Big]_1^\ell = \lim_{\ell\to+\infty}(\pi - \pi/\ell) = \pi$

$A = \displaystyle\lim_{\ell\to+\infty}\int_1^\ell 2\pi(1/x)\sqrt{1+1/x^4}\,dx;$

use Exercise 46(a) with $f(x) = 2\pi/x$, $g(x) = (2\pi/x)\sqrt{1+1/x^4}$
and $a = 1$ to see that the area is infinite.

51. The area under the curve $y = \dfrac{1}{1+x^2}$, above the x-axis, and to the
right of the y-axis is given by $\displaystyle\int_0^\infty \dfrac{1}{1+x^2}$. Solving for
$y = \sqrt{\dfrac{1-y}{y}}$, the area is also given by the improper integral

$\displaystyle\int_0^1 \sqrt{\dfrac{1-y}{y}}\,dy.$

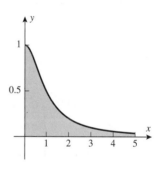

53. Let $x = r\tan\theta$ to get $\displaystyle\int \dfrac{dx}{(r^2+x^2)^{3/2}} = \dfrac{1}{r^2}\int \cos\theta\,d\theta = \dfrac{1}{r^2}\sin\theta + C = \dfrac{x}{r^2\sqrt{r^2+x^2}} + C$

so $u = \dfrac{2\pi NIr}{k}\displaystyle\lim_{\ell\to+\infty}\dfrac{x}{r^2\sqrt{r^2+x^2}}\Big]_a^\ell = \dfrac{2\pi NI}{kr}\lim_{\ell\to+\infty}\left(\ell/\sqrt{r^2+\ell^2} - a/\sqrt{r^2+a^2}\right)$

$= \dfrac{2\pi NI}{kr}\left(1 - a/\sqrt{r^2+a^2}\right).$

55. (a) Satellite's weight $= w(x) = k/x^2$ lb when $x =$ distance from center of Earth; $w(4000) = 6000$
so $k = 9.6 \times 10^{10}$ and $W = \displaystyle\int_{4000}^{4000+\ell} 9.6 \times 10^{10}x^{-2}\,dx$ mi·lb.

(b) $\displaystyle\int_{4000}^{+\infty} 9.6 \times 10^{10}x^{-2}\,dx = \lim_{\ell\to+\infty} -9.6 \times 10^{10}/x\Big]_{4000}^\ell = 2.4 \times 10^7$ mi·lb

57. (a) $\mathcal{L}\{f(t)\} = \displaystyle\int_0^{+\infty} te^{-st}\,dt = \lim_{\ell\to+\infty} -(t/s+1/s^2)e^{-st}\Big]_0^\ell = \dfrac{1}{s^2}$

(b) $\mathcal{L}\{f(t)\} = \displaystyle\int_0^{+\infty} t^2 e^{-st}\,dt = \lim_{\ell\to+\infty} -(t^2/s + 2t/s^2 + 2/s^3)e^{-st}\Big]_0^{\ell} = \dfrac{2}{s^3}$

(c) $\mathcal{L}\{f(t)\} = \displaystyle\int_3^{+\infty} e^{-st}\,dt = \lim_{\ell\to+\infty} -\dfrac{1}{s}e^{-st}\Big]_3^{\ell} = \dfrac{e^{-3s}}{s}$

59. (a) $u = \sqrt{a}\,x,\, du = \sqrt{a}\,dx,\, 2\displaystyle\int_0^{+\infty} e^{-ax^2}\,dx = \dfrac{2}{\sqrt{a}}\int_0^{+\infty} e^{-u^2}\,du = \sqrt{\pi/a}$

(b) $x = \sqrt{2}\sigma u,\, dx = \sqrt{2}\sigma\,du,\, \dfrac{2}{\sqrt{2\pi}\sigma}\displaystyle\int_0^{+\infty} e^{-x^2/2\sigma^2}\,dx = \dfrac{2}{\sqrt{\pi}}\int_0^{+\infty} e^{-u^2}\,du = 1$

61. (a) $\displaystyle\int_0^4 \dfrac{1}{x^6+1}\,dx \approx 1.047;\ \pi/3 \approx 1.047$

(b) $\displaystyle\int_0^{+\infty} \dfrac{1}{x^6+1}\,dx = \int_0^4 \dfrac{1}{x^6+1}\,dx + \int_4^{+\infty} \dfrac{1}{x^6+1}\,dx$ so

$E = \displaystyle\int_4^{+\infty} \dfrac{1}{x^6+1}\,dx < \int_4^{+\infty} \dfrac{1}{x^6}\,dx = \dfrac{1}{5(4)^5} < 2\times 10^{-4}$

63. If $p = 1$, then $\displaystyle\int_0^1 \dfrac{dx}{x} = \lim_{\ell\to 0^+} \ln x\Big]_{\ell}^1 = +\infty$;

if $p \neq 1$, then $\displaystyle\int_0^1 \dfrac{dx}{x^p} = \lim_{\ell\to 0^+} \dfrac{x^{1-p}}{1-p}\Big]_{\ell}^1 = \lim_{\ell\to 0^+}[(1-\ell^{1-p})/(1-p)] = \begin{cases} 1/(1-p), & p < 1 \\ +\infty, & p > 1 \end{cases}$.

65. $2\displaystyle\int_0^1 \cos(u^2)\,du \approx 1.809$

67. (a) $\Gamma(1) = \displaystyle\int_0^{+\infty} e^{-t}\,dt = \lim_{\ell\to+\infty} -e^{-t}\Big]_0^{\ell} = \lim_{\ell\to+\infty}(-e^{-\ell}+1) = 1$

(b) $\Gamma(x+1) = \displaystyle\int_0^{+\infty} t^x e^{-t}\,dt$; let $u = t^x$, $dv = e^{-t}\,dt$ to get

$\Gamma(x+1) = -t^x e^{-t}\Big]_0^{+\infty} + x\displaystyle\int_0^{+\infty} t^{x-1}e^{-t}\,dt = -t^x e^{-t}\Big]_0^{+\infty} + x\Gamma(x)$

$\displaystyle\lim_{t\to+\infty} t^x e^{-t} = \lim_{t\to+\infty} \dfrac{t^x}{e^t} = 0$ (by multiple applications of L'Hôpital's rule)

so $\Gamma(x+1) = x\Gamma(x)$

(c) $\Gamma(2) = (1)\Gamma(1) = (1)(1) = 1,\ \Gamma(3) = 2\Gamma(2) = (2)(1) = 2,\ \Gamma(4) = 3\Gamma(3) = (3)(2) = 6$

It appears that $\Gamma(n) = (n-1)!$ if n is a positive integer.

(d) $\Gamma\left(\dfrac{1}{2}\right) = \displaystyle\int_0^{+\infty} t^{-1/2}e^{-t}\,dt = 2\int_0^{+\infty} e^{-u^2}\,du$ (with $u = \sqrt{t}$) $= 2(\sqrt{\pi}/2) = \sqrt{\pi}$

(e) $\Gamma\left(\dfrac{3}{2}\right) = \dfrac{1}{2}\Gamma\left(\dfrac{1}{2}\right) = \dfrac{1}{2}\sqrt{\pi},\ \Gamma\left(\dfrac{5}{2}\right) = \dfrac{3}{2}\Gamma\left(\dfrac{3}{2}\right) = \dfrac{3}{4}\sqrt{\pi}$

69. (a) $\sqrt{\cos\theta - \cos\theta_0} = \sqrt{2\left[\sin^2(\theta_0/2) - \sin^2(\theta/2)\right]} = \sqrt{2(k^2 - k^2\sin^2\phi)} = \sqrt{2k^2\cos^2\phi}$

$= \sqrt{2}\,k\cos\phi;\ k\sin\phi = \sin(\theta/2)$ so $k\cos\phi\,d\phi = \dfrac{1}{2}\cos(\theta/2)\,d\theta = \dfrac{1}{2}\sqrt{1 - \sin^2(\theta/2)}\,d\theta$

$$= \frac{1}{2}\sqrt{1 - k^2 \sin^2 \phi}\, d\theta, \text{ thus } d\theta = \frac{2k\cos\phi}{\sqrt{1 - k^2 \sin^2 \phi}}\, d\phi \text{ and hence}$$

$$T = \sqrt{\frac{8L}{g}} \int_0^{\pi/2} \frac{1}{\sqrt{2}k\cos\phi} \cdot \frac{2k\cos\phi}{\sqrt{1 - k^2 \sin^2 \phi}}\, d\phi = 4\sqrt{\frac{L}{g}} \int_0^{\pi/2} \frac{1}{\sqrt{1 - k^2 \sin^2 \phi}}\, d\phi$$

(b) If $L = 1.5$ ft and $\theta_0 = (\pi/180)(20) = \pi/9$, then

$$T = \frac{\sqrt{3}}{2} \int_0^{\pi/2} \frac{d\phi}{\sqrt{1 - \sin^2(\pi/18)\sin^2 \phi}} \approx 1.37 \text{ s.}$$

REVIEW EXERCISES, CHAPTER 8

1. $u = 4 + 9x, du = 9\, dx, \quad \frac{1}{9}\int u^{1/2}\, du = \frac{2}{27}(4 + 9x)^{3/2} + C$

3. $u = \cos\theta, -\int u^{1/2} du = -\frac{2}{3}\cos^{3/2}\theta + C$

5. $u = \tan(x^2), \frac{1}{2}\int u^2 du = \frac{1}{6}\tan^3(x^2) + C$

7. (a) With $u = \sqrt{x}$:

$$\int \frac{1}{\sqrt{x}\sqrt{2 - x}}\, dx = 2\int \frac{1}{\sqrt{2 - u^2}}\, du = 2\sin^{-1}(u/\sqrt{2}) + C = 2\sin^{-1}(\sqrt{x/2}) + C;$$

with $u = \sqrt{2 - x}$:

$$\int \frac{1}{\sqrt{x}\sqrt{2 - x}}\, dx = -2\int \frac{1}{\sqrt{2 - u^2}}\, du = -2\sin^{-1}(u/\sqrt{2}) + C = -2\sin^{-1}(\sqrt{2 - x}/\sqrt{2}) + C_1;$$

completing the square:

$$\int \frac{1}{\sqrt{1 - (x - 1)^2}}\, dx = \sin^{-1}(x - 1) + C.$$

(b) In the three results in Part (a) the antiderivatives differ by a constant, in particular
$2\sin^{-1}(\sqrt{x/2}) = \pi - 2\sin^{-1}(\sqrt{2 - x}/\sqrt{2}) = \pi/2 + \sin^{-1}(x - 1).$

9. $u = x, dv = e^{-x}dx, du = dx, v = -e^{-x};$

$$\int xe^{-x}dx = -xe^{-x} + \int e^{-x}dx = -xe^{-x} - e^{-x} + C$$

11. $u = \ln(2x + 3), dv = dx, du = \frac{2}{2x + 3}dx, v = x; \int \ln(2x + 3)dx = x\ln(2x + 3) - \int \frac{2x}{2x + 3}dx$

but $\int \frac{2x}{2x + 3}dx = \int \left(1 - \frac{3}{2x + 3}\right)dx = x - \frac{3}{2}\ln(2x + 3) + C_1$ so

$$\int \ln(2x + 3)dx = x\ln(2x + 3) - x + \frac{3}{2}\ln(2x + 3) + C$$

13.

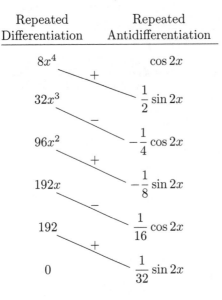

Repeated Differentiation	Repeated Antidifferentiation

$8x^4$ $\cos 2x$

$+$

$32x^3$ $\dfrac{1}{2}\sin 2x$

$-$

$96x^2$ $-\dfrac{1}{4}\cos 2x$

$+$

$192x$ $-\dfrac{1}{8}\sin 2x$

$-$

192 $\dfrac{1}{16}\cos 2x$

$+$

0 $\dfrac{1}{32}\sin 2x$

$$\int 8x^4 \cos 2x\, dx = (4x^4 - 12x^2 + 6)\sin 2x + (8x^3 - 12x)\cos 2x + C$$

15. $\displaystyle\int \sin^2 5\theta\, d\theta = \frac{1}{2}\int (1 - \cos 10\theta)d\theta = \frac{1}{2}\theta - \frac{1}{20}\sin 10\theta + C$

17. $\displaystyle\int \sin x \cos 2x\, dx = \frac{1}{2}\int (\sin 3x - \sin x)dx = -\frac{1}{6}\cos 3x + \frac{1}{2}\cos x + C$

19. $u = 2x$,

$$\int \sin^4 2x\, dx = \frac{1}{2}\int \sin^4 u\, du = \frac{1}{2}\left[-\frac{1}{4}\sin^3 u \cos u + \frac{3}{4}\int \sin^2 u\, du\right]$$

$$= -\frac{1}{8}\sin^3 u \cos u + \frac{3}{8}\left[-\frac{1}{2}\sin u \cos u + \frac{1}{2}\int du\right]$$

$$= -\frac{1}{8}\sin^3 u \cos u - \frac{3}{16}\sin u \cos u + \frac{3}{16}u + C$$

$$= -\frac{1}{8}\sin^3 2x \cos 2x - \frac{3}{16}\sin 2x \cos 2x + \frac{3}{8}x + C$$

21. $x = 3\sin\theta$, $dx = 3\cos\theta\, d\theta$,

$$9\int \sin^2\theta\, d\theta = \frac{9}{2}\int (1 - \cos 2\theta)d\theta = \frac{9}{2}\theta - \frac{9}{4}\sin 2\theta + C = \frac{9}{2}\theta - \frac{9}{2}\sin\theta \cos\theta + C$$

$$= \frac{9}{2}\sin^{-1}(x/3) - \frac{1}{2}x\sqrt{9 - x^2} + C$$

23. $x = \sec\theta$, $dx = \sec\theta \tan\theta\, d\theta$, $\displaystyle\int \sec\theta\, d\theta = \ln|\sec\theta + \tan\theta| + C = \ln\left|x + \sqrt{x^2 - 1}\right| + C$

25. $x = 3\tan\theta$, $dx = 3\sec^2\theta\, d\theta$,

$$9 \int \tan^2 \theta \sec \theta \, d\theta = 9 \int \sec^3 \theta \, d\theta - 9 \int \sec \theta \, d\theta$$

$$= \frac{9}{2} \sec \theta \tan \theta - \frac{9}{2} \ln|\sec \theta + \tan \theta| + C$$

$$= \frac{1}{2} x \sqrt{9 + x^2} - \frac{9}{2} \ln|\frac{1}{3} \sqrt{9 + x^2} + \frac{1}{3} x| + C$$

27. $\dfrac{1}{(x + 4)(x - 1)} = \dfrac{A}{x + 4} + \dfrac{B}{x - 1}; \ A = -\dfrac{1}{5}, \ B = \dfrac{1}{5}$ so

$$-\frac{1}{5} \int \frac{1}{x + 4} dx + \frac{1}{5} \int \frac{1}{x - 1} dx = -\frac{1}{5} \ln|x + 4| + \frac{1}{5} \ln|x - 1| + C = \frac{1}{5} \ln\left|\frac{x - 1}{x + 4}\right| + C$$

29. $\dfrac{x^2 + 2}{x + 2} = x - 2 + \dfrac{6}{x + 2}, \ \displaystyle\int \left(x - 2 + \dfrac{6}{x + 2}\right) dx = \dfrac{1}{2} x^2 - 2x + 6 \ \ln|x + 2| + C$

31. $\dfrac{x^2}{(x + 2)^3} = \dfrac{A}{x + 2} + \dfrac{B}{(x + 2)^2} + \dfrac{C}{(x + 2)^3}; \ A = 1, \ B = -4, \ C = 4$ so

$$\int \frac{1}{x + 2} dx - 4 \int \frac{1}{(x + 2)^2} dx + 4 \int \frac{1}{(x + 2)^3} dx = \ln|x + 2| + \frac{4}{x + 2} - \frac{2}{(x + 2)^2} + C$$

33. **(a)** With $x = \sec \theta$:

$$\int \frac{1}{x^3 - x} dx = \int \cot \theta \, d\theta = \ln|\sin \theta| + C = \ln \frac{\sqrt{x^2 - 1}}{|x|} + C; \text{ valid for } |x| > 1.$$

(b) With $x = \sin \theta$:

$$\int \frac{1}{x^3 - x} dx = -\int \frac{1}{\sin \theta \cos \theta} d\theta = -\int 2 \csc 2\theta \, d\theta$$

$$= -\ln|\csc 2\theta - \cot 2\theta| + C = \ln|\cot \theta| + C = \ln \frac{\sqrt{1 - x^2}}{|x|} + C, \ \ 0 < |x| < 1.$$

35. #40 **37.** #113 **39.** #28

41. exact value $= 14/3 \approx 4.666666667$
  (a) $4.667600663, \ |E_M| \approx 0.000933996$
  (b) $4.664795679, \ |E_T| \approx 0.001870988$
  (c) $4.666651630, \ |E_S| \approx 0.000015037$

43. $f(x) = \sqrt{x + 1}, \ f''(x) = -\dfrac{1}{4}(x + 1)^{-3/2}, \ f^{(4)}(x) = -\dfrac{15}{16}(x + 1)^{-7/2}; \ K_2 = 1/4, \ K_4 = 15/16$

  (a) $|E_M| \leq \dfrac{27}{2400}(1/4) = 0.002812500$ **(b)** $|E_T| \leq \dfrac{27}{1200}(1/4) = 0.005625000$

  (c) $|E_S| \leq \dfrac{243}{180 \times 10^4}(15/16) \approx 0.000126563$

45. **(a)** $n > \left[\dfrac{(27)(1/4)}{(24)(5 \times 10^{-4})}\right]^{1/2} \approx 23.7; \ n = 24$ **(b)** $n > \left[\dfrac{(27)(1/4)}{(12)(5 \times 10^{-4})}\right]^{1/2} \approx 33.5; \ n = 34$

  (c) $n > \left[\dfrac{(243)(15/16)}{(180)(5 \times 10^{-4})}\right]^{1/4} \approx 7.1; \ n = 8$

47. $\displaystyle\lim_{\ell \to +\infty} (-e^{-x})\Big]_0^\ell = \lim_{\ell \to +\infty} (-e^{-\ell} + 1) = 1$

49. $\displaystyle\lim_{\ell \to 9^-} -2\sqrt{9-x}\,\Big]_0^\ell = \lim_{\ell \to 9^-} 2(-\sqrt{9-\ell} + 3) = 6$

51. $A = \displaystyle\int_e^{+\infty} \frac{\ln x - 1}{x^2}\, dx = \lim_{\ell \to +\infty} c - \frac{\ln x}{x}\Big]_e^\ell = 1/e$

53. $\displaystyle\int_0^{+\infty} \frac{dx}{x^2 + a^2} = \lim_{\ell \to +\infty} \frac{1}{a} \tan^{-1}(x/a)\Big]_0^\ell = \lim_{\ell \to +\infty} \frac{1}{a} \tan^{-1}(\ell/a) = \frac{\pi}{2a} = 1, a = \pi/2$

55. $x = \sqrt{3}\tan\theta$, $dx = \sqrt{3}\sec^2\theta\, d\theta$,

$\displaystyle\frac{1}{3}\int \frac{1}{\sec\theta}\, d\theta = \frac{1}{3}\int \cos\theta\, d\theta = \frac{1}{3}\sin\theta + C = \frac{x}{3\sqrt{3+x^2}} + C$

57. Use Endpaper Formula (31) to get $\displaystyle\int \tan^7\theta\, d\theta = \frac{1}{6}\tan^6\theta - \frac{1}{4}\tan^4\theta + \frac{1}{2}\tan^2\theta + \ln|\cos\theta| + C$.

59. $\displaystyle\int \sin^2 2t \cos^3 2t\, dt = \int \sin^2 2t(1 - \sin^2 2t)\cos 2t\, dt = \int (\sin^2 2t - \sin^4 2t)\cos 2t\, dt$

$\displaystyle\qquad\qquad = \frac{1}{6}\sin^3 2t - \frac{1}{10}\sin^5 2t + C$

61. $u = e^{2x}$, $dv = \cos 3x\, dx$, $du = 2e^{2x}dx$, $v = \frac{1}{3}\sin 3x$;

$\displaystyle\int e^{2x}\cos 3x\, dx = \frac{1}{3}e^{2x}\sin 3x - \frac{2}{3}\int e^{2x}\sin 3x\, dx.$ Use $u = e^{2x}$, $dv = \sin 3x\, dx$ to get

$\displaystyle\int e^{2x}\sin 3x\, dx = -\frac{1}{3}e^{2x}\cos 3x + \frac{2}{3}\int e^{2x}\cos 3x\, dx$ so

$\displaystyle\int e^{2x}\cos 3x\, dx = \frac{1}{3}e^{2x}\sin 3x + \frac{2}{9}e^{2x}\cos 3x - \frac{4}{9}\int e^{2x}\cos 3x\, dx,$

$\displaystyle\frac{13}{9}\int e^{2x}\cos 3x\, dx = \frac{1}{9}e^{2x}(3\sin 3x + 2\cos 3x) + C_1, \int e^{2x}\cos 3x\, dx = \frac{1}{13}e^{2x}(3\sin 3x + 2\cos 3x) + C$

63. $\displaystyle\frac{1}{(x-1)(x+2)(x-3)} = \frac{A}{x-1} + \frac{B}{x+2} + \frac{C}{x-3}; A = -\frac{1}{6}, B = \frac{1}{15}, C = \frac{1}{10}$ so

$\displaystyle -\frac{1}{6}\int \frac{1}{x-1}dx + \frac{1}{15}\int \frac{1}{x+2}dx + \frac{1}{10}\int \frac{1}{x-3}dx$

$\displaystyle\qquad = -\frac{1}{6}\ln|x-1| + \frac{1}{15}\ln|x+2| + \frac{1}{10}\ln|x-3| + C$

65. $u = \sqrt{x-4}$, $x = u^2 + 4$, $dx = 2u\, du$,

$\displaystyle\int_0^2 \frac{2u^2}{u^2+4}du = 2\int_0^2 \left[1 - \frac{4}{u^2+4}\right]du = \left[2u - 4\tan^{-1}(u/2)\right]_0^2 = 4 - \pi$

67. $u = \sqrt{e^x + 1}$, $e^x = u^2 - 1$, $x = \ln(u^2 - 1)$, $dx = \frac{2u}{u^2-1}du$,

$\displaystyle\int \frac{2}{u^2-1}du = \int \left[\frac{1}{u-1} - \frac{1}{u+1}\right]du = \ln|u-1| - \ln|u+1| + C = \ln\frac{\sqrt{e^x+1}-1}{\sqrt{e^x+1}+1} + C$

69. $u = \sin^{-1} x$, $dv = dx$, $du = \dfrac{1}{\sqrt{1 - x^2}} dx$, $v = x$;

$$\int_0^{1/2} \sin^{-1} x \, dx = x \sin^{-1} x \Big]_0^{1/2} - \int_0^{1/2} \frac{x}{\sqrt{1 - x^2}} dx = \frac{1}{2} \sin^{-1} \frac{1}{2} + \sqrt{1 - x^2} \Big]_0^{1/2}$$

$$= \frac{1}{2}\left(\frac{\pi}{6}\right) + \sqrt{\frac{3}{4}} - 1 = \frac{\pi}{12} + \frac{\sqrt{3}}{2} - 1$$

71. $\displaystyle\int \frac{x + 3}{\sqrt{(x + 1)^2 + 1}} dx$, let $u = x + 1$,

$$\int \frac{u + 2}{\sqrt{u^2 + 1}} du = \int \left[u(u^2 + 1)^{-1/2} + \frac{2}{\sqrt{u^2 + 1}} \right] du = \sqrt{u^2 + 1} + 2\sinh^{-1} u + C$$

$$= \sqrt{x^2 + 2x + 2} + 2\sinh^{-1}(x + 1) + C$$

<u>Alternate solution</u>: let $x + 1 = \tan\theta$,

$$\int (\tan\theta + 2) \sec\theta \, d\theta = \int \sec\theta \tan\theta \, d\theta + 2 \int \sec\theta \, d\theta = \sec\theta + 2\ln|\sec\theta + \tan\theta| + C$$

$$= \sqrt{x^2 + 2x + 2} + 2\ln(\sqrt{x^2 + 2x + 2} + x + 1) + C.$$

73. $\displaystyle\lim_{\ell \to +\infty} -\frac{1}{2(x^2 + 1)} \Big]_a^\ell = \lim_{\ell \to +\infty} \left[-\frac{1}{2(\ell^2 + 1)} + \frac{1}{2(a^2 + 1)} \right] = \frac{1}{2(a^2 + 1)}$

CHAPTER 9
Mathematical Modeling with Differential Equations

EXERCISE SET 9.1

1. $y' = 9x^2 e^{x^3} = 3x^2 y$ and $y(0) = 3$ by inspection.

3. **(a)** first order; $\dfrac{dy}{dx} = c$; $(1+x)\dfrac{dy}{dx} = (1+x)c = y$

   (b) second order; $y' = c_1 \cos t - c_2 \sin t$, $y'' + y = -c_1 \sin t - c_2 \cos t + (c_1 \sin t + c_2 \cos t) = 0$

5. $\dfrac{1}{y}\dfrac{dy}{dx} = y^2 + 2xy\dfrac{dy}{dx}$, $\dfrac{dy}{dx}(1 - 2xy^2) = y^3$, $\dfrac{dy}{dx} = \dfrac{y^3}{1 - 2xy^2}$

7. **(a)** IF: $\mu = e^{3\int dx} = e^{3x}$, $\dfrac{d}{dx}\left[ye^{3x}\right] = 0$, $ye^{3x} = C$, $y = Ce^{-3x}$

   separation of variables: $\dfrac{dy}{y} = -3dx$, $\ln|y| = -3x + C_1$, $y = \pm e^{-3x}e^{C_1} = Ce^{-3x}$

   including $C = 0$ by inspection

   (b) IF: $\mu = e^{-2\int dt} = e^{-2t}$, $\dfrac{d}{dt}[ye^{-2t}] = 0$, $ye^{-2t} = C$, $y = Ce^{2t}$

   separation of variables: $\dfrac{dy}{y} = 2dt$, $\ln|y| = 2t + C_1$, $y = \pm e^{C_1}e^{2t} = Ce^{2t}$

   including $C = 0$ by inspection

9. $\mu = e^{\int 4dx} = e^{4x}$, $e^{4x}y = \displaystyle\int e^x \, dx = e^x + C$, $y = e^{-3x} + Ce^{-4x}$

11. $\mu = e^{\int dx} = e^x$, $e^x y = \displaystyle\int e^x \cos(e^x)dx = \sin(e^x) + C$, $y = e^{-x}\sin(e^x) + Ce^{-x}$

13. $\dfrac{dy}{dx} + \dfrac{x}{x^2+1}y = 0$, $\mu = e^{\int (x/(x^2+1))dx} = e^{\frac{1}{2}\ln(x^2+1)} = \sqrt{x^2+1}$,

   $\dfrac{d}{dx}\left[y\sqrt{x^2+1}\right] = 0$, $y\sqrt{x^2+1} = C$, $y = \dfrac{C}{\sqrt{x^2+1}}$

15. $\dfrac{1}{y}dy = \dfrac{1}{x}dx$, $\ln|y| = \ln|x| + C_1$, $\ln\left|\dfrac{y}{x}\right| = C_1$, $\dfrac{y}{x} = \pm e^{C_1} = C$, $y = Cx$

   including $C = 0$ by inspection

17. $\dfrac{dy}{1+y} = -\dfrac{x}{\sqrt{1+x^2}}dx$, $\ln|1+y| = -\sqrt{1+x^2} + C_1$, $1+y = \pm e^{-\sqrt{1+x^2}}e^{C_1} = Ce^{-\sqrt{1+x^2}}$,

   $y = Ce^{-\sqrt{1+x^2}} - 1, C \neq 0$

19. $\left(\dfrac{2(1+y^2)}{y}\right) dy = e^x dx$, $2\ln|y| + y^2 = e^x + C$; by inspection, $y = 0$ is also a solution

21. $e^y dy = \dfrac{\sin x}{\cos^2 x}dx = \sec x \tan x \, dx$, $e^y = \sec x + C$, $y = \ln(\sec x + C)$

23. $\dfrac{dy}{y^2 - y} = \dfrac{dx}{\sin x}$, $\displaystyle\int \left[-\dfrac{1}{y} + \dfrac{1}{y-1} \right] dy = \int \csc x\, dx$, $\ln \left| \dfrac{y-1}{y} \right| = \ln | \csc x - \cot x | + C_1$,

$\dfrac{y-1}{y} = \pm e^{C_1}(\csc x - \cot x) = C(\csc x - \cot x)$, $y = \dfrac{1}{1 - C(\csc x - \cot x)}$, $C \neq 0$;

by inspection, $y = 0$ is also a solution, as is $y = 1$.

25. $\dfrac{dy}{dx} + \dfrac{1}{x}y = 1$, $\mu = e^{\int (1/x)dx} = e^{\ln x} = x$, $\dfrac{d}{dx}[xy] = x$, $xy = \dfrac{1}{2}x^2 + C$, $y = x/2 + C/x$

   (a) $2 = y(1) = \dfrac{1}{2} + C, C = \dfrac{3}{2}, y = x/2 + 3/(2x)$

   (b) $2 = y(-1) = -1/2 - C, C = -5/2, y = x/2 - 5/(2x)$

27. $\mu = e^{-2 \int x\, dx} = e^{-x^2}$, $e^{-x^2} y = \displaystyle\int 2xe^{-x^2}\, dx = -e^{-x^2} + C$,

$y = -1 + Ce^{x^2}$, $3 = -1 + C$, $C = 4$, $y = -1 + 4e^{x^2}$

29. $(2y + \cos y)\, dy = 3x^2\, dx$, $y^2 + \sin y = x^3 + C$, $\pi^2 + \sin \pi = C, C = \pi^2$,

$y^2 + \sin y = x^3 + \pi^2$

31. $2(y - 1)\, dy = (2t + 1)\, dt$, $y^2 - 2y = t^2 + t + C$, $1 + 2 = C, C = 3$, $y^2 - 2y = t^2 + t + 3$

33. (a) $\dfrac{dy}{y} = \dfrac{dx}{2x}$, $\ln |y| = \dfrac{1}{2} \ln |x| + C_1$,

$|y| = C|x|^{1/2}$, $y^2 = Cx$;

by inspection $y = 0$ is also a solution.

(b) $1 = C(2)^2, C = 1/4, y^2 = x/4$

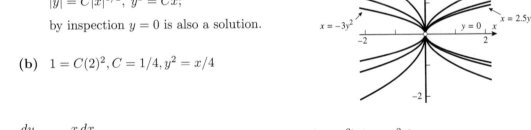

35. $\dfrac{dy}{y} = -\dfrac{x\, dx}{x^2 + 4}$,

$\ln |y| = -\dfrac{1}{2} \ln(x^2 + 4) + C_1$,

$y = \dfrac{C}{\sqrt{x^2 + 4}}$

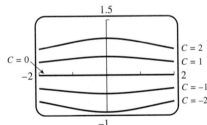

37. $(1 - y^2)\, dy = x^2\, dx$,

$y - \dfrac{y^3}{3} = \dfrac{x^3}{3} + C_1$, $x^3 + y^3 - 3y = C$

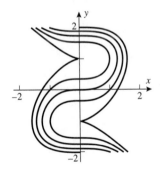

39. Of the solutions $y = \dfrac{1}{2x^2 - C}$, all pass through the point $\left(0, -\dfrac{1}{C}\right)$ and thus never through $(0, 0)$.

A solution of the initial value problem with $y(0) = 0$ is (by inspection) $y = 0$. The methods of Example 4 fail because the integrals there become divergent when the point $x = 0$ is included in the integral.

41. $\dfrac{dy}{dx} = xe^{-y}, e^y \, dy = x \, dx, e^y = \dfrac{x^2}{2} + C, x = 2$ when $y = 0$ so $1 = 2 + C, C = -1, e^y = x^2/2 - 1$

43. $\dfrac{dy}{dt} = $ rate in $-$ rate out, where y is the amount of salt at time t,

$$\dfrac{dy}{dt} = (4)(2) - \left(\dfrac{y}{50}\right)(2) = 8 - \dfrac{1}{25}y, \text{ so } \dfrac{dy}{dt} + \dfrac{1}{25}y = 8 \text{ and } y(0) = 25.$$

$$\mu = e^{\int (1/25)dt} = e^{t/25}, \; e^{t/25}y = \int 8e^{t/25} dt = 200e^{t/25} + C,$$

$$y = 200 + Ce^{-t/25}, \; 25 = 200 + C, \; C = -175,$$

(a) $y = 200 - 175e^{-t/25}$ oz   **(b)** when $t = 25, y = 200 - 175e^{-1} \approx 136$ oz

45. The volume V of the (polluted) water is $V(t) = 500 + (20 - 10)t = 500 + 10t$; if $y(t)$ is the number of pounds of particulate matter in the water,

then $y(0) = 50$, and $\dfrac{dy}{dt} = 0 - 10\dfrac{y}{V} = -\dfrac{1}{50 + t}y, \dfrac{dy}{dt} + \dfrac{1}{50 + t}y = 0; \mu = e^{\int \frac{dt}{50+t}} = 50 + t;$

$$\dfrac{d}{dt}[(50 + t)y] = 0, \; (50 + t)y = C, \; 2500 = 50y(0) = C, \; y(t) = 2500/(50 + t).$$

The tank reaches the point of overflowing when $V = 500 + 10t = 1000, t = 50$ min, so $y = 2500/(50 + 50) = 25$ lb.

47. (a) $\dfrac{dv}{dt} + \dfrac{c}{m}v = -g, \mu = e^{(c/m)\int dt} = e^{ct/m}, \dfrac{d}{dt}\left[ve^{ct/m}\right] = -ge^{ct/m}, ve^{ct/m} = -\dfrac{gm}{c}e^{ct/m} + C,$

$v = -\dfrac{gm}{c} + Ce^{-ct/m}$, but $v_0 = v(0) = -\dfrac{gm}{c} + C, C = v_0 + \dfrac{gm}{c}, v = -\dfrac{gm}{c} + \left(v_0 + \dfrac{gm}{c}\right)e^{-ct/m}$

(b) Replace $\dfrac{mg}{c}$ with v_τ and $-ct/m$ with $-gt/v_\tau$ in (23).

(c) From Part (b), $s(t) = C - v_\tau t - (v_0 + v_\tau)\dfrac{v_\tau}{g}e^{-gt/v_\tau}$;

$$s_0 = s(0) = C - (v_0 + v_\tau)\dfrac{v_\tau}{g}, \; C = s_0 + (v_0 + v_\tau)\dfrac{v_\tau}{g}, \; s(t) = s_0 - v_\tau t + \dfrac{v_\tau}{g}(v_0 + v_\tau)\left(1 - e^{-gt/v_\tau}\right)$$

49. $\dfrac{dI}{dt} + \dfrac{R}{L}I = \dfrac{V(t)}{L}, \mu = e^{(R/L)\int dt} = e^{Rt/L}, \dfrac{d}{dt}(e^{Rt/L}I) = \dfrac{V(t)}{L}e^{Rt/L},$

$$Ie^{Rt/L} = I(0) + \dfrac{1}{L}\int_0^t V(u)e^{Ru/L}du, I(t) = I(0)e^{-Rt/L} + \dfrac{1}{L}e^{-Rt/L}\int_0^t V(u)e^{Ru/L}du.$$

(a) $I(t) = \dfrac{1}{5}e^{-2t}\int_0^t 20e^{2u}du = 2e^{-2t}e^{2u}\Big]_0^t = 2\left(1 - e^{-2t}\right)$ A.

(b) $\lim\limits_{t \to +\infty} I(t) = 2$ A

51. **(a)** $\dfrac{dv}{dt} = \dfrac{ck}{m_0 - kt} - g, v = -c\ln(m_0 - kt) - gt + C; v = 0$ when $t = 0$ so $0 = -c\ln m_0 + C$,

$C = c\ln m_0, v = c\ln m_0 - c\ln(m_0 - kt) - gt = c\ln \dfrac{m_0}{m_0 - kt} - gt.$

(b) $m_0 - kt = 0.2m_0$ when $t = 100$ so

$v = 2500\ln \dfrac{m_0}{0.2m_0} - 9.8(100) = 2500\ln 5 - 980 \approx 3044\,\text{m/s}.$

53. **(a)** $A(h) = \pi(1)^2 = \pi, \pi\dfrac{dh}{dt} = -0.025\sqrt{h}, \dfrac{\pi}{\sqrt{h}}dh = -0.025dt, 2\pi\sqrt{h} = -0.025t + C; h = 4$ when

$t = 0$, so $4\pi = C, 2\pi\sqrt{h} = -0.025t + 4\pi, \sqrt{h} = 2 - \dfrac{0.025}{2\pi}t, h \approx (2 - 0.003979\,t)^2.$

(b) $h = 0$ when $t \approx 2/0.003979 \approx 502.6$ s ≈ 8.4 min.

55. $\dfrac{dv}{dt} = -\dfrac{1}{32}v^2, \dfrac{1}{v^2}dv = -\dfrac{1}{32}dt, -\dfrac{1}{v} = -\dfrac{1}{32}t + C; v = 128$ when $t = 0$ so $-\dfrac{1}{128} = C$,

$-\dfrac{1}{v} = -\dfrac{1}{32}t - \dfrac{1}{128}, v = \dfrac{128}{4t + 1}$ cm/s. But $v = \dfrac{dx}{dt}$ so $\dfrac{dx}{dt} = \dfrac{128}{4t + 1}, x = 32\ln(4t + 1) + C_1;$

$x = 0$ when $t = 0$ so $C_1 = 0, x = 32\ln(4t + 1)$ cm.

57. Differentiate to get $\dfrac{dy}{dx} = -\sin x + e^{-x^2}, y(0) = 1.$

59. Suppose that $H(y) = G(x) + C.$ Then $\dfrac{dH}{dy}\dfrac{dy}{dx} = G'(x).$ But $\dfrac{dH}{dy} = h(y)$ and $\dfrac{dG}{dx} = g(x),$ hence $y(x)$ is a solution of (10).

EXERCISE SET 9.2

1. $y^1 = xy/4$
$-2 \le x \le +2$
$-2 \le y \le +2$

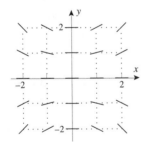

3.

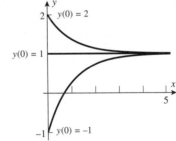

5.

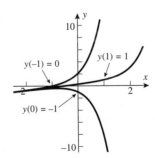

7. $\displaystyle\lim_{x \to +\infty} y = 1$

9. (a) IV, since the slope is positive for $x > 0$ and negative for $x < 0$.

(b) VI, since the slope is positive for $y > 0$ and negative for $y < 0$.

(c) V, since the slope is always positive.

(d) II, since the slope changes sign when crossing the lines $y = \pm 1$.

(e) I, since the slope can be positive or negative in each quadrant but is not periodic.

(f) III, since the slope is periodic in both x and y.

11. (a) $y_0 = 1$,

$y_{n+1} = y_n + (x_n + y_n)(0.2) = (x_n + 6y_n)/5$

n	0	1	2	3	4	5
x_n	0	0.2	0.4	0.6	0.8	1.0
y_n	1	1.20	1.48	1.86	2.35	2.98

(b) $y' - y = x$, $\mu = e^{-x}$, $\dfrac{d}{dx}\left[ye^{-x}\right] = xe^{-x}$,

$ye^{-x} = -(x+1)e^{-x} + C$, $1 = -1 + C$,

$C = 2$, $y = -(x+1) + 2e^x$

x_n	0	0.2	0.4	0.6	0.8	1.0
$y(x_n)$	1	1.24	1.58	2.04	2.65	3.44
abs. error	0	0.04	0.10	0.19	0.30	0.46
perc. error	0	3	6	9	11	13

(c)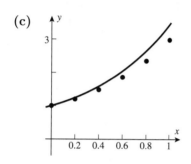

13. $y_0 = 1, y_{n+1} = y_n + \frac{1}{2}y_n^{1/3}$

n	0	1	2	3	4	5	6	7	8
x_n	0	0.5	1	1.5	2	2.5	3	3.5	4
y_n	1	1.50	2.11	2.84	3.68	4.64	5.72	6.91	8.23

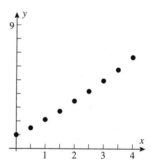

15. $y_0 = 1, y_{n+1} = y_n + \dfrac{1}{2}\cos y_n$

n	0	1	2	3	4
t_n	0	0.5	1	1.5	2
y_n	1	1.27	1.42	1.49	1.53

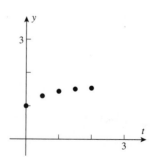

17. $h = 1/5$, $y_0 = 1$, $y_{n+1} = y_n + \dfrac{1}{5}\sin(\pi n/5)$

n	0	1	2	3	4	5
t_n	0	0.2	0.4	0.6	0.8	1.0
y_n	1	1	1.12	1.31	1.50	1.62

19. **(b)** $y\,dy = -x\,dx$, $y^2/2 = -x^2/2 + C_1$, $x^2 + y^2 = C$; if $y(0) = 1$ then $C = 1$ so $y(1/2) = \sqrt{3}/2$.

21. **(a)** The slope field does not vary with t, hence along a given parallel line all values are equal since they only depend on the height y.

    (b) As in part (a), the slope field does not vary with t; it is independent of t.

    (c) From $G(y) - x = C$ we obtain $\dfrac{d}{dx}(G(y) - x) = \dfrac{1}{f(y)}\dfrac{dy}{dx} - 1 = \dfrac{d}{dx}C = 0$, i.e. $\dfrac{dy}{dx} = f(y)$.

23. **(a)** By implicit differentiation, $y^3 + 3xy^2\dfrac{dy}{dx} - 2xy - x^2\dfrac{dy}{dx} = 0$, $\dfrac{dy}{dx} = \dfrac{y^3 - 2xy}{x^2 - 3xy^2}$.

    (b) If $y(x)$ is an integral curve of the slope field in part (a), then

$$\frac{d}{dx}\{x[y(x)]^3 - x^2 y(x)\} = [y(x)]^3 + 3xy(x)^2 y'(x) - 2xy(x) - x^2 y'(x) = 0,$$ so the integral curve must be of the form $x[y(x)]^3 - x^2 y(x) = C$.

    (c) $x[y(x)]^3 - x^2 y(x) = 2$

25. Euler's Method is repeated application of local linear approximation, each step dependent on the previous step.

EXERCISE SET 9.3

1. **(a)** $\dfrac{dy}{dt} = ky^2$, $y(0) = y_0, k > 0$ **(b)** $\dfrac{dy}{dt} = -ky^2$, $y(0) = y_0, k > 0$

3. **(a)** $\dfrac{ds}{dt} = \dfrac{1}{2}s$ **(b)** $\dfrac{d^2 s}{dt^2} = 2\dfrac{ds}{dt}$

5. **(a)** $\dfrac{dy}{dt} = 0.02y$, $y_0 = 10{,}000$ **(b)** $y = 10{,}000 e^{2t/100}$

    (c) $T = \dfrac{1}{0.02}\ln 2 \approx 34.657$ h   **(d)** $45{,}000 = 10{,}000 e^{2t/100}$,

$$t = 50\ln\frac{45{,}000}{10{,}000} \approx 75.20 \text{ h}$$

7. **(a)** $\dfrac{dy}{dt} = -ky$, $y(0) = 5.0 \times 10^7$; $3.83 = T = \dfrac{1}{k}\ln 2$, so $k = \dfrac{\ln 2}{3.83} \approx 0.1810$

    (b) $y = 5.0 \times 10^7 e^{-0.181t}$

    (c) $y(30) = 5.0 \times 10^7 e^{-0.1810(30)} \approx 219{,}000$

    (d) $y(t) = (0.1)y_0 = y_0 e^{-kt}$, $-kt = \ln 0.1$, $t = -\dfrac{\ln 0.1}{0.1810} = 12.72$ days

9. $100 e^{0.02t} = 10{,}000$, $e^{0.02t} = 100$, $t = \dfrac{1}{0.02}\ln 100 \approx 230$ days

11. $y(t) = y_0 e^{-kt} = 10.0e^{-kt}$, $3.5 = 10.0e^{-k(5)}$, $k = -\dfrac{1}{5}\ln\dfrac{3.5}{10.0} \approx 0.2100$, $T = \dfrac{1}{k}\ln 2 \approx 3.30$ days

13. **(a)** $k = \dfrac{\ln 2}{6} \approx 0.1155$; $y \approx 3e^{0.1155t}$ **(b)** $y(t) = 4e^{0.02t}$

     (c) $y = y_0 e^{kt}$, $1 = y_0 e^k$, $200 = y_0 e^{10k}$. Divide: $200 = e^{9k}$, $k = \dfrac{1}{9}\ln 200 \approx 0.5887$,

         $y \approx y_0 e^{0.5887t}$; also $y(1) = 1$, so $y_0 = e^{-0.5887} \approx 0.5550$, $y \approx 0.5550 e^{0.5887t}$.

     (d) $k = \dfrac{\ln 2}{T} \approx 0.1155$, $2 = y(1) \approx y_0 e^{0.1155}$, $y_0 \approx 2e^{-0.1155} \approx 1.7818$, $y \approx 1.7818 e^{0.1155t}$

17. **(a)** $T = \dfrac{\ln 2}{k}$; and $\ln 2 \approx 0.6931$. If k is measured in percent, $k' = 100k$,

         then $T = \dfrac{\ln 2}{k} \approx \dfrac{69.31}{k'} \approx \dfrac{70}{k'}$.

     (b) 70 yr   **(c)** 20 yr   **(d)** 7%

19. From (11), $y(t) = y_0 e^{-0.000121t}$. If $0.27 = \dfrac{y(t)}{y_0} = e^{-0.000121t}$ then $t = -\dfrac{\ln 0.27}{0.000121} \approx 10{,}820$ yr, and

     if $0.30 = \dfrac{y(t)}{y_0}$ then $t = -\dfrac{\ln 0.30}{0.000121} \approx 9950$, or roughly between 9000 B.C. and 8000 B.C.

21. **(a)** Let $T_1 = 5730 - 40 = 5690$, $k_1 = \dfrac{\ln 2}{T_1} \approx 0.00012182$; $T_2 = 5730 + 40 = 5770$, $k_2 \approx 0.00012013$.

         With $y/y_0 = 0.92, 0.93$, $t_1 = -\dfrac{1}{k_1}\ln\dfrac{y}{y_0} = 684.5, 595.7$; $t_2 = -\dfrac{1}{k_2}\ln(y/y_0) = 694.1, 604.1$; in

         1988 the shroud was at most 695 years old, which places its creation in or after the year 1293.

     (b) Suppose T is the true half-life of carbon-14 and $T_1 = T(1 + r/100)$ is the false half-life. Then

         with $k = \dfrac{\ln 2}{T}$, $k_1 = \dfrac{\ln 2}{T_1}$ we have the formulae $y(t) = y_0 e^{-kt}$, $y_1(t) = y_0 e^{-k_1 t}$. At a certain

         point in time a reading of the carbon-14 is taken resulting in a certain value y, which in the

         case of the true formula is given by $y = y(t)$ for some t, and in the case of the false formula

         is given by $y = y_1(t_1)$ for some t_1.

         If the true formula is used then the time t since the beginning is given by $t = -\dfrac{1}{k}\ln\dfrac{y}{y_0}$. If

         the false formula is used we get a false value $t_1 = -\dfrac{1}{k_1}\ln\dfrac{y}{y_0}$; note that in both cases the

         value y/y_0 is the same. Thus $t_1/t = k/k_1 = T_1/T = 1 + r/100$, so the percentage error in

         the time to be measured is the same as the percentage error in the half-life.

23. **(a)** $y = y_0 b^t = y_0 e^{t\ln b} = y_0 e^{kt}$ with $k = \ln b > 0$ since $b > 1$.

     (b) $y = y_0 b^t = y_0 e^{t\ln b} = y_0 e^{-kt}$ with $k = -\ln b > 0$ since $0 < b < 1$.

     (c) $y = 4(2^t) = 4e^{t\ln 2}$ **(d)** $y = 4(0.5^t) = 4e^{t\ln 0.5} = 4e^{-t\ln 2}$

25. **(a)** If $y = y_0 e^{kt}$, then $y_1 = y_0 e^{kt_1}$, $y_2 = y_0 e^{kt_2}$, divide: $y_2/y_1 = e^{k(t_2 - t_1)}$, $k = \dfrac{1}{t_2 - t_1}\ln(y_2/y_1)$,

         $T = \dfrac{\ln 2}{k} = \dfrac{(t_2 - t_1)\ln 2}{\ln(y_2/y_1)}$. If $y = y_0 e^{-kt}$, then $y_1 = y_0 e^{-kt_1}$, $y_2 = y_0 e^{-kt_2}$,

         $y_2/y_1 = e^{-k(t_2 - t_1)}$, $k = -\dfrac{1}{t_2 - t_1}\ln(y_2/y_1)$, $T = \dfrac{\ln 2}{k} = -\dfrac{(t_2 - t_1)\ln 2}{\ln(y_2/y_1)}$.

In either case, T is positive, so $T = \left| \dfrac{(t_2 - t_1)\ln 2}{\ln(y_2/y_1)} \right|$.

(b) In Part (a) assume $t_2 = t_1 + 1$ and $y_2 = 1.25 y_1$. Then $T = \dfrac{\ln 2}{\ln 1.25} \approx 3.1$ h.

27. (a) $A = 1000 e^{(0.08)(5)} = 1000 e^{0.4} \approx \$1,491.82$

(b) $P e^{(0.08)(10)} = 10,000$, $P e^{0.8} = 10,000$, $P = 10,000 e^{-0.8} \approx \$4,493.29$

(c) From (11), with $k = r = 0.08$, $T = (\ln 2)/0.08 \approx 8.7$ years.

29. (a) $\dfrac{dT}{dt} = -k(T - 21)$, $T(0) = 95$, $\dfrac{dT}{T - 21} = -k\,dt$, $\ln(T - 21) = -kt + C_1$,

$T = 21 + e^{C_1} e^{-kt} = 21 + C e^{-kt}$, $95 = T(0) = 21 + C$, $C = 74$, $T = 21 + 74 e^{-kt}$

(b) $85 = T(1) = 21 + 74 e^{-k}$, $k = -\ln \dfrac{64}{74} = -\ln \dfrac{32}{37}$, $T = 21 + 74 e^{t\ln(32/37)} = 21 + 74 \left(\dfrac{32}{37}\right)^t$,

$T = 51$ when $\dfrac{30}{74} = \left(\dfrac{32}{37}\right)^t$, $t = \dfrac{\ln(30/74)}{\ln(32/37)} \approx 6.22$ min

31. Let T denote the body temperature of McHam's body at time t, the number of hours elapsed after 10:06 P.M.; then $\dfrac{dT}{dt} = -k(T - 72)$, $\dfrac{dT}{T - 72} = -k\,dt$, $\ln(T - 72) = -kt + C$, $T = 72 + e^C e^{-kt}$,

$77.9 = 72 + e^C$, $e^C = 5.9$, $T = 72 + 5.9 e^{-kt}$, $75.6 = 72 + 5.9 e^{-k}$, $k = -\ln \dfrac{3.6}{5.9} \approx 0.4940$,

$T = 72 + 5.9 e^{-0.4940 t}$. McHam's body temperature was last 98.6° when $t = -\dfrac{\ln(26.6/5.9)}{0.4940} \approx -3.05$,

so around 3 hours and 3 minutes before 10:06; the death took place at approximately 7:03 P.M., while Moore was on stage.

33. (a) Both $y(t) = 0$ and $y(t) = L$ are solutions of the logistic equation $\dfrac{dy}{dt} = k\left(1 - \dfrac{y}{L}\right) y$ as both sides of the equation are then zero.

(b) If y is very small relative to L then $y/L \approx 0$, and the logistic equation becomes $\dfrac{dy}{dt} \approx ky$, which is a form of the equation for exponential growth.

(c) All the terms on the right-hand-side of the logistic equation are positive, except perhaps $1 - \dfrac{y}{L}$, which is positive if $y < L$ and negative if $y > L$.

(d) The rate of change of y is a function of only one variable, y itself. The right-hand-side of the differential equation is a quadratic equation in y, which can be thought of as a parabola in y which opens down and crosses the y-axis at $y = 0$ and $y = L$. The parabola thus takes its maximum midway between the two y-intercepts, namely at $y = L/2$.

35. (a) $k = L = 1$, $y_0 = 2$ **(b)** $k = L = y_0 = 1$

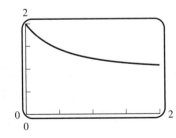   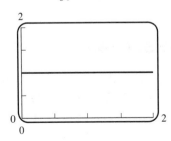

(c) $k = y_0 = 1, L = 2$

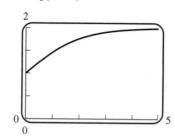

(d) $k = y_0 = 1, L = 4$

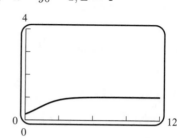

37. $y_0 \approx 2$, $L \approx 8$; since the curve $y = \dfrac{2 \cdot 8}{2 + 6e^{-kt}}$ passes through the point $(2, 4)$, $4 = \dfrac{16}{2 + 6e^{-2k}}$,

$6e^{-2k} = 2$, $k = \dfrac{1}{2} \ln 3 \approx 0.5493$.

39. (a) $y_0 = 5$ **(b)** $L = 12$ **(c)** $k = 1$

(d) $L/2 = 6 = \dfrac{60}{5 + 7e^{-t}}$, $5 + 7e^{-t} = 10$, $t = -\ln(5/7) \approx 0.3365$

(e) $\dfrac{dy}{dt} = \dfrac{1}{12} y(12 - y)$, $y(0) = 5$

41. Assume $y(t)$ students have had the flu t days after semester break. Then $y(0) = 20$, $y(5) = 35$.

(a) $\dfrac{dy}{dt} = ky(L - y) = ky(1000 - y)$, $y_0 = 20$

(b) Part (a) has solution $y = \dfrac{20000}{20 + 980e^{-kt}} = \dfrac{1000}{1 + 49e^{-kt}}$;

$35 = \dfrac{1000}{1 + 49e^{-5k}}$, $k = 0.115$, $y \approx \dfrac{1000}{1 + 49e^{-0.115t}}$.

(c)

t	0	1	2	3	4	5	6	7	8	9	10	11	12	13	14
$y(t)$	20	22	25	28	31	35	39	44	49	54	61	67	75	83	93

(d)

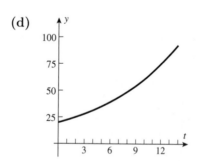

EXERCISE SET 9.4

1. (a) $y = e^{-2x}$, $y' = -2e^{-2x}$, $y'' = 4e^{-2x}$; $y'' + y' - 2y = 0$
$y = e^x$, $y' = e^x$, $y'' = e^x$; $y'' + y' - 2y = 0$.

(b) $y = c_1 e^{-2x} + c_2 e^x$, $y' = -2c_1 e^{-2x} + c_2 e^x$, $y'' = 4c_1 e^{-2x} + c_2 e^x$; $y'' + y' - 2y = 0$

3. $m^2 + 3m - 4 = 0$, $(m-1)(m+4) = 0$; $m = 1, -4$ so $y = c_1 e^x + c_2 e^{-4x}$.

5. $m^2 - 2m + 1 = 0$, $(m-1)^2 = 0$; $m = 1$, so $y = c_1 e^x + c_2 x e^x$.

7. $m^2 + 1 = 0$, $m = \pm i$ so $y = c_1 \cos x + c_2 \sin x$.

9. $m^2 - m = 0$, $m(m-1) = 0$; $m = 0, 1$ so $y = c_1 + c_2 e^x$.

11. $m^2 - 4m + 4 = 0$, $(m-2)^2 = 0$; $m = 2$ so $y = c_1 e^{2t} + c_2 t e^{2t}$.

13. $m^2 + 4m + 13 = 0$, $m = -2 \pm 3i$ so $y = e^{-2x}(c_1 \cos 3x + c_2 \sin 3x)$.

15. $8m^2 - 2m - 1 = 0$, $(4m+1)(2m-1) = 0$; $m = -1/4, 1/2$ so $y = c_1 e^{-x/4} + c_2 e^{x/2}$.

17. $m^2 + 2m - 3 = 0$, $(m+3)(m-1) = 0$; $m = -3, 1$ so $y = c_1 e^{-3x} + c_2 e^x$ and $y' = -3c_1 e^{-3x} + c_2 e^x$. Solve the system $c_1 + c_2 = 1$, $-3c_1 + c_2 = 9$ to get $c_1 = -2$, $c_2 = 3$ so $y = -2e^{-3x} + 3e^x$.

19. $m^2 + 6m + 9 = 0$, $(m+3)^2 = 0$; $m = -3$ so $y = (c_1 + c_2 x)e^{-3x}$ and $y' = (-3c_1 + c_2 - 3c_2 x)e^{-3x}$. Solve the system $c_1 = 2$, $-3c_1 + c_2 = -5$ to get $c_1 = 2$, $c_2 = 1$ so $y = (2 + x)e^{-3x}$.

21. $m^2 + 4m + 5 = 0$, $m = -2 \pm i$ so $y = e^{-2x}(c_1 \cos x + c_2 \sin x)$,
$y' = e^{-2x}[(c_2 - 2c_1)\cos x - (c_1 + 2c_2)\sin x]$. Solve the system $c_1 = -3$, $c_2 - 2c_1 = 0$ to get $c_1 = -3$, $c_2 = -6$ so $y = -e^{-2x}(3\cos x + 6\sin x)$.

23. **(a)** $m = 5, -2$ so $(m-5)(m+2) = 0$, $m^2 - 3m - 10 = 0$; $y'' - 3y' - 10y = 0$.

(b) $m = 4, 4$ so $(m-4)^2 = 0$, $m^2 - 8m + 16 = 0$; $y'' - 8y' + 16y = 0$.

(c) $m = -1 \pm 4i$ so $(m+1-4i)(m+1+4i) = 0$, $m^2 + 2m + 17 = 0$; $y'' + 2y' + 17y = 0$.

25. $m^2 + km + k = 0$, $m = \left(-k \pm \sqrt{k^2 - 4k}\right)/2$

(a) $k^2 - 4k > 0$, $k(k-4) > 0$; $k < 0$ or $k > 4$

(b) $k^2 - 4k = 0$; $k = 0, 4$ 　　　　　　　**(c)** $k^2 - 4k < 0$, $k(k-4) < 0$; $0 < k < 4$

27. **(a)** $\dfrac{d^2 y}{dz^2} + 2\dfrac{dy}{dz} + 2y = 0$, $m^2 + 2m + 2 = 0$; $m = -1 \pm i$ so

$$y = e^{-z}(c_1 \cos z + c_2 \sin z) = \frac{1}{x}[c_1 \cos(\ln x) + c_2 \sin(\ln x)].$$

(b) $\dfrac{d^2 y}{dz^2} - 2\dfrac{dy}{dz} - 2y = 0$, $m^2 - 2m - 2 = 0$; $m = 1 \pm \sqrt{3}$ so

$$y = c_1 e^{(1+\sqrt{3})z} + c_2 e^{(1-\sqrt{3})z} = c_1 x^{1+\sqrt{3}} + c_2 x^{1-\sqrt{3}}$$

29. **(a)** Neither is a constant multiple of the other, since, e.g. if $y_1 = ky_2$ then $e^{m_1 x} = ke^{m_2 x}$, $e^{(m_1 - m_2)x} = k$. But the right hand side is constant, and the left hand side is constant only if $m_1 = m_2$, which is false.

(b) If $y_1 = ky_2$ then $e^{mx} = kxe^{mx}$, $kx = 1$ which is impossible. If $y_2 = y_1$ then $xe^{mx} = ke^{mx}$, $x = k$ which is impossible.

31. **(a)** The general solution is $c_1 e^{\mu x} + c_2 e^{mx}$; let $c_1 = 1/(\mu - m)$, $c_2 = -1/(\mu - m)$.

(b) $\displaystyle\lim_{\mu \to m} \frac{e^{\mu x} - e^{mx}}{\mu - m} = \lim_{\mu \to m} x e^{\mu x} = x e^{mx}$.

33. $k/M = 0.5/2 = 0.25$

    (a) From (20), $y = 0.4\cos(t/2)$

    (b) $T = 2\pi \cdot 2 = 4\pi$ s, $f = 1/T = 1/(4\pi)$ Hz

    (c)

    (d) $y = 0$ at the equilibrium position, so $t/2 = \pi/2, t = \pi$ s.

    (e) $t/2 = 2\pi$ at the maximum position below the equilibrium position, so $t = 4\pi$ s.

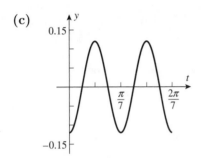

35. $l = 0.05$, $k/M = g/l = 9.8/0.05 = 196$ s^{-2}

    (a) From (20), $y = -0.12\cos 14t$.

    (b) $T = 2\pi\sqrt{M/k} = 2\pi/14 = \pi/7$ s, $f = 7/\pi$ Hz

    (c)

    (d) $14t = \pi/2$, $t = \pi/28$ s

    (e) $14t = \pi$, $t = \pi/14$ s

37. **(a)** From the graph it appears that the maximum velocity occurs at the equilibrium position, $t = \pi/8$.

    To show this mathematically, $v_{\max}$ occurs when $\dfrac{dv}{dt} = 0$. But $v = \dfrac{dy}{dt} = 8\sin 4t$, $\dfrac{dv}{dt} = 32\cos 4t$, $\dfrac{dv}{dt} = 0$ when $4t = \pi/2, t = \pi/8$.

    (b) From the graph it appears that the minimum velocity occurs at the equilibrium position as the block is falling, i.e. $t = 3\pi/8$. Mathematically, $\dfrac{dv}{dt} = 32\cos 4t = 0$ when $4t = \pi/2, 3\pi/2, \ldots$. When $4t = 3\pi/2$ the block is falling, so $t = 3\pi/8$.

39. By Hooke's Law, $F(t) = -kx(t)$, since the only force is the restoring force of the spring. Newton's Second Law gives $F(t) = Mx''(t)$, so $Mx''(t) + kx(t) = 0$, $x(0) = x_0, x'(0) = 0$.

41. $y = y_0\cos\sqrt{\dfrac{k}{M}}\,t$, $T = 2\pi\sqrt{\dfrac{M}{k}}$, $y = y_0\cos\dfrac{2\pi t}{T}$

    (a) $v = y'(t) = -\dfrac{2\pi}{T}y_0\sin\dfrac{2\pi t}{T}$ has maximum magnitude $2\pi|y_0|/T$ and occurs when $2\pi t/T = n\pi + \pi/2$, $y = y_0\cos(n\pi + \pi/2) = 0$.

    (b) $a = y''(t) = -\dfrac{4\pi^2}{T^2}y_0\cos\dfrac{2\pi t}{T}$ has maximum magnitude $4\pi^2|y_0|/T^2$ and occurs when $2\pi t/T = j\pi$, $y = y_0\cos j\pi = \pm y_0$.

43. **(a)** $m^2 + 2.4m + 1.44 = 0, (m + 1.2)^2 = 0, m = -1.2$, $y = C_1e^{-6t/5} + C_2te^{-6t/5}$,

    $C_1 = 1$, $2 = y'(0) = -\dfrac{6}{5}C_1 + C_2, C_2 = \dfrac{16}{5}$, $y = e^{-6t/5} + \dfrac{16}{5}te^{-6t/5}$

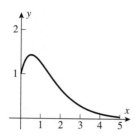

(b) $y'(t) = 0$ when $t = t_1 = 25/48 \approx 0.520833$, $y(t_1) = 1.427364$ cm

(c) $y = \dfrac{16}{5} e^{-6t/5}(t + 5/16) = 0$ only if $t = -5/16$, so $y \neq 0$ for $t \geq 0$.

45. **(a)** $m^2 + m + 5 = 0, m = -1/2 \pm (\sqrt{19}/2)i$, $y = e^{-t/2}\left[C_1 \cos(\sqrt{19}t/2) + C_2 \sin(\sqrt{19}t/2)\right]$,

$1 = y(0) = C_1, -3.5 = y'(0) = -(1/2)C_1 + (\sqrt{19}/2)C_2$, $C_2 = -6/\sqrt{19}$,

$y = e^{-t/2}\cos(\sqrt{19}\,t/2) - (6/\sqrt{19}\,)e^{-t/2}\sin(\sqrt{19}\,t/2)$

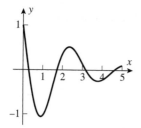

(b) $y'(t) = 0$ for the first time when $t = t_1 = 0.905533$, $y(t_1) = -1.054466$ cm so the maximum distance below the equilibrium position is 1.054466 cm.

(c) $y(t) = 0$ for the first time when $t = t_2 = 0.288274$, $y'(t_2) = -3.210357$ cm/s.

(d) The acceleration is $y''(t)$ so from the differential equation $y'' = -y' - 5y$. But $y = 0$ when the object passes through the equilibrium position, thus $y'' = -y' = 3.210357$ cm/s^2.

47. **(a)** $m^2 + 3.5m + 3 = (m + 1.5)(m + 2), y = C_1e^{-3t/2} + C_2e^{-2t}$,

$1 = y(0) = C_1 + C_2, v_0 = y'(0) = -(3/2)C_1 - 2C_2$, $C_1 = 4 + 2v_0, C_2 = -3 - 2v_0$,

$y(t) = (4 + 2v_0)e^{-3t/2} - (3 + 2v_0)e^{-2t}$

(b) $v_0 = 2, y(t) = 8e^{-3t/2} - 7e^{-2t}$, $v_0 = -1, y(t) = 2e^{-3t/2} - e^{-2t}$,

$v_0 = -4, y(t) = -4e^{-3t/2} + 5e^{-2t}$

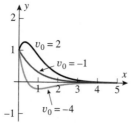

49. $\dfrac{dy}{dt} + p(x)y = c\dfrac{dy_1}{dt} + p(x)(cy_1) = c\left[\dfrac{dy_1}{dt} + p(x)y_1\right] = c \cdot 0 = 0$

REVIEW EXERCISES, CHAPTER 9

3. **(a)** linear   **(b)** both   **(c)** separable   **(d)** neither

5. $\dfrac{dy}{dx} - 4xy = x$

   (a) IF: $e^{\int(-4x)dx} = e^{-2x^2}$, $\quad \dfrac{d}{dx}[ye^{-2x^2}] = xe^{-2x^2}$, $\quad ye^{-2x^2} = \displaystyle\int xe^{-2x^2}\,dx = -\dfrac{1}{4}e^{-2x^2} + C$,

      $y = -\dfrac{1}{4} + Ce^{2x^2}$

   (b) $\dfrac{dy}{dx} = 4xy + x$, $\dfrac{dy}{4y+1} = x\,dx$, $\dfrac{1}{4}\ln(4y+1) = \dfrac{1}{2}x^2 + C$, $\ln(4y+1) = 2x^2 + C_1$,

      $4y + 1 = C_2e^{2x^2}$, $y = \frac{1}{4}(C_2e^{2x^2} - 1) = C_3e^{2x^2} - \frac{1}{4}$, same as in part (a)

7. $\dfrac{dy}{1+y^2} = x^2\,dx$, $\tan^{-1}y = \dfrac{1}{3}x^3 + C$, $y = \tan\left(\dfrac{1}{3}x^3 + C\right)$

9. $\left(\dfrac{1}{y} + y\right)dy = e^x dx$, $\ln|y| + y^2/2 = e^x + C$; by inspection, $y = 0$ is also a solution

11. $\mu = e^{-\int x\,dx} = e^{-x^2/2}$, $e^{-x^2/2}y = \displaystyle\int xe^{-x^2/2}dx = -e^{-x^2/2} + C$,

   $y = -1 + Ce^{x^2/2}$, $3 = -1 + C$, $C = 4$, $y = -1 + 4e^{x^2/2}$

13. IF: $\dfrac{d}{dx}(y\cosh x) = \cosh^2 x = \dfrac{1}{2}(1 + \cosh 2x)$, $y\cosh x = \dfrac{1}{2}x + \dfrac{1}{4}\sinh 2x + C$. When $x = 0, y = 2$ so

   $2 = C$, and $y = \dfrac{1}{2}x\,\mathrm{sech}x + \dfrac{1}{4}\sinh 2x\,\mathrm{sech}x + 2\,\mathrm{sech}x$.

15. $\left(\dfrac{1}{y^5} + \dfrac{1}{y}\right)dy = \dfrac{dx}{x}$, $-\dfrac{1}{4}y^{-4} + \ln|y| = \ln|x| + C$; $-\dfrac{1}{4} = C$, $y^{-4} + 4\ln(x/y) = 1$

17. **(a)** $\mu = e^{-\int dx} = e^{-x}$, $\dfrac{d}{dx}\left[ye^{-x}\right] = xe^{-x}\sin 3x$,

      $ye^{-x} = \displaystyle\int xe^{-x}\sin 3x\,dx = \left(-\dfrac{3}{10}x - \dfrac{3}{50}\right)e^{-x}\cos 3x + \left(-\dfrac{1}{10}x + \dfrac{2}{25}\right)e^{-x}\sin 3x + C$;

      $1 = y(0) = -\dfrac{3}{50} + C$, $C = \dfrac{53}{50}$, $y = \left(-\dfrac{3}{10}x - \dfrac{3}{50}\right)\cos 3x + \left(-\dfrac{1}{10}x + \dfrac{2}{25}\right)\sin 3x + \dfrac{53}{50}e^x$

   (c)

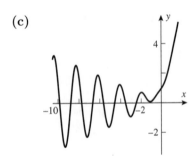

19. Assume the tank contains $y(t)$ oz of salt at time t. Then $y_0 = 0$ and for $0 < t < 15$,

   $\dfrac{dy}{dt} = 5 \cdot 10 - \dfrac{y}{1000}10 = (50 - y/100)$ oz/min, with solution $y = 5000 + Ce^{-t/100}$. But $y(0) = 0$ so

$C = -5000$, $y = 5000(1 - e^{-t/100})$ for $0 \le t \le 15$, and $y(15) = 5000(1 - e^{-0.15})$. For $15 < t < 30$,

$\dfrac{dy}{dt} = 0 - \dfrac{y}{1000}5$, $y = C_1 e^{-t/200}$, $C_1 e^{-0.075} = y(15) = 5000(1 - e^{-0.15})$, $C_1 = 5000(e^{0.075} - e^{-0.075})$,

$y = 5000(e^{0.075} - e^{-0.075})e^{-t/100}$, $y(30) = 5000(e^{0.075} - e^{-0.075})e^{-0.3} \approx 556.13$ oz.

21.

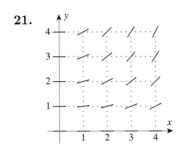

23. $y_0 = 1, y_{n+1} = y_n + \sqrt{y_n}/2$

n	0	1	2	3	4	5	6	7	8
x_n	0	0.5	1	1.5	2	2.5	3	3.5	4
y_n	1	1.50	2.11	2.84	3.68	4.64	5.72	6.91	8.23

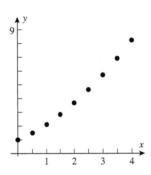

25. $h = 1/5$, $y_0 = 1$, $y_{n+1} = y_n + \dfrac{1}{5}\cos(2\pi n/5)$

n	0	1	2	3	4	5
t_n	0	0.2	0.4	0.6	0.8	1.0
y_n	1.00	1.06	0.90	0.74	0.80	1.00

27. **(a)** $k = \dfrac{\ln 2}{5} \approx 0.1386$; $y \approx 2e^{0.1386t}$ **(b)** $y(t) = 5e^{0.015t}$

(c) $y = y_0 e^{kt}$, $1 = y_0 e^k$, $100 = y_0 e^{10k}$. Divide: $100 = e^{9k}$, $k = \dfrac{1}{9}\ln 100 \approx 0.5117$,

$y \approx y_0 e^{0.5117t}$; also $y(1) = 1$, so $y_0 = e^{-0.5117} \approx 0.5995$, $y \approx 0.5995 e^{0.5117t}$.

(d) $k = \dfrac{\ln 2}{T} \approx 0.1386$, $1 = y(1) \approx y_0 e^{0.1386}$, $y_0 \approx e^{-0.1386} \approx 0.8706$, $y \approx 0.8706 e^{0.1386t}$

29. From section 9.3 formula (11), $y(t) = y_0 e^{-0.000121t}$, so $0.785 y_0 = y_0 e^{-0.000121t}$, $t = -\ln 0.785/0.000121 \approx 2000.6$ yr

31. **(a)** $y = C_1 e^x + C_2 e^{2x}$ **(b)** $y = C_1 e^{x/2} + C_2 x e^{x/2}$

(c) $y = e^{-x/2}\left[C_1 \cos \dfrac{\sqrt{7}}{2}x + C_2 \sin \dfrac{\sqrt{7}}{2}x \right]$

33. $k/M = 0.25/1 = 0.25$

(a) From (20) in Section 9.4, $y = 0.3\cos(t/2)$ **(b)** $T = 2\pi \cdot 2 = 4\pi$ s, $f = 1/T = 1/(4\pi)$ Hz

(c)

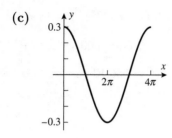

(d) $y = 0$ at the equilibrium position, so $t/2 = \pi/2, t = \pi$ s.

(e) $t/2 = \pi$ at the maximum position below the equlibrium position, so $t = 2\pi$ s.

CHAPTER 10
Infinite Series

EXERCISE SET 10.1

1. (a) $\dfrac{1}{3^{n-1}}$ (b) $\dfrac{(-1)^{n-1}}{3^{n-1}}$ (c) $\dfrac{2n-1}{2n}$ (d) $\dfrac{n^2}{\pi^{1/(n+1)}}$

3. (a) $2, 0, 2, 0$ (b) $1, -1, 1, -1$ (c) $2(1+(-1)^n); 2 + 2\cos n\pi$

5. $1/3, 2/4, 3/5, 4/6, 5/7, \ldots;\ \lim\limits_{n\to+\infty} \dfrac{n}{n+2} = 1,$ converges

7. $2, 2, 2, 2, 2, \ldots;\ \lim\limits_{n\to+\infty} 2 = 2,$ converges

9. $\dfrac{\ln 1}{1}, \dfrac{\ln 2}{2}, \dfrac{\ln 3}{3}, \dfrac{\ln 4}{4}, \dfrac{\ln 5}{5}, \ldots;$

    $\lim\limits_{n\to+\infty} \dfrac{\ln n}{n} = \lim\limits_{n\to+\infty} \dfrac{1}{n} = 0 \left(\text{apply L'Hôpital's Rule to } \dfrac{\ln x}{x}\right),$ converges

11. $0, 2, 0, 2, 0, \ldots;$ diverges

13. $-1, 16/9, -54/28, 128/65, -250/126, \ldots;$ diverges because odd-numbered terms approach -2, even-numbered terms approach 2.

15. $6/2, 12/8, 20/18, 30/32, 42/50, \ldots;\ \lim\limits_{n\to+\infty} \dfrac{1}{2}(1 + 1/n)(1 + 2/n) = 1/2,$ converges

17. $\cos(3), \cos(3/2), \cos(1), \cos(3/4), \cos(3/5), \ldots;\ \lim\limits_{n\to+\infty} \cos(3/n) = 1,$ converges

19. $e^{-1}, 4e^{-2}, 9e^{-3}, 16e^{-4}, 25e^{-5}, \ldots;\ \lim\limits_{x\to+\infty} x^2 e^{-x} = \lim\limits_{x\to+\infty} \dfrac{x^2}{e^x} = 0,$ so $\lim\limits_{n\to+\infty} n^2 e^{-n} = 0,$ converges

21. $2, (5/3)^2, (6/4)^3, (7/5)^4, (8/6)^5, \ldots;$ let $y = \left[\dfrac{x+3}{x+1}\right]^x,$ converges because

    $\lim\limits_{x\to+\infty} \ln y = \lim\limits_{x\to+\infty} \dfrac{\ln \dfrac{x+3}{x+1}}{1/x} = \lim\limits_{x\to+\infty} \dfrac{2x^2}{(x+1)(x+3)} = 2,$ so $\lim\limits_{n\to+\infty} \left[\dfrac{n+3}{n+1}\right]^n = e^2$

23. $\left\{\dfrac{2n-1}{2n}\right\}_{n=1}^{+\infty};\ \lim\limits_{n\to+\infty} \dfrac{2n-1}{2n} = 1,$ converges

25. $\left\{(-1)^{n-1}\dfrac{1}{3^n}\right\}_{n=1}^{+\infty};\ \lim\limits_{n\to+\infty} \dfrac{(-1)^{n-1}}{3^n} = 0,$ converges

27. $\left\{\dfrac{1}{n} - \dfrac{1}{n+1}\right\}_{n=1}^{+\infty};\ \lim\limits_{n\to+\infty} \left(\dfrac{1}{n} - \dfrac{1}{n+1}\right) = 0,$ converges

29. $\left\{\sqrt{n+1} - \sqrt{n+2}\right\}_{n=1}^{+\infty};$ converges because

    $\lim\limits_{n\to+\infty} (\sqrt{n+1} - \sqrt{n+2}) = \lim\limits_{n\to+\infty} \dfrac{(n+1) - (n+2)}{\sqrt{n+1} + \sqrt{n+2}} = \lim\limits_{n\to+\infty} \dfrac{-1}{\sqrt{n+1} + \sqrt{n+2}} = 0$

31. $a_n = \begin{cases} +1 & k \text{ even} \\ -1 & k \text{ odd} \end{cases}$ oscillates; there is no limit point which attracts all of the a_n.

$b_n = \cos n$; the terms lie all over the interval $[-1, 1]$ without any limit.

33. (a) $1, 2, 1, 4, 1, 6$ **(b)** $a_n = \begin{cases} n, & n \text{ odd} \\ 1/2^n, & n \text{ even} \end{cases}$ **(c)** $a_n = \begin{cases} 1/n, & n \text{ odd} \\ 1/(n+1), & n \text{ even} \end{cases}$

(d) In Part (a) the sequence diverges, since the even terms diverge to $+\infty$ and the odd terms equal 1; in Part (b) the sequence diverges, since the odd terms diverge to $+\infty$ and the even terms tend to zero; in Part (c) $\lim\limits_{n \to +\infty} a_n = 0$.

35. $\lim\limits_{n \to +\infty} \sqrt[n]{n} = 1$, so $\lim\limits_{n \to +\infty} \sqrt[n]{n^3} = 1^3 = 1$

37. $\lim\limits_{n \to +\infty} x_{n+1} = \dfrac{1}{2} \lim\limits_{n \to +\infty} \left(x_n + \dfrac{a}{x_n} \right)$ or $L = \dfrac{1}{2}\left(L + \dfrac{a}{L} \right), 2L^2 - L^2 - a = 0, L = \sqrt{a}$ (we reject $-\sqrt{a}$

because $x_n > 0$, thus $L \geq 0$.)

39. (a) $1, \dfrac{1}{4} + \dfrac{2}{4}, \dfrac{1}{9} + \dfrac{2}{9} + \dfrac{3}{9}, \dfrac{1}{16} + \dfrac{2}{16} + \dfrac{3}{16} + \dfrac{4}{16} = 1, \dfrac{3}{4}, \dfrac{2}{3}, \dfrac{5}{8}$

(c) $a_n = \dfrac{1}{n^2}(1 + 2 + \cdots + n) = \dfrac{1}{n^2}\dfrac{1}{2}n(n+1) = \dfrac{1}{2}\dfrac{n+1}{n}, \lim\limits_{n \to +\infty} a_n = 1/2$

41. Let $a_n = 0, b_n = \dfrac{\sin^2 n}{n}, c_n = \dfrac{1}{n}$; then $a_n \leq b_n \leq c_n, \lim\limits_{n \to +\infty} a_n = \lim\limits_{n \to +\infty} c_n = 0$, so $\lim\limits_{n \to +\infty} b_n = 0$.

43. (a) $a_1 = (0.5)^2, a_2 = a_1^2 = (0.5)^4, \ldots, a_n = (0.5)^{2^n}$

(c) $\lim\limits_{n \to +\infty} a_n = \lim\limits_{n \to +\infty} e^{2^n \ln(0.5)} = 0$, since $\ln(0.5) < 0$.

(d) Replace 0.5 in Part (a) with a_0; then the sequence converges for $-1 \leq a_0 \leq 1$, because if $a_0 = \pm 1$, then $a_n = 1$ for $n \geq 1$; if $a_0 = 0$ then $a_n = 0$ for $n \geq 1$; and if $0 < |a_0| < 1$ then $a_1 = a_0^2 > 0$ and $\lim\limits_{n \to +\infty} a_n = \lim\limits_{n \to +\infty} e^{2^{n-1} \ln a_1} = 0$ since $0 < a_1 < 1$. This same argument proves divergence to $+\infty$ for $|a| > 1$ since then $\ln a_1 > 0$.

45. (a)

(b) Let $y = (2^x + 3^x)^{1/x}, \lim\limits_{x \to +\infty} \ln y = \lim\limits_{x \to +\infty} \dfrac{\ln(2^x + 3^x)}{x} = \lim\limits_{x \to +\infty} \dfrac{2^x \ln 2 + 3^x \ln 3}{2^x + 3^x}$

$= \lim\limits_{x \to +\infty} \dfrac{(2/3)^x \ln 2 + \ln 3}{(2/3)^x + 1} = \ln 3$, so $\lim\limits_{n \to +\infty} (2^n + 3^n)^{1/n} = e^{\ln 3} = 3$

Alternate proof: $3 = (3^n)^{1/n} < (2^n + 3^n)^{1/n} < (2 \cdot 3^n)^{1/n} = 3 \cdot 2^{1/n}$. Then apply the Squeezing Theorem.

47. $a_n = \dfrac{1}{n-1} \displaystyle\int_1^n \dfrac{1}{x}\,dx = \dfrac{\ln n}{n-1}$, $\displaystyle\lim_{n \to +\infty} a_n = \lim_{n \to +\infty} \dfrac{\ln n}{n-1} = \lim_{n \to +\infty} \dfrac{1}{n} = 0$,

$\left(\text{apply L'Hôpital's Rule to } \dfrac{\ln n}{n-1}\right)$, converges

49. $\left|\dfrac{1}{n} - 0\right| = \dfrac{1}{n} < \epsilon$ if $n > 1/\epsilon$

    (a) $1/\epsilon = 1/0.5 = 2$, $N = 3$ **(b)** $1/\epsilon = 1/0.1 = 10$, $N = 11$

    (c) $1/\epsilon = 1/0.001 = 1000$, $N = 1001$

51. **(a)** $\left|\dfrac{1}{n} - 0\right| = \dfrac{1}{n} < \epsilon$ if $n > 1/\epsilon$, choose any $N > 1/\epsilon$.

    (b) $\left|\dfrac{n}{n+1} - 1\right| = \dfrac{1}{n+1} < \epsilon$ if $n > 1/\epsilon - 1$, choose any $N > 1/\epsilon - 1$.

EXERCISE SET 10.2

1. $a_{n+1} - a_n = \dfrac{1}{n+1} - \dfrac{1}{n} = -\dfrac{1}{n(n+1)} < 0$ for $n \geq 1$, so strictly decreasing.

3. $a_{n+1} - a_n = \dfrac{n+1}{2n+3} - \dfrac{n}{2n+1} = \dfrac{1}{(2n+1)(2n+3)} > 0$ for $n \geq 1$, so strictly increasing.

5. $a_{n+1} - a_n = (n+1-2^{n+1}) - (n-2^n) = 1 - 2^n < 0$ for $n \geq 1$, so strictly decreasing.

7. $\dfrac{a_{n+1}}{a_n} = \dfrac{(n+1)/(2n+3)}{n/(2n+1)} = \dfrac{(n+1)(2n+1)}{n(2n+3)} = \dfrac{2n^2+3n+1}{2n^2+3n} > 1$ for $n \geq 1$, so strictly increasing.

9. $\dfrac{a_{n+1}}{a_n} = \dfrac{(n+1)e^{-(n+1)}}{ne^{-n}} = (1+1/n)e^{-1} < 1$ for $n \geq 1$, so strictly decreasing.

11. $\dfrac{a_{n+1}}{a_n} = \dfrac{(n+1)^{n+1}}{(n+1)!} \cdot \dfrac{n!}{n^n} = \dfrac{(n+1)^n}{n^n} = (1+1/n)^n > 1$ for $n \geq 1$, so strictly increasing.

13. $f(x) = x/(2x+1)$, $f'(x) = 1/(2x+1)^2 > 0$ for $x \geq 1$, so strictly increasing.

15. $f(x) = 1/(x + \ln x)$, $f'(x) = -\dfrac{1+1/x}{(x+\ln x)^2} < 0$ for $x \geq 1$, so strictly decreasing.

17. $f(x) = \dfrac{\ln(x+2)}{x+2}$, $f'(x) = \dfrac{1 - \ln(x+2)}{(x+2)^2} < 0$ for $x \geq 1$, so strictly decreasing.

19. $f(x) = 2x^2 - 7x$, $f'(x) = 4x - 7 > 0$ for $x \geq 2$, so eventually strictly increasing.

21. $f(x) = \dfrac{x}{x^2+10}$, $f'(x) = \dfrac{10 - x^2}{(x^2+10)^2} < 0$ for $x \geq 4$, so eventually strictly decreasing.

23. $\dfrac{a_{n+1}}{a_n} = \dfrac{(n+1)!}{3^{n+1}} \cdot \dfrac{3^n}{n!} = \dfrac{n+1}{3} > 1$ for $n \geq 3$, so eventually strictly increasing.

25. (a) Yes: a monotone sequence is increasing or decreasing; if it is increasing, then it is increasing and bounded above, so by Theorem 10.2.3 it converges; if decreasing, then use Theorem 10.2.4. The limit lies in the interval $[1,2]$.

(b) Such a sequence may converge, in which case, by the argument in Part (a), its limit is ≤ 2. But convergence may not happen: for example, the sequence $\{-n\}_{n=1}^{+\infty}$ diverges.

27. (a) $a_{n+1} = \dfrac{|x|^{n+1}}{(n+1)!} = \dfrac{|x|}{n+1}\dfrac{|x|^n}{n!} = \dfrac{|x|}{n+1}a_n$

(b) $a_{n+1}/a_n = |x|/(n+1) < 1$ if $n > |x| - 1$.

(c) From Part (b) the sequence is eventually decreasing, and it is bounded below by 0, so by Theorem 10.2.4 it converges.

(d) If $\lim\limits_{n\to+\infty} a_n = L$ then from Part (a), $L = \dfrac{|x|}{\lim\limits_{n\to+\infty}(n+1)}L = 0$.

(e) $\lim\limits_{n\to+\infty}\dfrac{|x|^n}{n!} = \lim\limits_{n\to+\infty} a_n = 0$

29. (a) If $f(x) = \frac{1}{2}(x + 3/x)$, then $f'(x) = (x^2 - 3)/(2x^2)$ and $f'(x) = 0$ for $x = \sqrt{3}$; the minimum value of $f(x)$ for $x > 0$ is $f(\sqrt{3}) = \sqrt{3}$. Thus $f(x) \geq \sqrt{3}$ for $x > 0$ and hence $a_n \geq \sqrt{3}$ for $n \geq 2$.

(b) $a_{n+1} - a_n = (3 - a_n^2)/(2a_n) \leq 0$ for $n \geq 2$ since $a_n \geq \sqrt{3}$ for $n \geq 2$; $\{a_n\}$ is eventually decreasing.

(c) $\sqrt{3}$ is a lower bound for a_n so $\{a_n\}$ converges; $\lim\limits_{n\to+\infty} a_{n+1} = \lim\limits_{n\to+\infty}\frac{1}{2}(a_n + 3/a_n)$, $L = \frac{1}{2}(L + 3/L)$, $L^2 - 3 = 0$, $L = \sqrt{3}$.

31. $n! > \dfrac{n^n}{e^{n-1}}$, $\sqrt[n]{n!} > \dfrac{n}{e^{1-1/n}}$, $\lim\limits_{n\to+\infty}\dfrac{n}{e^{1-1/n}} = +\infty$ so $\lim\limits_{n\to+\infty}\sqrt[n]{n!} = +\infty$.

EXERCISE SET 10.3

1. (a) $s_1 = 2$, $s_2 = 12/5$, $s_3 = \dfrac{62}{25}$, $s_4 = \dfrac{312}{125}$ $s_n = \dfrac{2 - 2(1/5)^n}{1 - 1/5} = \dfrac{5}{2} - \dfrac{5}{2}(1/5)^n$,

$\lim\limits_{n\to+\infty} s_n = \dfrac{5}{2}$, converges

(b) $s_1 = \dfrac{1}{4}$, $s_2 = \dfrac{3}{4}$, $s_3 = \dfrac{7}{4}$, $s_4 = \dfrac{15}{4}$ $s_n = \dfrac{(1/4) - (1/4)2^n}{1 - 2} = -\dfrac{1}{4} + \dfrac{1}{4}(2^n)$,

$\lim\limits_{n\to+\infty} s_n = +\infty$, diverges

(c) $\dfrac{1}{(k+1)(k+2)} = \dfrac{1}{k+1} - \dfrac{1}{k+2}$, $s_1 = \dfrac{1}{6}$, $s_2 = \dfrac{1}{4}$, $s_3 = \dfrac{3}{10}$, $s_4 = \dfrac{1}{3}$;

$s_n = \dfrac{1}{2} - \dfrac{1}{n+2}$, $\lim\limits_{n\to+\infty} s_n = \dfrac{1}{2}$, converges

3. geometric, $a = 1$, $r = -3/4$, sum $= \dfrac{1}{1 - (-3/4)} = 4/7$

5. geometric, $a = 7$, $r = -1/6$, sum $= \dfrac{7}{1 + 1/6} = 6$

7. $s_n = \sum\limits_{k=1}^{n} \left(\dfrac{1}{k+2} - \dfrac{1}{k+3} \right) = \dfrac{1}{3} - \dfrac{1}{n+3}$, $\lim\limits_{n \to +\infty} s_n = 1/3$

9. $s_n = \sum\limits_{k=1}^{n} \left(\dfrac{1/3}{3k-1} - \dfrac{1/3}{3k+2} \right) = \dfrac{1}{6} - \dfrac{1/3}{3n+2}$, $\lim\limits_{n \to +\infty} s_n = 1/6$

11. $\sum\limits_{k=3}^{\infty} \dfrac{1}{k-2} = \sum\limits_{k=1}^{\infty} 1/k$, the harmonic series, so the series diverges.

13. $\sum\limits_{k=1}^{\infty} \dfrac{4^{k+2}}{7^{k-1}} = \sum\limits_{k=1}^{\infty} 64 \left(\dfrac{4}{7} \right)^{k-1}$; geometric, $a = 64$, $r = 4/7$, sum $= \dfrac{64}{1 - 4/7} = 448/3$

15. (a) Exercise 5   **(b)** Exercise 3   **(c)** Exercise 7   **(d)** Exercise 9

17. $0.4444 \cdots = 0.4 + 0.04 + 0.004 + \cdots = \dfrac{0.4}{1 - 0.1} = 4/9$

19. $5.373737 \cdots = 5 + 0.37 + 0.0037 + 0.000037 + \cdots = 5 + \dfrac{0.37}{1 - 0.01} = 5 + 37/99 = 532/99$

21. $0.a_1 a_2 \cdots a_n 9999 \cdots = 0.a_1 a_2 \cdots a_n + 0.9 \left(10^{-n} \right) + 0.09 \left(10^{-n} \right) + \cdots$

$$= 0.a_1 a_2 \cdots a_n + \dfrac{0.9 \left(10^{-n} \right)}{1 - 0.1} = 0.a_1 a_2 \cdots a_n + 10^{-n}$$

$$= 0.a_1 a_2 \cdots (a_n + 1) = 0.a_1 a_2 \cdots (a_n + 1) 0000 \cdots$$

23. $d = 10 + 2 \cdot \dfrac{3}{4} \cdot 10 + 2 \cdot \dfrac{3}{4} \cdot \dfrac{3}{4} \cdot 10 + 2 \cdot \dfrac{3}{4} \cdot \dfrac{3}{4} \cdot \dfrac{3}{4} \cdot 10 + \cdots$

$$= 10 + 20 \left(\dfrac{3}{4} \right) + 20 \left(\dfrac{3}{4} \right)^2 + 20 \left(\dfrac{3}{4} \right)^3 + \cdots = 10 + \dfrac{20(3/4)}{1 - 3/4} = 10 + 60 = 70 \text{ meters}$$

25. (a) $s_n = \ln \dfrac{1}{2} + \ln \dfrac{2}{3} + \ln \dfrac{3}{4} + \cdots + \ln \dfrac{n}{n+1} = \ln \left(\dfrac{1}{2} \cdot \dfrac{2}{3} \cdot \dfrac{3}{4} \cdots \dfrac{n}{n+1} \right) = \ln \dfrac{1}{n+1} = -\ln(n+1)$,

$\lim\limits_{n \to +\infty} s_n = -\infty$, series diverges.

(b) $\ln(1 - 1/k^2) = \ln \dfrac{k^2 - 1}{k^2} = \ln \dfrac{(k-1)(k+1)}{k^2} = \ln \dfrac{k-1}{k} + \ln \dfrac{k+1}{k} = \ln \dfrac{k-1}{k} - \ln \dfrac{k}{k+1}$,

$$s_n = \sum\limits_{k=2}^{n+1} \left[\ln \dfrac{k-1}{k} - \ln \dfrac{k}{k+1} \right]$$

$$= \left(\ln \dfrac{1}{2} - \ln \dfrac{2}{3} \right) + \left(\ln \dfrac{2}{3} - \ln \dfrac{3}{4} \right) + \left(\ln \dfrac{3}{4} - \ln \dfrac{4}{5} \right) + \cdots + \left(\ln \dfrac{n}{n+1} - \ln \dfrac{n+1}{n+2} \right)$$

$$= \ln \dfrac{1}{2} - \ln \dfrac{n+1}{n+2}, \; \lim\limits_{n \to +\infty} s_n = \ln \dfrac{1}{2} = -\ln 2$$

27. (a) Geometric series, $a = x$, $r = -x^2$. Converges for $|-x^2| < 1$, $|x| < 1$;

$S = \dfrac{x}{1 - (-x^2)} = \dfrac{x}{1 + x^2}$.

(b) Geometric series, $a = 1/x^2$, $r = 2/x$. Converges for $|2/x| < 1$, $|x| > 2$;

$$S = \frac{1/x^2}{1 - 2/x} = \frac{1}{x^2 - 2x}.$$

(c) Geometric series, $a = e^{-x}$, $r = e^{-x}$. Converges for $|e^{-x}| < 1$, $e^{-x} < 1$, $e^x > 1$, $x > 0$;

$$S = \frac{e^{-x}}{1 - e^{-x}} = \frac{1}{e^x - 1}.$$

29. $a_2 = \frac{1}{2}a_1 + \frac{1}{2}$, $a_3 = \frac{1}{2}a_2 + \frac{1}{2} = \frac{1}{2^2}a_1 + \frac{1}{2^2} + \frac{1}{2}$, $a_4 = \frac{1}{2}a_3 + \frac{1}{2} = \frac{1}{2^3}a_1 + \frac{1}{2^3} + \frac{1}{2^2} + \frac{1}{2}$,

$a_5 = \frac{1}{2}a_4 + \frac{1}{2} = \frac{1}{2^4}a_1 + \frac{1}{2^4} + \frac{1}{2^3} + \frac{1}{2^2} + \frac{1}{2}, \ldots, a_n = \frac{1}{2^{n-1}}a_1 + \frac{1}{2^{n-1}} + \frac{1}{2^{n-2}} + \cdots + \frac{1}{2}$,

$$\lim_{n \to +\infty} a_n = \lim_{n \to +\infty} \frac{a_1}{2^{n-1}} + \sum_{n=1}^{\infty} \left(\frac{1}{2}\right)^n = 0 + \frac{1/2}{1 - 1/2} = 1$$

31. $s_n = (1 - 1/3) + (1/2 - 1/4) + (1/3 - 1/5) + (1/4 - 1/6) + \cdots + [1/n - 1/(n+2)]$

$= (1 + 1/2 + 1/3 + \cdots + 1/n) - (1/3 + 1/4 + 1/5 + \cdots + 1/(n+2))$

$= 3/2 - 1/(n+1) - 1/(n+2)$, $\displaystyle\lim_{n \to +\infty} s_n = 3/2$

33. $s_n = \displaystyle\sum_{k=1}^{n} \frac{1}{(2k-1)(2k+1)} = \sum_{k=1}^{n} \left[\frac{1/2}{2k-1} - \frac{1/2}{2k+1}\right] = \frac{1}{2}\left[\sum_{k=1}^{n} \frac{1}{2k-1} - \sum_{k=1}^{n} \frac{1}{2k+1}\right]$

$= \frac{1}{2}\left[\displaystyle\sum_{k=1}^{n} \frac{1}{2k-1} - \sum_{k=2}^{n+1} \frac{1}{2k-1}\right] = \frac{1}{2}\left[1 - \frac{1}{2n+1}\right]$; $\displaystyle\lim_{n \to +\infty} s_n = \frac{1}{2}$

35. By inspection, $\dfrac{\theta}{2} - \dfrac{\theta}{4} + \dfrac{\theta}{8} - \dfrac{\theta}{16} + \cdots = \dfrac{\theta/2}{1 - (-1/2)} = \theta/3$

37. (b) $\dfrac{2^k A}{3^k - 2^k} + \dfrac{2^k B}{3^{k+1} - 2^{k+1}} = \dfrac{2^k\left(3^{k+1} - 2^{k+1}\right)A + 2^k\left(3^k - 2^k\right)B}{\left(3^k - 2^k\right)\left(3^{k+1} - 2^{k+1}\right)}$

$$= \frac{\left(3 \cdot 6^k - 2 \cdot 2^{2k}\right)A + \left(6^k - 2^{2k}\right)B}{\left(3^k - 2^k\right)\left(3^{k+1} - 2^{k+1}\right)} = \frac{(3A+B)6^k - (2A+B)2^{2k}}{\left(3^k - 2^k\right)\left(3^{k+1} - 2^{k+1}\right)}$$

so $3A + B = 1$ and $2A + B = 0$, $A = 1$ and $B = -2$.

(c) $s_n = \displaystyle\sum_{k=1}^{n} \left[\frac{2^k}{3^k - 2^k} - \frac{2^{k+1}}{3^{k+1} - 2^{k+1}}\right] = \sum_{k=1}^{n} (a_k - a_{k+1})$ where $a_k = \dfrac{2^k}{3^k - 2^k}$.

But $s_n = (a_1 - a_2) + (a_2 - a_3) + (a_3 - a_4) + \cdots + (a_n - a_{n+1})$ which is a telescoping sum,

$$s_n = a_1 - a_{n+1} = 2 - \frac{2^{n+1}}{3^{n+1} - 2^{n+1}}, \quad \lim_{n \to +\infty} s_n = \lim_{n \to +\infty} \left[2 - \frac{(2/3)^{n+1}}{1 - (2/3)^{n+1}}\right] = 2.$$

EXERCISE SET 10.4

1. (a) $\displaystyle\sum_{k=1}^{\infty} \frac{1}{2^k} = \frac{1/2}{1 - 1/2} = 1$; $\displaystyle\sum_{k=1}^{\infty} \frac{1}{4^k} = \frac{1/4}{1 - 1/4} = 1/3$; $\displaystyle\sum_{k=1}^{\infty} \left(\frac{1}{2^k} + \frac{1}{4^k}\right) = 1 + 1/3 = 4/3$

(b) $\displaystyle\sum_{k=1}^{\infty} \frac{1}{5^k} = \frac{1/5}{1-1/5} = 1/4;$ $\displaystyle\sum_{k=1}^{\infty} \frac{1}{k(k+1)} = 1$ (Example 5, Section 10.4);

$$\sum_{k=1}^{\infty} \left[\frac{1}{5^k} - \frac{1}{k(k+1)}\right] = 1/4 - 1 = -3/4$$

3. (a) $p=3$, converges   **(b)** $p=1/2$, diverges   **(c)** $p=1$, diverges   **(d)** $p=2/3$, diverges

5. (a) $\displaystyle\lim_{k\to+\infty} \frac{k^2+k+3}{2k^2+1} = \frac{1}{2}$; the series diverges.   **(b)** $\displaystyle\lim_{k\to+\infty} \left(1+\frac{1}{k}\right)^k = e$; the series diverges.

(c) $\displaystyle\lim_{k\to+\infty} \cos k\pi$ does not exist; the series diverges.   **(d)** $\displaystyle\lim_{k\to+\infty} \frac{1}{k!} = 0$; no information

7. (a) $\displaystyle\int_1^{+\infty} \frac{1}{5x+2} = \lim_{\ell\to+\infty} \frac{1}{5}\ln(5x+2)\Big]_1^{\ell} = +\infty$, the series diverges by the Integral Test.

(b) $\displaystyle\int_1^{+\infty} \frac{1}{1+9x^2}dx = \lim_{\ell\to+\infty} \frac{1}{3}\tan^{-1}3x\Big]_1^{\ell} = \frac{1}{3}\left(\pi/2 - \tan^{-1}3\right),$
the series converges by the Integral Test.

9. $\displaystyle\sum_{k=1}^{\infty} \frac{1}{k+6} = \sum_{k=7}^{\infty} \frac{1}{k}$, diverges because the harmonic series diverges.

11. $\displaystyle\sum_{k=1}^{\infty} \frac{1}{\sqrt{k+5}} = \sum_{k=6}^{\infty} \frac{1}{\sqrt{k}}$, diverges because the p-series with $p = 1/2 \leq 1$ diverges.

13. $\displaystyle\int_1^{+\infty} (2x-1)^{-1/3}dx = \lim_{\ell\to+\infty} \frac{3}{4}(2x-1)^{2/3}\Big]_1^{\ell} = +\infty$, the series diverges by the Integral Test.

15. $\displaystyle\lim_{k\to+\infty} \frac{k}{\ln(k+1)} = \lim_{k\to+\infty} \frac{1}{1/(k+1)} = +\infty$, the series diverges because $\displaystyle\lim_{k\to+\infty} u_k \neq 0$.

17. $\displaystyle\lim_{k\to+\infty} (1+1/k)^{-k} = 1/e \neq 0$, the series diverges.

19. $\displaystyle\int_1^{+\infty} \frac{\tan^{-1}x}{1+x^2}dx = \lim_{\ell\to+\infty} \frac{1}{2}\left(\tan^{-1}x\right)^2\Big]_1^{\ell} = 3\pi^2/32$, the series converges by the Integral Test, since

$$\frac{d}{dx}\frac{\tan^{-1}x}{1+x^2} = \frac{1-2x\tan^{-1}x}{(1+x^2)^2} < 0 \text{ for } x \geq 1.$$

21. $\displaystyle\lim_{k\to+\infty} k^2\sin^2(1/k) = 1 \neq 0$, the series diverges.

23. $7\displaystyle\sum_{k=5}^{\infty} k^{-1.01}$, p-series with $p > 1$, converges

25. $\dfrac{1}{x(\ln x)^p}$ is decreasing for $x \geq e^p$, so use the Integral Test with $\displaystyle\int_{e^p}^{+\infty} \dfrac{dx}{x(\ln x)^p}$ to get

$$\lim_{\ell \to +\infty} \ln(\ln x)\Big]_{e^p}^{\ell} = +\infty \text{ if } p = 1, \qquad \lim_{\ell \to +\infty} \dfrac{(\ln x)^{1-p}}{1-p}\Big]_{e^p}^{\ell} = \begin{cases} +\infty & \text{if } p < 1 \\ \dfrac{p^{1-p}}{p-1} & \text{if } p > 1 \end{cases}$$

Thus the series converges for $p > 1$.

27. Suppose $\Sigma(u_k + v_k)$ converges; then so does $\Sigma[(u_k + v_k) - u_k]$, but $\Sigma[(u_k + v_k) - u_k] = \Sigma v_k$, so Σv_k converges which contradicts the assumption that Σv_k diverges. Suppose $\Sigma(u_k - v_k)$ converges; then so does $\Sigma[u_k - (u_k - v_k)] = \Sigma v_k$ which leads to the same contradiction as before.

29. (a) diverges because $\displaystyle\sum_{k=1}^{\infty}(2/3)^{k-1}$ converges and $\displaystyle\sum_{k=1}^{\infty} 1/k$ diverges.

(b) diverges because $\displaystyle\sum_{k=1}^{\infty} 1/(3k+2)$ diverges and $\displaystyle\sum_{k=1}^{\infty} 1/k^{3/2}$ converges.

31. (a) $3\displaystyle\sum_{k=1}^{\infty} \dfrac{1}{k^2} - \sum_{k=1}^{\infty} \dfrac{1}{k^4} = \pi^2/2 - \pi^4/90$ **(b)** $\displaystyle\sum_{k=1}^{\infty} \dfrac{1}{k^2} - 1 - \dfrac{1}{2^2} = \pi^2/6 - 5/4$

(c) $\displaystyle\sum_{k=2}^{\infty} \dfrac{1}{(k-1)^4} = \sum_{k=1}^{\infty} \dfrac{1}{k^4} = \pi^4/90$

33. (a) In Exercise 32 above let $f(x) = \dfrac{1}{x^2}$. Then $\displaystyle\int_n^{+\infty} f(x)\,dx = -\dfrac{1}{x}\Big]_n^{+\infty} = \dfrac{1}{n}$; use this result and the same result with $n+1$ replacing n to obtain the desired result.

(b) $s_3 = 1 + 1/4 + 1/9 = 49/36$; $58/36 = s_3 + \dfrac{1}{4} < \dfrac{1}{6}\pi^2 < s_3 + \dfrac{1}{3} = 61/36$

(d) $1/11 < \dfrac{1}{6}\pi^2 - s_{10} < 1/10$

35. (a) $\displaystyle\int_n^{+\infty} \dfrac{1}{x^3}\,dx = \dfrac{1}{2n^2}$; use Exercise 32(b) sw

(b) $\dfrac{1}{2n^2} - \dfrac{1}{2(n+1)^2} < 0.01$ for $n = 5$.

(c) From Part (a) with $n = 5$ obtain $1.200 < S < 1.206$, so $S \approx 1.203$.

37. (a) Let $F(x) = \dfrac{1}{x}$, then $\displaystyle\int_1^n \dfrac{1}{x}\,dx = \ln n$ and $\displaystyle\int_1^{n+1} \dfrac{1}{x}\,dx = \ln(n+1)$, $u_1 = 1$ so $\ln(n+1) < s_n < 1 + \ln n$.

(b) $\ln(1,000,001) < s_{1,000,000} < 1 + \ln(1,000,000)$, $13 < s_{1,000,000} < 15$

(c) $s_{10^9} < 1 + \ln 10^9 = 1 + 9\ln 10 < 22$

(d) $s_n > \ln(n+1) \geq 100$, $n \geq e^{100} - 1 \approx 2.688 \times 10^{43}$; $n = 2.69 \times 10^{43}$

39. $x^2 e^{-x}$ is decreasing and positive for $x > 2$ so the Integral Test applies:
$$\int_1^{\infty} x^2 e^{-x}\,dx = -(x^2 + 2x + 2)e^{-x}\Big]_1^{\infty} = 5e^{-1} \text{ so the series converges.}$$

EXERCISE SET 10.5

1. **(a)** $\dfrac{1}{5k^2 - k} \le \dfrac{1}{5k^2 - k^2} = \dfrac{1}{4k^2}, \displaystyle\sum_{k=1}^{\infty} \dfrac{1}{4k^2}$ converges

   (b) $\dfrac{3}{k - 1/4} > \dfrac{3}{k}, \displaystyle\sum_{k=1}^{\infty} 3/k$ diverges

3. **(a)** $\dfrac{1}{3^k + 5} < \dfrac{1}{3^k}, \displaystyle\sum_{k=1}^{\infty} \dfrac{1}{3^k}$ converges   **(b)** $\dfrac{5\sin^2 k}{k!} < \dfrac{5}{k!}, \displaystyle\sum_{k=1}^{\infty} \dfrac{5}{k!}$ converges

5. compare with the convergent series $\displaystyle\sum_{k=1}^{\infty} 1/k^5$, $\rho = \lim_{k \to +\infty} \dfrac{4k^7 - 2k^6 + 6k^5}{8k^7 + k - 8} = 1/2$, converges

7. compare with the convergent series $\displaystyle\sum_{k=1}^{\infty} 5/3^k$, $\rho = \lim_{k \to +\infty} \dfrac{3^k}{3^k + 1} = 1$, converges

9. compare with the divergent series $\displaystyle\sum_{k=1}^{\infty} \dfrac{1}{k^{2/3}}$,

   $\rho = \lim_{k \to +\infty} \dfrac{k^{2/3}}{(8k^2 - 3k)^{1/3}} = \lim_{k \to +\infty} \dfrac{1}{(8 - 3/k)^{1/3}} = 1/2$, diverges

11. $\rho = \lim_{k \to +\infty} \dfrac{3^{k+1}/(k+1)!}{3^k/k!} = \lim_{k \to +\infty} \dfrac{3}{k+1} = 0$, the series converges

13. $\rho = \lim_{k \to +\infty} \dfrac{k}{k+1} = 1$, the result is inconclusive

15. $\rho = \lim_{k \to +\infty} \dfrac{(k+1)!/(k+1)^3}{k!/k^3} = \lim_{k \to +\infty} \dfrac{k^3}{(k+1)^2} = +\infty$, the series diverges

17. $\rho = \lim_{k \to +\infty} \dfrac{3k + 2}{2k - 1} = 3/2$, the series diverges

19. $\rho = \lim_{k \to +\infty} \dfrac{k^{1/k}}{5} = 1/5$, the series converges

21. Ratio Test, $\rho = \lim_{k \to +\infty} 7/(k+1) = 0$, converges

23. Ratio Test, $\rho = \lim_{k \to +\infty} \dfrac{(k+1)^2}{5k^2} = 1/5$, converges

25. Ratio Test, $\rho = \lim_{k \to +\infty} e^{-1}(k+1)^{50}/k^{50} = e^{-1} < 1$, converges

27. Limit Comparison Test, compare with the convergent series $\displaystyle\sum_{k=1}^{\infty} 1/k^{5/2}$, $\rho = \lim_{k \to +\infty} \dfrac{k^3}{k^3 + 1} = 1$, converges

29. Limit Comparison Test, compare with the divergent series $\displaystyle\sum_{k=1}^{\infty} 1/k$, $\rho = \lim_{k \to +\infty} \dfrac{k}{\sqrt{k^2 + k}} = 1$, diverges

31. Limit Comparison Test, compare with the convergent series $\sum\limits_{k=1}^{\infty} 1/k^{5/2}$,

$$\rho = \lim_{k \to +\infty} \frac{k^3 + 2k^{5/2}}{k^3 + 3k^2 + 3k} = 1, \text{ converges}$$

33. Limit Comparison Test, compare with the divergent series $\sum\limits_{k=1}^{\infty} 1/\sqrt{k}$

35. Ratio Test, $\rho = \lim\limits_{k \to +\infty} \dfrac{\ln(k+1)}{e \ln k} = \lim\limits_{k \to +\infty} \dfrac{k}{e(k+1)} = 1/e < 1$, converges

37. Ratio Test, $\rho = \lim\limits_{k \to +\infty} \dfrac{k+5}{4(k+1)} = 1/4$, converges

39. diverges because $\lim\limits_{k \to +\infty} \dfrac{1}{4 + 2^{-k}} = 1/4 \neq 0$

41. $\dfrac{\tan^{-1} k}{k^2} < \dfrac{\pi/2}{k^2}, \sum\limits_{k=1}^{\infty} \dfrac{\pi/2}{k^2}$ converges so $\sum\limits_{k=1}^{\infty} \dfrac{\tan^{-1} k}{k^2}$ converges

43. Ratio Test, $\rho = \lim\limits_{k \to +\infty} \dfrac{(k+1)^2}{(2k+2)(2k+1)} = 1/4$, converges

45. $u_k = \dfrac{k!}{1 \cdot 3 \cdot 5 \cdots (2k-1)}$, by the Ratio Test $\rho = \lim\limits_{k \to +\infty} \dfrac{k+1}{2k+1} = 1/2$; converges

47. Root Test: $\rho = \lim\limits_{k \to +\infty} \dfrac{1}{3}(\ln k)^{1/k} = 1/3$, converges

49. **(b)** $\rho = \lim\limits_{k \to +\infty} \dfrac{\sin(\pi/k)}{\pi/k} = 1$ and $\sum\limits_{k=1}^{\infty} \pi/k$ diverges

51. Set $g(x) = \sqrt{x} - \ln x$; $\dfrac{d}{dx} g(x) = \dfrac{1}{2\sqrt{x}} - \dfrac{1}{x} = 0$ when $x = 4$. Since $\lim\limits_{x \to 0^+} g(x) = \lim\limits_{x \to +\infty} g(x) = +\infty$ it follows that $g(x)$ has its minimum at $x = 4$, $g(4) = \sqrt{4} - \ln 4 > 0$, and thus $\sqrt{x} - \ln x > 0$ for $x > 0$.

(a) $\dfrac{\ln k}{k^2} < \dfrac{\sqrt{k}}{k^2} = \dfrac{1}{k^{3/2}}, \sum\limits_{k=1}^{\infty} \dfrac{1}{k^{3/2}}$ converges so $\sum\limits_{k=1}^{\infty} \dfrac{\ln k}{k^2}$ converges.

(b) $\dfrac{1}{(\ln k)^2} > \dfrac{1}{k}, \sum\limits_{k=2}^{\infty} \dfrac{1}{k}$ diverges so $\sum\limits_{k=2}^{\infty} \dfrac{1}{(\ln k)^2}$ diverges.

53. **(a)** If $\sum b_k$ converges, then set $M = \sum b_k$. Then $a_1 + a_2 + \cdots + a_n \leq b_1 + b_2 + \cdots + b_n \leq M$; apply Theorem 10.4.6 to get convergence of $\sum a_k$.

(b) Assume the contrary, that $\sum b_k$ converges; then use Part (a) of the Theorem to show that $\sum a_k$ converges, a contradiction.

EXERCISE SET 10.6

1. $a_{k+1} < a_k, \displaystyle\lim_{k \to +\infty} a_k = 0, a_k > 0$

3. diverges because $\displaystyle\lim_{k \to +\infty} a_k = \lim_{k \to +\infty} \frac{k+1}{3k+1} = 1/3 \neq 0$

5. $\{e^{-k}\}$ is decreasing and $\displaystyle\lim_{k \to +\infty} e^{-k} = 0$, converges

7. $\rho = \displaystyle\lim_{k \to +\infty} \frac{(3/5)^{k+1}}{(3/5)^k} = 3/5$, converges absolutely

9. $\rho = \displaystyle\lim_{k \to +\infty} \frac{3k^2}{(k+1)^2} = 3$, diverges

11. $\rho = \displaystyle\lim_{k \to +\infty} \frac{(k+1)^3}{ek^3} = 1/e$, converges absolutely

13. conditionally convergent, $\displaystyle\sum_{k=1}^{\infty} \frac{(-1)^{k+1}}{3k}$ converges by the Alternating Series Test but $\displaystyle\sum_{k=1}^{\infty} \frac{1}{3k}$ diverges

15. divergent, $\displaystyle\lim_{k \to +\infty} a_k \neq 0$

17. $\displaystyle\sum_{k=1}^{\infty} \frac{\cos k\pi}{k} = \sum_{k=1}^{\infty} \frac{(-1)^k}{k}$ is conditionally convergent, $\displaystyle\sum_{k=1}^{\infty} \frac{(-1)^k}{k}$ converges by the Alternating Series Test but $\displaystyle\sum_{k=1}^{\infty} 1/k$ diverges.

19. conditionally convergent, $\displaystyle\sum_{k=1}^{\infty} (-1)^{k+1} \frac{k+2}{k(k+3)}$ converges by the

Alternating Series Test but $\displaystyle\sum_{k=1}^{\infty} \frac{k+2}{k(k+3)}$ diverges (Limit Comparison Test with $\sum 1/k$)

21. $\displaystyle\sum_{k=1}^{\infty} \sin(k\pi/2) = 1 + 0 - 1 + 0 + 1 + 0 - 1 + 0 + \cdots$, divergent ($\displaystyle\lim_{k \to +\infty} \sin(k\pi/2)$ does not exist)

23. conditionally convergent, $\displaystyle\sum_{k=2}^{\infty} \frac{(-1)^k}{k \ln k}$ converges by the Alternating Series Test but $\displaystyle\sum_{k=2}^{\infty} \frac{1}{k \ln k}$ diverges (Integral Test)

25. absolutely convergent, $\displaystyle\sum_{k=2}^{\infty} (1/\ln k)^k$ converges by the Root Test

27. conditionally convergent, let $f(x) = \dfrac{x^2+1}{x^3+2}$ then $f'(x) = \dfrac{x(4 - 3x - x^3)}{(x^3+2)^2} \leq 0$ for $x \geq 1$ so

$\{a_k\}_{k=2}^{+\infty} = \left\{\dfrac{k^2+1}{k^3+2}\right\}_{k=2}^{+\infty}$ is decreasing, $\displaystyle\lim_{k \to +\infty} a_k = 0$; the series converges by the

Alternating Series Test but $\displaystyle\sum_{k=2}^{\infty} \frac{k^2+1}{k^3+2}$ diverges (Limit Comparison Test with $\sum 1/k$)

29. absolutely convergent by the Ratio Test, $\rho = \lim\limits_{k \to +\infty} \dfrac{k+1}{(2k+1)(2k)} = 0$

31. $|\text{error}| < a_8 = 1/8 = 0.125$

33. $|\text{error}| < a_{100} = 1/\sqrt{100} = 0.1$

35. $|\text{error}| < 0.0001$ if $a_{n+1} \le 0.0001$, $1/(n+1) \le 0.0001$, $n+1 \ge 10,000$, $n \ge 9,999$, $n = 9,999$

37. $|\text{error}| < 0.005$ if $a_{n+1} \le 0.005$, $1/\sqrt{n+1} \le 0.005$, $\sqrt{n+1} \ge 200$, $n+1 \ge 40,000$, $n \ge 39,999$, $n = 39,999$

39. $a_k = \dfrac{3}{2^{k+1}}$, $|\text{error}| < a_{11} = \dfrac{3}{2^{12}} < 0.00074$; $s_{10} \approx 0.4995$; $S = \dfrac{3/4}{1-(-1/2)} = 0.5$

41. $a_k = \dfrac{1}{(2k-1)!}$, $a_{n+1} = \dfrac{1}{(2n+1)!} \le 0.005$, $(2n+1)! \ge 200$, $2n+1 \ge 6$, $n \ge 2.5$; $n = 3$, $s_3 = 1 - 1/6 + 1/120 \approx 0.84$

43. $a_k = \dfrac{1}{k2^k}$, $a_{n+1} = \dfrac{1}{(n+1)2^{n+1}} \le 0.005$, $(n+1)2^{n+1} \ge 200$, $n+1 \ge 6$, $n \ge 5$; $n = 5$, $s_5 \approx 0.41$

45. **(c)** $a_k = \dfrac{1}{2k-1}$, $a_{n+1} = \dfrac{1}{2n+1} \le 10^{-2}$, $2n+1 \ge 100$, $n \ge 49.5$; $n = 50$

47. **(a)** $\sum (-1)^k/k$ diverges, but $\sum 1/k^2$ converges; $\sum (-1)^k/\sqrt{k}$ converges, but $\sum 1/k$ diverges

(b) Let $a_k = \dfrac{(-1)^k}{k}$, then $\sum a_k^2$ converges but $\sum |a_k|$ diverges, $\sum a_k$ converges.

49. $1 + \dfrac{1}{3^2} + \dfrac{1}{5^2} + \cdots = \left[1 + \dfrac{1}{2^2} + \dfrac{1}{3^2} + \cdots \right] - \left[\dfrac{1}{2^2} + \dfrac{1}{4^2} + \dfrac{1}{6^2} + \cdots \right]$

$= \dfrac{\pi^2}{6} - \dfrac{1}{2^2} \left[1 + \dfrac{1}{2^2} + \dfrac{1}{3^2} + \cdots \right] = \dfrac{\pi^2}{6} - \dfrac{1}{4}\dfrac{\pi^2}{6} = \dfrac{\pi^2}{8}$

51. $1 + \dfrac{1}{3^4} + \dfrac{1}{5^4} + \cdots = \left[1 + \dfrac{1}{2^4} + \dfrac{1}{3^4} + \cdots \right] - \left[\dfrac{1}{2^4} + \dfrac{1}{4^4} + \dfrac{1}{6^4} + \cdots \right]$

$= \dfrac{\pi^4}{90} - \dfrac{1}{2^4} \left[1 + \dfrac{1}{2^4} + \dfrac{1}{3^4} + \cdots \right] = \dfrac{\pi^4}{90} - \dfrac{1}{16}\dfrac{\pi^4}{90} = \dfrac{\pi^4}{96}$

53. **(a)** Write the series in the form $\left(1 - \dfrac{1}{2} \right) + \left(\dfrac{2}{3} - \dfrac{1}{3} \right) + \left(\dfrac{2}{4} - \dfrac{1}{4} \right) + \ldots = \sum\limits_{k=2}^{+\infty} \dfrac{1}{k}$, which diverges.

(b) The alternating series test requires a sequence that i) alternates in sign, which this does, and ii) decreases monotonely to 0, which this does not.

55. **(a)** The distance d from the starting point is

$d = 180 - \dfrac{180}{2} + \dfrac{180}{3} - \cdots - \dfrac{180}{1000} = 180 \left[1 - \dfrac{1}{2} + \dfrac{1}{3} - \cdots - \dfrac{1}{1000} \right].$

From Theorem 10.6.2, $1 - \dfrac{1}{2} + \dfrac{1}{3} - \cdots - \dfrac{1}{1000}$ differs from $\ln 2$ by less than $1/1001$ so $180(\ln 2 - 1/1001) < d < 180 \ln 2$, $124.58 < d < 124.77$.

(b) The total distance traveled is $s = 180 + \dfrac{180}{2} + \dfrac{180}{3} + \cdots + \dfrac{180}{1000}$, and from inequality (2) in Section 10.4,

$$\int_1^{1001} \frac{180}{x} dx < s < 180 + \int_1^{1000} \frac{180}{x} dx$$

$$180 \ln 1001 < s < 180(1 + \ln 1000)$$

$$1243 < s < 1424$$

EXERCISE SET 10.7

1. **(a)** $f^{(k)}(x) = (-1)^k e^{-x}$, $f^{(k)}(0) = (-1)^k$; $e^{-x} \approx 1 - x + x^2/2$ (quadratic), $e^{-x} \approx 1 - x$ (linear)

   (b) $f'(x) = -\sin x$, $f''(x) = -\cos x$, $f(0) = 1$, $f'(0) = 0$, $f''(0) = -1$,

   $\cos x \approx 1 - x^2/2$ (quadratic), $\cos x \approx 1$ (linear)

   (c) $f'(x) = \cos x$, $f''(x) = -\sin x$, $f(\pi/2) = 1$, $f'(\pi/2) = 0$, $f''(\pi/2) = -1$,

   $\sin x \approx 1 - (x - \pi/2)^2/2$ (quadratic), $\sin x \approx 1$ (linear)

   (d) $f(1) = 1$, $f'(1) = 1/2$, $f''(1) = -1/4$;

   $\sqrt{x} = 1 + \dfrac{1}{2}(x - 1) - \dfrac{1}{8}(x - 1)^2$ (quadratic), $\sqrt{x} \approx 1 + \dfrac{1}{2}(x - 1)$ (linear)

3. **(a)** $f'(x) = \dfrac{1}{2}x^{-1/2}$, $f''(x) = -\dfrac{1}{4}x^{-3/2}$; $f(1) = 1$, $f'(1) = \dfrac{1}{2}$, $f''(1) = -\dfrac{1}{4}$;

   $\sqrt{x} \approx 1 + \dfrac{1}{2}(x - 1) - \dfrac{1}{8}(x - 1)^2$

   (b) $x = 1.1, x_0 = 1$, $\sqrt{1.1} \approx 1 + \dfrac{1}{2}(0.1) - \dfrac{1}{8}(0.1)^2 = 1.04875$, calculator value ≈ 1.0488088

5. $f(x) = \tan x$, $61° = \pi/3 + \pi/180$ rad; $x_0 = \pi/3$, $f'(x) = \sec^2 x$, $f''(x) = 2\sec^2 x \tan x$;
   $f(\pi/3) = \sqrt{3}, f'(\pi/3) = 4, f''(x) = 8\sqrt{3}$; $\tan x \approx \sqrt{3} + 4(x - \pi/3) + 4\sqrt{3}(x - \pi/3)^2$,
   $\tan 61° = \tan(\pi/3 + \pi/180) \approx \sqrt{3} + 4\pi/180 + 4\sqrt{3}(\pi/180)^2 \approx 1.80397443$,
   calculator value ≈ 1.80404776

7. $f^{(k)}(x) = (-1)^k e^{-x}$, $f^{(k)}(0) = (-1)^k$; $p_0(x) = 1$, $p_1(x) = 1 - x$, $p_2(x) = 1 - x + \dfrac{1}{2}x^2$,

   $p_3(x) = 1 - x + \dfrac{1}{2}x^2 - \dfrac{1}{3!}x^3$, $p_4(x) = 1 - x + \dfrac{1}{2}x^2 - \dfrac{1}{3!}x^3 + \dfrac{1}{4!}x^4$; $\displaystyle\sum_{k=0}^{n} \frac{(-1)^k}{k!} x^k$

9. $f^{(k)}(0) = 0$ if k is odd, $f^{(k)}(0)$ is alternately π^k and $-\pi^k$ if k is even; $p_0(x) = 1$, $p_1(x) = 1$,

   $p_2(x) = 1 - \dfrac{\pi^2}{2!}x^2$; $p_3(x) = 1 - \dfrac{\pi^2}{2!}x^2$, $p_4(x) = 1 - \dfrac{\pi^2}{2!}x^2 + \dfrac{\pi^4}{4!}x^4$; $\displaystyle\sum_{k=0}^{[\frac{n}{2}]} \frac{(-1)^k \pi^{2k}}{(2k)!} x^{2k}$

   NB: The function $[x]$ defined for real x indicates the greatest integer which is $\leq x$.

11. $f^{(0)}(0) = 0$; for $k \geq 1$, $f^{(k)}(x) = \dfrac{(-1)^{k+1}(k-1)!}{(1+x)^k}$, $f^{(k)}(0) = (-1)^{k+1}(k-1)!$; $p_0(x) = 0$,

   $p_1(x) = x$, $p_2(x) = x - \dfrac{1}{2}x^2$, $p_3(x) = x - \dfrac{1}{2}x^2 + \dfrac{1}{3}x^3$, $p_4(x) = x - \dfrac{1}{2}x^2 + \dfrac{1}{3}x^3 - \dfrac{1}{4}x^4$; $\displaystyle\sum_{k=1}^{n} \frac{(-1)^{k+1}}{k} x^k$

13. $f^{(k)}(0) = 0$ if k is odd, $f^{(k)}(0) = 1$ if k is even; $p_0(x) = 1, p_1(x) = 1,$

$$p_2(x) = 1 + x^2/2, \ p_3(x) = 1 + x^2/2, \ p_4(x) = 1 + x^2/2 + x^4/4!; \ \sum_{k=0}^{[\frac{n}{2}]} \frac{1}{(2k)!} x^{2k}$$

15. $f^{(k)}(x) = \begin{cases} (-1)^{k/2}(x \sin x - k \cos x) & k \text{ even} \\ (-1)^{(k-1)/2}(x \cos x + k \sin x) & k \text{ odd} \end{cases}$, $\quad f^{(k)}(0) = \begin{cases} (-1)^{1+k/2}k & k \text{ even} \\ 0 & k \text{ odd} \end{cases}$

$$p_0(x) = 0, \ p_1(x) = 0, \ p_2(x) = x^2, p_3(x) = x^2, \ p_4(x) = x^2 - \frac{1}{6}x^4; \ \sum_{k=0}^{[\frac{n}{2}]-1} \frac{(-1)^k}{(2k+1)!} x^{2k+2}$$

17. $f^{(k)}(x_0) = e; \ p_0(x) = e, \ p_1(x) = e + e(x - 1),$

$$p_2(x) = e + e(x - 1) + \frac{e}{2}(x - 1)^2, \ p_3(x) = e + e(x - 1) + \frac{e}{2}(x - 1)^2 + \frac{e}{3!}(x - 1)^3,$$

$$p_4(x) = e + e(x - 1) + \frac{e}{2}(x - 1)^2 + \frac{e}{3!}(x - 1)^3 + \frac{e}{4!}(x - 1)^4; \ \sum_{k=0}^{n} \frac{e}{k!}(x - 1)^k$$

19. $f^{(k)}(x) = \dfrac{(-1)^k k!}{x^{k+1}}, \ f^{(k)}(-1) = -k!; \ p_0(x) = -1; \ p_1(x) = -1 - (x + 1);$

$$p_2(x) = -1 - (x + 1) - (x + 1)^2; \ p_3(x) = -1 - (x + 1) - (x + 1)^2 - (x + 1)^3;$$

$$p_4(x) = -1 - (x + 1) - (x + 1)^2 - (x + 1)^3 - (x + 1)^4; \ \sum_{k=0}^{n}(-1)(x + 1)^k$$

21. $f^{(k)}(1/2) = 0$ if k is odd, $f^{(k)}(1/2)$ is alternately π^k and $-\pi^k$ if k is even;

$$p_0(x) = p_1(x) = 1, p_2(x) = p_3(x) = 1 - \frac{\pi^2}{2}(x - 1/2)^2,$$

$$p_4(x) = 1 - \frac{\pi^2}{2}(x - 1/2)^2 + \frac{\pi^4}{4!}(x - 1/2)^4; \ \sum_{k=0}^{[\frac{n}{2}]} \frac{(-1)^k \pi^{2k}}{(2k)!}(x - 1/2)^{2k}$$

23. $f(1) = 0$, for $k \geq 1, f^{(k)}(x) = \dfrac{(-1)^{k-1}(k-1)!}{x^k}; \ f^{(k)}(1) = (-1)^{k-1}(k-1)!;$

$$p_0(x) = 0, \ p_1(x) = (x - 1); \ p_2(x) = (x - 1) - \frac{1}{2}(x - 1)^2; \ p_3(x) = (x - 1) - \frac{1}{2}(x - 1)^2 + \frac{1}{3}(x - 1)^3,$$

$$p_4(x) = (x - 1) - \frac{1}{2}(x - 1)^2 + \frac{1}{3}(x - 1)^3 - \frac{1}{4}(x - 1)^4; \ \sum_{k=1}^{n} \frac{(-1)^{k-1}}{k}(x - 1)^k$$

25. **(a)** $f(0) = 1, f'(0) = 2, f''(0) = -2, f'''(0) = 6$, the third MacLaurin polynomial for $f(x)$ is $f(x)$.
   (b) $f(1) = 1, f'(1) = 2, f''(1) = -2, f'''(1) = 6$, the third Taylor polynomial for $f(x)$ is $f(x)$.

27. $f^{(k)}(0) = (-2)^k; \ p_0(x) = 1, \ p_1(x) = 1 - 2x,$

$$p_2(x) = 1 - 2x + 2x^2, \ p_3(x) = 1 - 2x + 2x^2 - \frac{4}{3}x^3$$

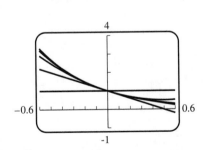

29. $f^{(k)}(\pi) = 0$ if k is odd, $f^{(k)}(\pi)$ is alternately -1

and 1 if k is even; $p_0(x) = -1$, $p_2(x) = -1 + \dfrac{1}{2}(x-\pi)^2$,

$p_4(x) = -1 + \dfrac{1}{2}(x-\pi)^2 - \dfrac{1}{24}(x-\pi)^4$,

$p_6(x) = -1 + \dfrac{1}{2}(x-\pi)^2 - \dfrac{1}{24}(x-\pi)^4 + \dfrac{1}{720}(x-\pi)^6$

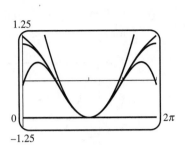

31. $f^{(k)}(x) = e^x$, $|f^{(k)}(x)| \le e^{1/2} < 2$ on $[0, 1/2]$, let $M = 2$,

$e^{1/2} = 1 + \dfrac{1}{2} + \dfrac{1}{8} + \dfrac{1}{48} + \dfrac{1}{24 \cdot 16} + \cdots + \dfrac{1}{n! 2^n} + R_n(1/2)$;

$|R_n(1/2)| \le \dfrac{M}{(n+1)!}(1/2)^{n+1} \le \dfrac{2}{(n+1)!}(1/2)^{n+1} \le 0.00005$ for $n = 5$;

$e^{1/2} \approx 1 + \dfrac{1}{2} + \dfrac{1}{8} + \dfrac{1}{48} + \dfrac{1}{24 \cdot 16} + \dfrac{1}{120 \cdot 32} \approx 1.64870$, calculator value 1.64872

33. $f^{(k)}(\ln 4) = 15/8$ for k even, $f^{(k)}(\ln 4) = 17/8$ for k odd, which can be written as

$f^{(k)}(\ln 4) = \dfrac{16 - (-1)^k}{8}$; $\displaystyle\sum_{k=0}^{n} \dfrac{16 - (-1)^k}{8k!}(x - \ln 4)^k$

35. $p(0) = 1$, $p(x)$ has slope -1 at $x = 0$, and $p(x)$ is concave up at $x = 0$, eliminating I, II and III respectively and leaving IV.

37. From Exercise 2(a), $p_1(x) = 1 + x$, $p_2(x) = 1 + x + x^2/2$

(a)

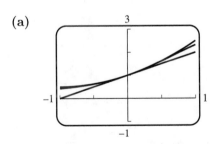

(b)

x	-1.000	-0.750	-0.500	-0.250	0.000	0.250	0.500	0.750	1.000
$f(x)$	0.431	0.506	0.619	0.781	1.000	1.281	1.615	1.977	2.320
$p_1(x)$	0.000	0.250	0.500	0.750	1.000	1.250	1.500	1.750	2.000
$p_2(x)$	0.500	0.531	0.625	0.781	1.000	1.281	1.625	2.031	2.500

(c) $|e^{\sin x} - (1+x)| < 0.01$
for $-0.14 < x < 0.14$

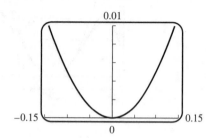

(d) $|e^{\sin x} - (1 + x + x^2/2)| < 0.01$
for $-0.50 < x < 0.50$

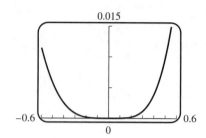

39. (a) $\sin x = x - \dfrac{x^3}{3!} + 0 \cdot x^4 + R_4(x),$

$$|R_4(x)| \le \frac{|x|^5}{5!} < 0.5 \times 10^{-3} \text{ if } |x|^5 < 0.06,$$

$$|x| < (0.06)^{1/5} \approx 0.569, (-0.569, 0.569)$$

(b)

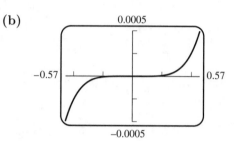

41. $f^{(5)}(x) = -\dfrac{3840}{(1+x^2)^7} + \dfrac{3840x^3}{(1+x^2)^5} - \dfrac{720x}{(1+x^2)^4},$ let $M = 8400,$

$$R_4(x) \le \frac{8400}{4!}|x|^4 < 0.0005 \text{ if } x < 0.0677$$

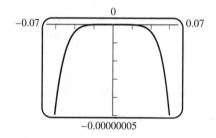

EXERCISE SET 10.8

1. $f^{(k)}(x) = (-1)^k e^{-x}, f^{(k)}(0) = (-1)^k;$ $\displaystyle\sum_{k=0}^{\infty} \frac{(-1)^k}{k!} x^k$

3. $f^{(k)}(0) = 0$ if k is odd, $f^{(k)}(0)$ is alternately π^k and $-\pi^k$ if k is even; $\displaystyle\sum_{k=0}^{\infty} \frac{(-1)^k \pi^{2k}}{(2k)!} x^{2k}$

5. $f^{(0)}(0) = 0;$ for $k \ge 1, f^{(k)}(x) = \dfrac{(-1)^{k+1}(k-1)!}{(1+x)^k}, f^{(k)}(0) = (-1)^{k+1}(k-1)!;$ $\displaystyle\sum_{k=1}^{\infty} \frac{(-1)^{k+1}}{k} x^k$

7. $f^{(k)}(0) = 0$ if k is odd, $f^{(k)}(0) = 1$ if k is even; $\displaystyle\sum_{k=0}^{\infty} \frac{1}{(2k)!} x^{2k}$

9. $f^{(k)}(x) = \begin{cases} (-1)^{k/2}(x\sin x - k\cos x) & k \text{ even} \\ (-1)^{(k-1)/2}(x\cos x + k\sin x) & k \text{ odd} \end{cases}$, $\quad f^{(k)}(0) = \begin{cases} (-1)^{1+k/2}k & k \text{ even} \\ 0 & k \text{ odd} \end{cases}$

$$\sum_{k=0}^{\infty} \frac{(-1)^k}{(2k+1)!} x^{2k+2}$$

11. $f^{(k)}(x_0) = e;\quad \displaystyle\sum_{k=0}^{\infty} \frac{e}{k!}(x-1)^k$

13. $f^{(k)}(x) = \dfrac{(-1)^k k!}{x^{k+1}}$, $f^{(k)}(-1) = -k!;\quad \displaystyle\sum_{k=0}^{\infty} (-1)(x+1)^k$

15. $f^{(k)}(1/2) = 0$ if k is odd, $f^{(k)}(1/2)$ is alternately π^k and $-\pi^k$ if k is even;

$$\sum_{k=0}^{\infty} \frac{(-1)^k \pi^{2k}}{(2k)!} (x-1/2)^{2k}$$

17. $f(1) = 0$, for $k \geq 1$, $f^{(k)}(x) = \dfrac{(-1)^{k-1}(k-1)!}{x^k}$; $f^{(k)}(1) = (-1)^{k-1}(k-1)!;$

$$\sum_{k=1}^{\infty} \frac{(-1)^{k-1}}{k} (x-1)^k$$

19. geometric series, $\rho = \displaystyle\lim_{k \to +\infty} \left|\frac{u_{k+1}}{u_k}\right| = |x|$, so the interval of convergence is $-1 < x < 1$, converges there to $\dfrac{1}{1+x}$ (the series diverges for $x = \pm 1$)

21. geometric series, $\rho = \displaystyle\lim_{k \to +\infty} \left|\frac{u_{k+1}}{u_k}\right| = |x-2|$, so the interval of convergence is $1 < x < 3$, converges there to $\dfrac{1}{1-(x-2)} = \dfrac{1}{3-x}$ (the series diverges for $x = 1, 3$)

23. **(a)** geometric series, $\rho = \displaystyle\lim_{k \to +\infty} \left|\frac{u_{k+1}}{u_k}\right| = |x/2|$, so the interval of convergence is $-2 < x < 2$, converges there to $\dfrac{1}{1+x/2} = \dfrac{2}{2+x}$; (the series diverges for $x = -2, 2$)

(b) $f(0) = 1$; $f(1) = 2/3$

25. $\rho = \displaystyle\lim_{k \to +\infty} \frac{k+1}{k+2}|x| = |x|$, the series converges if $|x| < 1$ and diverges if $|x| > 1$. If $x = -1$,

$$\sum_{k=0}^{\infty} \frac{(-1)^k}{k+1}$$ converges by the Alternating Series Test; if $x = 1$, $\displaystyle\sum_{k=0}^{\infty} \frac{1}{k+1}$ diverges. The radius of convergence is 1, the interval of convergence is $[-1, 1)$.

27. $\rho = \displaystyle\lim_{k \to +\infty} \frac{|x|}{k+1} = 0$, the radius of convergence is $+\infty$, the interval is $(-\infty, +\infty)$.

29. $\rho = \lim\limits_{k\to+\infty} \dfrac{5k^2|x|}{(k+1)^2} = 5|x|$, converges if $|x| < 1/5$ and diverges if $|x| > 1/5$. If $x = -1/5$, $\sum\limits_{k=1}^{\infty} \dfrac{(-1)^k}{k^2}$

converges; if $x = 1/5$, $\sum\limits_{k=1}^{\infty} 1/k^2$ converges. Radius of convergence is $1/5$, interval of convergence is

$[-1/5, 1/5]$.

31. $\rho = \lim\limits_{k\to+\infty} \dfrac{k|x|}{k+2} = |x|$, converges if $|x| < 1$, diverges if $|x| > 1$. If $x = -1$, $\sum\limits_{k=1}^{\infty} \dfrac{(-1)^k}{k(k+1)}$ converges;

if $x = 1$, $\sum\limits_{k=1}^{\infty} \dfrac{1}{k(k+1)}$ converges. Radius of convergence is 1, interval of convergence is $[-1, 1]$.

33. $\rho = \lim\limits_{k\to+\infty} \dfrac{\sqrt{k}}{\sqrt{k+1}}|x| = |x|$, converges if $|x| < 1$, diverges if $|x| > 1$. If $x = -1$, $\sum\limits_{k=1}^{\infty} \dfrac{-1}{\sqrt{k}}$ diverges; if

$x = 1$, $\sum\limits_{k=1}^{\infty} \dfrac{(-1)^{k-1}}{\sqrt{k}}$ converges. Radius of convergence is 1, interval of convergence is $(-1, 1]$.

35. $\rho = \lim\limits_{k\to+\infty} \dfrac{|x|^2}{(2k+3)(2k+2)} = 0$, radius of convergence is $+\infty$, interval of convergence is $(-\infty, +\infty)$.

37. $\rho = \lim\limits_{k\to+\infty} \dfrac{3|x|}{k+1} = 0$, radius of convergence is $+\infty$, interval of convergence is $(-\infty, +\infty)$.

39. $\rho = \lim\limits_{k\to+\infty} \dfrac{1+k^2}{1+(k+1)^2}|x| = |x|$, converges if $|x| < 1$, diverges if $|x| > 1$. If $x = -1$, $\sum\limits_{k=0}^{\infty} \dfrac{(-1)^k}{1+k^2}$

converges; if $x = 1$, $\sum\limits_{k=0}^{\infty} \dfrac{1}{1+k^2}$ converges. Radius of convergence is 1, interval of convergence is

$[-1, 1]$.

41. $\rho = \lim\limits_{k\to+\infty} \dfrac{k|x+1|}{k+1} = |x+1|$, converges if $|x+1| < 1$, diverges if $|x+1| > 1$. If $x = -2$, $\sum\limits_{k=1}^{\infty} \dfrac{-1}{k}$

diverges; if $x = 0$, $\sum\limits_{k=1}^{\infty} \dfrac{(-1)^{k+1}}{k}$ converges. Radius of convergence is 1, interval of convergence is

$(-2, 0]$.

43. $\rho = \lim\limits_{k\to+\infty} (3/4)|x+5| = \dfrac{3}{4}|x+5|$, converges if $|x+5| < 4/3$, diverges if $|x+5| > 4/3$. If

$x = -19/3$, $\sum\limits_{k=0}^{\infty}(-1)^k$ diverges; if $x = -11/3$, $\sum\limits_{k=0}^{\infty} 1$ diverges. Radius of convergence is $4/3$, interval

of convergence is $(-19/3, -11/3)$.

45. $\rho = \lim\limits_{k\to+\infty} \dfrac{k^2+4}{(k+1)^2+4}|x+1|^2 = |x+1|^2$, converges if $|x+1| < 1$, diverges if $|x+1| > 1$. If $x = -2$,

$\sum\limits_{k=1}^{\infty} \dfrac{(-1)^{3k+1}}{k^2+4}$ converges; if $x = 0$, $\sum\limits_{k=1}^{\infty} \dfrac{(-1)^k}{k^2+4}$ converges. Radius of convergence is 1, interval of

convergence is $[-2, 0]$.

47. $\rho = \lim\limits_{k \to +\infty} \dfrac{\pi|x-1|^2}{(2k+3)(2k+2)} = 0$, radius of convergence $+\infty$, interval of convergence $(-\infty, +\infty)$.

49. $\rho = \lim\limits_{k \to +\infty} \sqrt[k]{|u_k|} = \lim\limits_{k \to +\infty} \dfrac{|x|}{\ln k} = 0$, the series converges absolutely for all x so the interval of convergence is $(-\infty, +\infty)$.

51. If $x \geq 0$, then $\cos \sqrt{x} = 1 - \dfrac{(\sqrt{x})^2}{2!} + \dfrac{(\sqrt{x})^4}{4!} - \dfrac{(\sqrt{x})^6}{6!} + \cdots = 1 - \dfrac{x}{2!} + \dfrac{x^2}{4!} - \dfrac{x^3}{6!} + \cdots$; if $x \leq 0$, then

$\cosh(\sqrt{-x}) = 1 + \dfrac{(\sqrt{-x})^2}{2!} + \dfrac{(\sqrt{-x})^4}{4!} + \dfrac{(\sqrt{-x})^6}{6!} + \cdots = 1 - \dfrac{x}{2!} + \dfrac{x^2}{4!} - \dfrac{x^3}{6!} + \cdots$.

53. By Exercise 76 of Section 3.6, the derivative of an odd (even) function is even (odd); hence all odd-numbered derivatives of an odd function are even, all even-numbered derivatives of an odd function are odd; a similar statement holds for an even function.

   (a) If $f(x)$ is an even function, then $f^{(2k-1)}(x)$ is an odd function, so $f^{(2k-1)}(0) = 0$, and thus the MacLaurin series coefficients $a_{2k-1} = 0, k = 1, 2, \cdots$.

   (b) If $f(x)$ is an odd function, then $f^{(2k)}(x)$ is an odd function, so $f^{(2k)}(0) = 0$, and thus the MacLaurin series coefficients $a_{2k} = 0, k = 1, 2, \cdots$.

55. By Theorem 10.4.3(a) the series $\sum(c_k + d_k)(x - x_0)^k$ converges if $|x - x_0| < R$; if $|x - x_0| > R$ then $\sum(c_k + d_k)(x - x_0)^k$ cannot converge, as otherwise $\sum c_k(x - x_0)^k$ would converge by the same Theorem. Hence the radius of convergence of $\sum(c_k + d_k)(x - x_0)^k$ is R.

57. By the Ratio Test for absolute convergence,

$$\rho = \lim_{k \to +\infty} \frac{(pk+p)!(k!)^p}{(pk)![(k+1)!]^p}|x| = \lim_{k \to +\infty} \frac{(pk+p)(pk+p-1)(pk+p-2)\cdots(pk+p-[p-1])}{(k+1)^p}|x|$$

$$= \lim_{k \to +\infty} p\left(p - \frac{1}{k+1}\right)\left(p - \frac{2}{k+1}\right)\cdots\left(p - \frac{p-1}{k+1}\right)|x| = p^p|x|,$$

converges if $|x| < 1/p^p$, diverges if $|x| > 1/p^p$. Radius of convergence is $1/p^p$.

59. Ratio Test: $\rho = \lim\limits_{k \to +\infty} \dfrac{|x|^2}{4(k+1)(k+2)} = 0$, $R = +\infty$

61. By the Root Test for absolute convergence,

$\rho = \lim\limits_{k \to +\infty} |c_k|^{1/k}|x| = L|x|, L|x| < 1$ if $|x| < 1/L$ so the radius of convergence is $1/L$.

63. The assumption is that $\sum\limits_{k=0}^{\infty} c_k R^k$ is convergent and $\sum\limits_{k=0}^{\infty} c_k(-R)^k$ is divergent. Suppose that $\sum\limits_{k=0}^{\infty} c_k R^k$

is absolutely convergent then $\sum\limits_{k=0}^{\infty} c_k(-R)^k$ is also absolutely convergent and hence convergent

because $|c_k R^k| = |c_k(-R)^k|$, which contradicts the assumption that $\sum\limits_{k=0}^{\infty} c_k(-R)^k$ is divergent so

$\sum\limits_{k=0}^{\infty} c_k R^k$ must be conditionally convergent.

EXERCISE SET 10.9

1. $\sin 4° = \sin\left(\dfrac{\pi}{45}\right) = \dfrac{\pi}{45} - \dfrac{(\pi/45)^3}{3!} + \dfrac{(\pi/45)^5}{5!} - \cdots$

    (a) Method 1: $|R_n(\pi/45)| \le \dfrac{(\pi/45)^{n+1}}{(n+1)!} < 0.000005$ for $n+1 = 4, n = 3$;

    $\sin 4° \approx \dfrac{\pi}{45} - \dfrac{(\pi/45)^3}{3!} \approx 0.069756$

    (b) Method 2: The first term in the alternating series that is less than 0.000005 is $\dfrac{(\pi/45)^5}{5!}$, so the result is the same as in Part (a).

3. $|R_n(0.1)| \le \dfrac{(0.1)^{n+1}}{(n+1)!} \le 0.000005$ for $n = 3$; $\cos 0.1 \approx 1 - (0.1)^2/2 = 0.99500$, calculator value $0.995004\ldots$

5. Expand about $\pi/2$ to get $\sin x = 1 - \dfrac{1}{2!}(x - \pi/2)^2 + \dfrac{1}{4!}(x - \pi/2)^4 - \cdots$, $85° = 17\pi/36$ radians,

    $|R_n(x)| \le \dfrac{|x - \pi/2|^{n+1}}{(n+1)!}$, $|R_n(17\pi/36)| \le \dfrac{|17\pi/36 - \pi/2|^{n+1}}{(n+1)!} = \dfrac{(\pi/36)^{n+1}}{(n+1)!} < 0.5 \times 10^{-4}$

    if $n = 3$, $\sin 85° \approx 1 - \dfrac{1}{2}(-\pi/36)^2 \approx 0.99619$, calculator value $0.99619\ldots$

7. $f^{(k)}(x) = \cosh x$ or $\sinh x$, $|f^{(k)}(x)| \le \cosh x \le \cosh 0.5 = \dfrac{1}{2}\left(e^{0.5} + e^{-0.5}\right) < \dfrac{1}{2}(2 + 1) = 1.5$

    so $|R_n(x)| < \dfrac{1.5(0.5)^{n+1}}{(n+1)!} \le 0.5 \times 10^{-3}$ if $n = 4$, $\sinh 0.5 \approx 0.5 + \dfrac{(0.5)^3}{3!} \approx 0.5208$, calculator

    value $0.52109\ldots$

9. $f(x) = \sin x$, $f^{(n+1)}(x) = \pm \sin x$ or $\pm \cos x$, $|f^{(n+1)}(x)| \le 1$, $|R_n(x)| \le \dfrac{|x - \pi/4|^{n+1}}{(n+1)!}$,

    $\displaystyle\lim_{n \to +\infty} \dfrac{|x - \pi/4|^{n+1}}{(n+1)!} = 0$; by the Squeezing Theorem, $\displaystyle\lim_{n \to +\infty} |R_n(x)| = 0$

    so $\displaystyle\lim_{n \to +\infty} R_n(x) = 0$ for all x.

11. (a) Let $x = 1/9$ in series (12).

    (b) $\ln 1.25 \approx 2\left(1/9 + \dfrac{(1/9)^3}{3}\right) = 2(1/9 + 1/3^7) \approx 0.223$, which agrees with the calculator value $0.22314\ldots$ to three decimal places.

13. (a) $(1/2)^9/9 < 0.5 \times 10^{-3}$ and $(1/3)^7/7 < 0.5 \times 10^{-3}$ so

    $\tan^{-1}(1/2) \approx 1/2 - \dfrac{(1/2)^3}{3} + \dfrac{(1/2)^5}{5} - \dfrac{(1/2)^7}{7} \approx 0.4635$

    $\tan^{-1}(1/3) \approx 1/3 - \dfrac{(1/3)^3}{3} + \dfrac{(1/3)^5}{5} \approx 0.3218$

    (b) From Formula (16), $\pi \approx 4(0.4635 + 0.3218) = 3.1412$

    (c) Let $a = \tan^{-1}\dfrac{1}{2}$, $b = \tan^{-1}\dfrac{1}{3}$; then $|a - 0.4635| < 0.0005$ and $|b - 0.3218| < 0.0005$, so

    $|4(a + b) - 3.1412| \le 4|a - 0.4635| + 4|b - 0.3218| < 0.004$, so two decimal-place accuracy is guaranteed, but not three.

15. (a) $\cos x = 1 - \dfrac{x^2}{2!} + \dfrac{x^4}{4!} + (0)x^5 + R_5(x),$ **(b)**

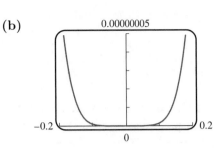

$$|R_5(x)| \le \dfrac{|x|^6}{6!} \le \dfrac{(0.2)^6}{6!} < 9 \times 10^{-8}$$

17. (a) $(1+x)^{-1} = 1 - x + \dfrac{-1(-2)}{2!}x^2 + \dfrac{-1(-2)(-3)}{3!}x^3 + \cdots + \dfrac{-1(-2)(-3)\cdots(-k)}{k!}x^k + \cdots$

$$= \sum_{k=0}^{\infty}(-1)^k x^k$$

(b) $(1+x)^{1/3} = 1 + (1/3)x + \dfrac{(1/3)(-2/3)}{2!}x^2 + \dfrac{(1/3)(-2/3)(-5/3)}{3!}x^3 + \cdots$

$$+ \dfrac{(1/3)(-2/3)\cdots(4-3k)/3}{k!}x^k + \cdots$$

$$= 1 + x/3 + \sum_{k=2}^{\infty}(-1)^{k-1}\dfrac{2\cdot 5\cdots(3k-4)}{3^k k!}x^k$$

(c) $(1+x)^{-3} = 1 - 3x + \dfrac{(-3)(-4)}{2!}x^2 + \dfrac{(-3)(-4)(-5)}{3!}x^3 + \cdots + \dfrac{(-3)(-4)\cdots(-2-k)}{k!}x^k + \cdots$

$$= \sum_{k=0}^{\infty}(-1)^k\dfrac{(k+2)!}{2\cdot k!}x^k = \sum_{k=0}^{\infty}(-1)^k\dfrac{(k+2)(k+1)}{2}x^k$$

19. (a) $\dfrac{d}{dx}\ln(1+x) = \dfrac{1}{1+x}, \dfrac{d^k}{dx^k}\ln(1+x) = (-1)^{k-1}\dfrac{(k-1)!}{(1+x)^k}$; similarly $\dfrac{d}{dx}\ln(1-x) = -\dfrac{(k-1)!}{(1-x)^k}$,

so $f^{(n+1)}(x) = n!\left[\dfrac{(-1)^n}{(1+x)^{n+1}} + \dfrac{1}{(1-x)^{n+1}}\right].$

(b) $\left|f^{(n+1)}(x)\right| \le n!\left|\dfrac{(-1)^n}{(1+x)^{n+1}}\right| + n!\left|\dfrac{1}{(1-x)^{n+1}}\right| = n!\left[\dfrac{1}{(1+x)^{n+1}} + \dfrac{1}{(1-x)^{n+1}}\right]$

(c) If $\left|f^{(n+1)}(x)\right| \le M$ on the interval $[0, 1/3]$ then $|R_n(1/3)| \le \dfrac{M}{(n+1)!}\left(\dfrac{1}{3}\right)^{n+1}.$

(d) If $0 \le x \le 1/3$ then $1+x \ge 1, 1-x \ge 2/3, \left|f^{(n+1)}(x)\right| \le M = n!\left[1 + \dfrac{1}{(2/3)^{n+1}}\right].$

(e) $0.000005 \ge \dfrac{M}{(n+1)!}\left(\dfrac{1}{3}\right)^{n+1} = \dfrac{1}{n+1}\left[\left(\dfrac{1}{3}\right)^{n+1} + \dfrac{(1/3)^{n+1}}{(2/3)^{n+1}}\right] = \dfrac{1}{n+1}\left[\left(\dfrac{1}{3}\right)^{n+1} + \left(\dfrac{1}{2}\right)^{n+1}\right]$

21. $f(x) = \cos x, f^{(n+1)}(x) = \pm\sin x$ or $\pm\cos x, |f^{(n+1)}(x)| \le 1$, set $M = 1$,

$$|R_n(x)| \le \dfrac{1}{(n+1)!}|x-a|^{n+1}, \lim_{n\to+\infty}\dfrac{|x-a|^{n+1}}{(n+1)!} = 0 \text{ so } \lim_{n\to+\infty}R_n(x) = 0 \text{ for all } x.$$

23. $e^{-x} = 1 - x + x^2/2! + \cdots$. Replace x with $-\left(\dfrac{x-100}{16}\right)^2/2$ to obtain

$$e^{-\left(\frac{x-100}{16}\right)^2/2} = 1 - \frac{(x-100)^2}{2 \cdot 16^2} + \frac{(x-100)^4}{8 \cdot 16^4} + \cdots, \text{ thus}$$

$$p \approx \frac{1}{16\sqrt{2\pi}} \int_{100}^{110} \left[1 - \frac{(x-100)^2}{2 \cdot 16^2} + \frac{(x-100)^4}{8 \cdot 16^4} \right] dx \approx 0.23406 \text{ or } 23.406\%.$$

25. (a) From Machin's formula and a CAS, $\frac{\pi}{4} \approx 0.7853981633974483096156608$, accurate to the 25th decimal place.

(b)

n	s_n
0	$0.318309878\ldots$
1	$0.3183098861837906067\ldots$
2	$0.31830988618379067153776695\ldots$
3	$0.3183098861837906715377675267450234\ldots$
$1/\pi$	$0.3183098861837906715377675267450287\ldots$

EXERCISE SET 10.10

1. (a) Replace x with $-x$: $\dfrac{1}{1+x} = 1 - x + x^2 - \cdots + (-1)^k x^k + \cdots$; $R = 1$.

(b) Replace x with x^2: $\dfrac{1}{1-x^2} = 1 + x^2 + x^4 + \cdots + x^{2k} + \cdots$; $R = 1$.

(c) Replace x with $2x$: $\dfrac{1}{1-2x} = 1 + 2x + 4x^2 + \cdots + 2^k x^k + \cdots$; $R = 1/2$.

(d) $\dfrac{1}{2-x} = \dfrac{1/2}{1-x/2}$; replace x with $x/2$: $\dfrac{1}{2-x} = \dfrac{1}{2} + \dfrac{1}{2^2}x + \dfrac{1}{2^3}x^2 + \cdots + \dfrac{1}{2^{k+1}}x^k + \cdots$; $R = 2$.

3. (a) From Section 10.9, Example 4(b), $\dfrac{1}{\sqrt{1+x}} = 1 - \dfrac{1}{2}x + \dfrac{1 \cdot 3}{2^2 \cdot 2!}x^2 - \dfrac{1 \cdot 3 \cdot 5}{2^3 \cdot 3!}x^3 + \cdots$, so

$$(2+x)^{-1/2} = \frac{1}{\sqrt{2}\sqrt{1+x/2}} = \frac{1}{2^{1/2}} - \frac{1}{2^{5/2}}x + \frac{1 \cdot 3}{2^{9/2} \cdot 2!}x^2 - \frac{1 \cdot 3 \cdot 5}{2^{13/2} \cdot 3!}x^3 + \cdots$$

(b) Example 4(a): $\dfrac{1}{(1+x)^2} = 1 - 2x + 3x^2 - 4x^3 + \cdots$, so $\dfrac{1}{(1-x^2)^2} = 1 + 2x^2 + 3x^4 + 4x^6 + \cdots$

5. (a) $2x - \dfrac{2^3}{3!}x^3 + \dfrac{2^5}{5!}x^5 - \dfrac{2^7}{7!}x^7 + \cdots$; $R = +\infty$

(b) $1 - 2x + 2x^2 - \dfrac{4}{3}x^3 + \cdots$; $R = +\infty$

(c) $1 + x^2 + \dfrac{1}{2!}x^4 + \dfrac{1}{3!}x^6 + \cdots$; $R = +\infty$

(d) $x^2 - \dfrac{\pi^2}{2}x^4 + \dfrac{\pi^4}{4!}x^6 - \dfrac{\pi^6}{6!}x^8 + \cdots$; $R = +\infty$

7. (a) $x^2\left(1 - 3x + 9x^2 - 27x^3 + \cdots\right) = x^2 - 3x^3 + 9x^4 - 27x^5 + \cdots$; $R = 1/3$

(b) $x\left(2x + \dfrac{2^3}{3!}x^3 + \dfrac{2^5}{5!}x^5 + \dfrac{2^7}{7!}x^7 + \cdots\right) = 2x^2 + \dfrac{2^3}{3!}x^4 + \dfrac{2^5}{5!}x^6 + \dfrac{2^7}{7!}x^8 + \cdots$; $R = +\infty$

(c) Substitute $3/2$ for m and $-x^2$ for x in Equation (17) of Section 10.9, then multiply by x:

$$x - \frac{3}{2}x^3 + \frac{3}{8}x^5 + \frac{1}{16}x^7 + \cdots; \; R = 1$$

9. (a) $\sin^2 x = \frac{1}{2}(1 - \cos 2x) = \frac{1}{2}\left[1 - \left(1 - \frac{2^2}{2!}x^2 + \frac{2^4}{4!}x^4 - \frac{2^6}{6!}x^6 + \cdots\right)\right]$

$$= x^2 - \frac{2^3}{4!}x^4 + \frac{2^5}{6!}x^6 - \frac{2^7}{8!}x^8 + \cdots$$

(b) $\ln\left[(1+x^3)^{12}\right] = 12\ln(1+x^3) = 12x^3 - 6x^6 + 4x^9 - 3x^{12} + \cdots$

11. (a) $\dfrac{1}{x} = \dfrac{1}{1-(1-x)} = 1 + (1-x) + (1-x)^2 + \cdots + (1-x)^k + \cdots$

$$= 1 - (x-1) + (x-1)^2 - \cdots + (-1)^k(x-1)^k + \cdots$$

(b) $(0, 2)$

13. (a) $(1 + x + x^2/2 + x^3/3! + x^4/4! + \cdots)(x - x^3/3! + x^5/5! - \cdots) = x + x^2 + x^3/3 - x^5/30 + \cdots$

(b) $(1 + x/2 - x^2/8 + x^3/16 - (5/128)x^4 + \cdots)(x - x^2/2 + x^3/3 - x^4/4 + x^5/5 - \cdots)$

$$= x - x^3/24 + x^4/24 - (71/1920)x^5 + \cdots$$

15. (a) $\dfrac{1}{\cos x} = 1 \Big/ \left(1 - \dfrac{1}{2!}x^2 + \dfrac{1}{4!}x^4 - \dfrac{1}{6!}x^6 + \cdots\right) = 1 + \dfrac{1}{2}x^2 + \dfrac{5}{24}x^4 + \dfrac{61}{720}x^6 + \cdots$

(b) $\dfrac{\sin x}{e^x} = \left(x - \dfrac{x^3}{3!} + \dfrac{x^5}{5!} - \cdots\right) \Big/ \left(1 + x + \dfrac{x^2}{2!} + \dfrac{x^3}{3!} + \dfrac{x^4}{4!} + \cdots\right) = x - x^2 + \dfrac{1}{3}x^3 - \dfrac{1}{30}x^5 + \cdots$

17. $e^x = 1 + x + x^2/2 + x^3/3! + \cdots + x^k/k! + \cdots$, $\quad e^{-x} = 1 - x + x^2/2 - x^3/3! + \cdots + (-1)^k x^k/k! + \cdots$;

$$\sinh x = \frac{1}{2}\left(e^x - e^{-x}\right) = x + x^3/3! + x^5/5! + \cdots + x^{2k+1}/(2k+1)! + \cdots, \; R = +\infty$$

$$\cosh x = \frac{1}{2}\left(e^x + e^{-x}\right) = 1 + x^2/2 + x^4/4! + \cdots + x^{2k}/(2k)! + \cdots, \; R = +\infty$$

19. $\dfrac{4x-2}{x^2-1} = \dfrac{-1}{1-x} + \dfrac{3}{1+x} = -\left(1 + x + x^2 + x^3 + x^4 + \cdots\right) + 3\left(1 - x + x^2 - x^3 + x^4 + \cdots\right)$

$$= 2 - 4x + 2x^2 - 4x^3 + 2x^4 + \cdots$$

21. (a) $\dfrac{d}{dx}\left(1 - x^2/2! + x^4/4! - x^6/6! + \cdots\right) = -x + x^3/3! - x^5/5! + \cdots = -\sin x$

(b) $\dfrac{d}{dx}\left(x - x^2/2 + x^3/3 - \cdots\right) = 1 - x + x^2 - \cdots = 1/(1+x)$

23. (a) $\displaystyle\int \left(1 + x + x^2/2! + \cdots\right) dx = \left(x + x^2/2! + x^3/3! + \cdots\right) + C_1$

$$= \left(1 + x + x^2/2! + x^3/3! + \cdots\right) + C_1 - 1 = e^x + C$$

(b) $\displaystyle\int \left(x + x^3/3! + x^5/5! + \cdots\right) = x^2/2! + x^4/4! + \cdots + C_1$

$$= 1 + x^2/2! + x^4/4! + \cdots + C_1 - 1 = \cosh x + C$$

25. **(a)** Substitute x^2 for x in the Maclaurin Series for $1/(1-x)$ (Table 10.9.1)

and then multiply by x: $\dfrac{x}{1-x^2} = x \displaystyle\sum_{k=0}^{\infty} (x^2)^k = \sum_{k=0}^{\infty} x^{2k+1}$

(b) $f^{(5)}(0) = 5!c_5 = 5!$, $f^{(6)}(0) = 6!c_6 = 0$ **(c)** $f^{(n)}(0) = n!c_n = \begin{cases} n! & \text{if } n \text{ odd} \\ 0 & \text{if } n \text{ even} \end{cases}$

27. **(a)** $\displaystyle\lim_{x\to 0} \frac{\sin x}{x} = \lim_{x\to 0} \left(1 - x^2/3! + x^4/5! - \cdots\right) = 1$

(b) $\displaystyle\lim_{x\to 0} \frac{\tan^{-1} x - x}{x^3} = \lim_{x\to 0} \frac{(x - x^3/3 + x^5/5 - x^7/7 + \cdots) - x}{x^3} = -1/3$

29. $\displaystyle\int_0^1 \sin\left(x^2\right) dx = \int_0^1 \left(x^2 - \frac{1}{3!}x^6 + \frac{1}{5!}x^{10} - \frac{1}{7!}x^{14} + \cdots\right) dx$

$= \dfrac{1}{3}x^3 - \dfrac{1}{7\cdot 3!}x^7 + \dfrac{1}{11\cdot 5!}x^{11} - \dfrac{1}{15\cdot 7!}x^{15} + \cdots \Big]_0^1$

$= \dfrac{1}{3} - \dfrac{1}{7\cdot 3!} + \dfrac{1}{11\cdot 5!} - \dfrac{1}{15\cdot 7!} + \cdots,$

but $\dfrac{1}{15\cdot 7!} < 0.5 \times 10^{-3}$ so $\displaystyle\int_0^1 \sin(x^2)dx \approx \dfrac{1}{3} - \dfrac{1}{7\cdot 3!} + \dfrac{1}{11\cdot 5!} \approx 0.3103$

31. $\displaystyle\int_0^{0.2} \left(1+x^4\right)^{1/3} dx = \int_0^{0.2} \left(1 + \frac{1}{3}x^4 - \frac{1}{9}x^8 + \cdots\right) dx$

$= x + \dfrac{1}{15}x^5 - \dfrac{1}{81}x^9 + \cdots \Big]_0^{0.2} = 0.2 + \dfrac{1}{15}(0.2)^5 - \dfrac{1}{81}(0.2)^9 + \cdots,$

but $\dfrac{1}{15}(0.2)^5 < 0.5 \times 10^{-3}$ so $\displaystyle\int_0^{0.2} (1+x^4)^{1/3}dx \approx 0.200$

33. **(a)** Substitute x^4 for x in the MacLaurin Series for e^x to obtain $\displaystyle\sum_{k=0}^{+\infty} \frac{x^{4k}}{k!}$. The radius of convergence is $R = +\infty$.

(b) The first method is to multiply the MacLaurin Series for e^{x^4} by x^3: $x^3 e^{x^4} = \displaystyle\sum_{k=0}^{+\infty} \frac{x^{4k+3}}{k!}$. The second method involves differentiation: $\dfrac{d}{dx} e^{x^4} = 4x^3 e^{x^4}$, so

$x^3 e^{x^4} = \dfrac{1}{4}\dfrac{d}{dx} e^{x^4} = \dfrac{1}{4}\dfrac{d}{dx} \displaystyle\sum_{k=0}^{+\infty} \frac{x^{4k}}{k!} = \dfrac{1}{4} \sum_{k=0}^{+\infty} \frac{4kx^{4k-1}}{k!} = \sum_{k=0}^{+\infty} \frac{x^{4k-1}}{(k-1)!}$. Use the change of variable $j = k-1$ to show equality of the two series.

35. **(a)** In Exercise 34(a), set $x = \dfrac{1}{3}$, $S = \dfrac{1/3}{(1-1/3)^2} = \dfrac{3}{4}$

(b) In Part (b) set $x = 1/4, S = \ln(4/3)$

37. **(a)** $\sinh^{-1} x = \displaystyle\int (1+x^2)^{-1/2} dx - C = \int \left(1 - \frac{1}{2}x^2 + \frac{3}{8}x^4 - \frac{5}{16}x^6 + \cdots\right) dx - C$

$= \left(x - \dfrac{1}{6}x^3 + \dfrac{3}{40}x^5 - \dfrac{5}{112}x^7 + \cdots\right) - C;\ \sinh^{-1} 0 = 0$ so $C = 0.$

(b) $\left(1+x^2\right)^{-1/2} = 1 + \displaystyle\sum_{k=1}^{\infty} \frac{(-1/2)(-3/2)(-5/2)\cdots(-1/2-k+1)}{k!}\left(x^2\right)^k$

$$= 1 + \sum_{k=1}^{\infty}(-1)^k\frac{1\cdot3\cdot5\cdots(2k-1)}{2^k k!}x^{2k},$$

$$\sinh^{-1}x = x + \sum_{k=1}^{\infty}(-1)^k\frac{1\cdot3\cdot5\cdots(2k-1)}{2^k k!(2k+1)}x^{2k+1}$$

(c) $R = 1$

39. (a) $y(t) = y_0\displaystyle\sum_{k=0}^{\infty}\frac{(-1)^k(0.000121)^k t^k}{k!}$

(b) $y(1) \approx y_0(1 - 0.000121t)\Big]_{t=1} = 0.999879y_0$

(c) $y_0 e^{-0.000121} \approx 0.9998790073y_0$

41. $\theta_0 = 5° = \pi/36$ rad, $k = \sin(\pi/72)$

(b) $T \approx 2\pi\sqrt{\dfrac{L}{g}} = 2\pi\sqrt{1/9.8} \approx 2.00709$ ⠀⠀⠀⠀ **(b)** $T \approx 2\pi\sqrt{\dfrac{L}{g}}\left(1 + \dfrac{k^2}{4}\right) \approx 2.008044621$

(c) 2.008045644

43. (a) $F = \dfrac{mgR^2}{(R+h)^2} = \dfrac{mg}{(1+h/R)^2} = mg\left(1 - 2h/R + 3h^2/R^2 - 4h^3/R^3 + \cdots\right)$

(b) If $h = 0$, then the binomial series converges to 1 and $F = mg$.

(c) Sum the series to the linear term, $F \approx mg - 2mgh/R$.

(d) $\dfrac{mg - 2mgh/R}{mg} = 1 - \dfrac{2h}{R} = 1 - \dfrac{2\cdot 29{,}028}{4000\cdot 5280} \approx 0.9973$, so about 0.27% less.

45. Suppose not, and suppose that k_0 is the first integer for which $a_k \neq b_k$. Then $a_{k_0}x^{k_0} + a_{k_0+1}x^{k_0+1} + \ldots = b_{k_0}x^{k_0} + b_{k_0+1}x^{k_0+1} + \ldots$. Divide by x^{k_0} and let $x \to 0$ to show that $a_{k_0} = b_{k_0}$ which contradicts the assumption that they were not equal. Thus $a_k = b_k$ for all k.

REVIEW EXERCISES, CHAPTER 10

9. (a) always true by Theorem 10.4.2

(b) sometimes false, for example the harmonic series diverges but $\sum(1/k^2)$ converges

(c) sometimes false, for example $f(x) = \sin\pi x, a_k = 0, L = 0$

(d) always true by the comments which follow Example 3(d) of Section 10.1

(e) sometimes false, for example $a_n = \dfrac{1}{2} + (-1)^n\dfrac{1}{4}$

(f) sometimes false, for example $u_k = 1/2$

(g) always false by Theorem 10.4.3

(h) sometimes false, for example $u_k = 1/k, v_k = 2/k$

(i) always true by the Comparison Test

(j) always true by the Comparison Test

(k) sometimes false, for example $\sum (-1)^k/k$

(l) sometimes false, for example $\sum (-1)^k/k$

11. **(a)** $a_n = \dfrac{n+2}{(n+1)^2 - n^2} = \dfrac{n+2}{((n+1)+n)((n+1)-n)} = \dfrac{n+2}{2n+1}$, limit $= 1/2$.

(b) $a_n = (-1)^{n-1}\dfrac{n}{2n+1}$, diverges by the Divergence Test (Theorem 10.4.1)

13. **(a)** $a_{n+1}/a_n = (n+1-10)^4/(n-10)^4 = (n-9)^4/(n-10)^4$. Since $n-9 > n-10$ for all n it follows that $(n-9)^4 > (n-10)^4$ and thus that $a_{n+1}/a_n > 1$ for all n, hence the sequence is strictly monotone increasing.

(b) $\dfrac{100^{n+1}}{(2(n+1))!(n+1)!} \times \dfrac{(2n)!n!}{100^n} = \dfrac{100}{(2n+2)(2n+1)(n+1)} < 1$ for $n \geq 3$, so the sequence is ultimately strictly monotone decreasing.

15. **(a)** geometric, $r = 1/5$, converges   **(b)** $1/(5^k + 1) < 1/5^k$, converges

17. **(a)** $\dfrac{1}{k^3 + 2k + 1} < \dfrac{1}{k^3}$, $\displaystyle\sum_{k=1}^{\infty} 1/k^3$ converges, so $\displaystyle\sum_{k=1}^{\infty} \dfrac{1}{k^3 + 2k + 1}$ converges by the Comparison Test

(b) Limit Comparison Test, compare with the divergent series $\displaystyle\sum_{k=1}^{\infty} \dfrac{1}{k^{2/5}}$, diverges

19. **(a)** $\dfrac{9}{\sqrt{k}+1} \geq \dfrac{9}{\sqrt{k}+\sqrt{k}} = \dfrac{9}{2\sqrt{k}}$, $\displaystyle\sum_{k=1}^{\infty} \dfrac{9}{2\sqrt{k}}$ diverges

(b) converges absolutely, because $\left|\dfrac{\cos(1/k)}{k^2}\right| \leq \dfrac{1}{k^2}$ and $\displaystyle\sum_{k=1}^{+\infty} \dfrac{1}{k^2}$ converges

21. $\displaystyle\sum_{k=0}^{\infty} \dfrac{1}{5^k} - \sum_{k=0}^{99} \dfrac{1}{5^k} = \sum_{k=100}^{\infty} \dfrac{1}{5^k} = \dfrac{1}{5^{100}} \sum_{k=0}^{\infty} \dfrac{1}{5^k} = \dfrac{1}{4 \cdot 5^{99}}$

23. **(a)** $\displaystyle\sum_{k=1}^{\infty} \left(\dfrac{3}{2^k} - \dfrac{2}{3^k}\right) = \sum_{k=1}^{\infty} \dfrac{3}{2^k} - \sum_{k=1}^{\infty} \dfrac{2}{3^k} = \left(\dfrac{3}{2}\right)\dfrac{1}{1-(1/2)} - \left(\dfrac{2}{3}\right)\dfrac{1}{1-(1/3)} = 2$ (geometric series)

(b) $\displaystyle\sum_{k=1}^{n} [\ln(k+1) - \ln k] = \ln(n+1)$, so $\displaystyle\sum_{k=1}^{\infty} [\ln(k+1) - \ln k] = \lim_{n\to+\infty} \ln(n+1) = +\infty$, diverges

(c) $\displaystyle\lim_{n\to+\infty} \sum_{k=1}^{n} \dfrac{1}{2}\left(\dfrac{1}{k} - \dfrac{1}{k+2}\right) = \lim_{n\to+\infty} \dfrac{1}{2}\left(1 + \dfrac{1}{2} - \dfrac{1}{n+1} - \dfrac{1}{n+2}\right) = \dfrac{3}{4}$

(d) $\displaystyle\lim_{n\to+\infty} \sum_{k=1}^{n} \left[\tan^{-1}(k+1) - \tan^{-1} k\right] = \lim_{n\to+\infty} \left[\tan^{-1}(n+1) - \tan^{-1}(1)\right] = \dfrac{\pi}{2} - \dfrac{\pi}{4} = \dfrac{\pi}{4}$

25. Compare with $1/k^p$: converges if $p > 1$, diverges otherwise.

27. **(a)** $1 \leq k, 2 \leq k, 3 \leq k, \ldots, k \leq k$, therefore $1 \cdot 2 \cdot 3 \cdots k \leq k \cdot k \cdot k \cdots k$, or $k! \leq k^k$.

(b) $\displaystyle\sum \frac{1}{k^k} \le \sum \frac{1}{k!}$, converges

(c) $\displaystyle\lim_{k \to +\infty} \left(\frac{1}{k^k}\right)^{1/k} = \lim_{k \to +\infty} \frac{1}{k} = 0$, converges

29. (a) $p_0(x) = 1, p_1(x) = 1 - 7x, p_2(x) = 1 - 7x + 5x^2, p_3(x) = 1 - 7x + 5x^2 + 4x^3,$
$p_4(x) = 1 - 7x + 5x^2 + 4x^3$

(b) If $f(x)$ is a polynomial of degree n and $k \ge n$ then the Maclaurin polynomial of degree k is the polynomial itself; if $k < n$ then it is the truncated polynomial.

31. $\ln(1 + x) = x - x^2/2 + \cdots$; so $|\ln(1 + x) - x| \le x^2/2$ by Theorem 10.6.2.

33. (a) $e^2 - 1$ **(b)** $\sin \pi = 0$ **(c)** $\cos e$ **(d)** $e^{-\ln 3} = 1/3$

35. $(27+x)^{1/3} = 3(1+x/3^3)^{1/3} = 3\left(1 + \dfrac{1}{3^4}x - \dfrac{1 \cdot 2}{3^8 2}x^2 + \dfrac{1 \cdot 2 \cdot 5}{3^{12} 3!}x^3 + \dots\right)$, alternates after first term,

$\dfrac{3 \cdot 2}{3^8 2} < 0.0005$, $\sqrt[3]{28} \approx 3\left(1 + \dfrac{1}{3^4}\right) \approx 3.0370$

37. Both (a) and (b): $x, -\dfrac{2}{3}x^3, \dfrac{2}{15}x^5, -\dfrac{4}{315}x^7$

CHAPTER 11
Analytic Geometry in Calculus

EXERCISE SET 11.1

1.

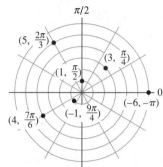

3. (a) $(3\sqrt{3}, 3)$ (b) $(-7/2, 7\sqrt{3}/2)$ (c) $(3\sqrt{3}, 3)$

    (d) $(0, 0)$ (e) $(-7\sqrt{3}/2, 7/2)$ (f) $(-5, 0)$

5. (a) $(5, \pi), (5, -\pi)$ (b) $(4, 11\pi/6), (4, -\pi/6)$ (c) $(2, 3\pi/2), (2, -\pi/2)$

    (d) $(8\sqrt{2}, 5\pi/4), (8\sqrt{2}, -3\pi/4)$ (e) $(6, 2\pi/3), (6, -4\pi/3)$ (f) $(\sqrt{2}, \pi/4), (\sqrt{2}, -7\pi/4)$

7. (a) $(5, 0.9273)$ (b) $(10, -0.92730)$ (c) $(1.27155, -0.66577)$

9. (a) $r^2 = x^2 + y^2 = 4$; circle   (b) $y = 4$; horizontal line

    (c) $r^2 = 3r\cos\theta$, $x^2 + y^2 = 3x$, $(x - 3/2)^2 + y^2 = 9/4$; circle

    (d) $3r\cos\theta + 2r\sin\theta = 6$, $3x + 2y = 6$; line

11. (a) $r\cos\theta = 3$ (b) $r = \sqrt{7}$

    (c) $r^2 + 6r\sin\theta = 0$, $r = -6\sin\theta$

    (d) $9(r\cos\theta)(r\sin\theta) = 4$, $9r^2\sin\theta\cos\theta = 4$, $r^2\sin 2\theta = 8/9$

13.

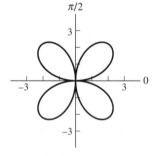

$r = 3\sin 2\theta$

15.

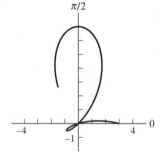

$r = 3 - 4\sin 3\theta$

17. (a) $r = 5$

    (b) $(x - 3)^2 + y^2 = 9$, $r = 6\cos\theta$

    (c) Example 6, $r = 1 - \cos\theta$

19. (a) Figure 11.1.18, $a = 3, n = 2, r = 3\sin 2\theta$

(b) From (8-9), symmetry about the y-axis and Theorem 11.1.1(b), the equation is of the form $r = a \pm b \sin \theta$. The cartesian points $(3,0)$ and $(0,5)$ give $a = 3$ and $5 = a + b$, so $b = 2$ and $r = 3 + 2 \sin \theta$.

(c) Example 8, $r^2 = 9 \cos 2\theta$

21.

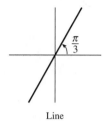

Line

23.

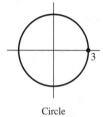

Circle

25.

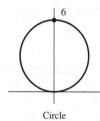

Circle

27.

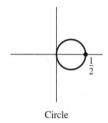

Circle

29.

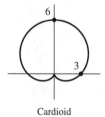

Cardioid

31.

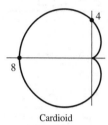

Cardioid

33.

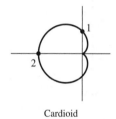

Cardioid

35.

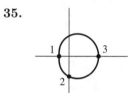

Limaçon

37.

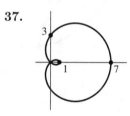

Limaçon

39.

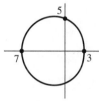

Limaçon

41.

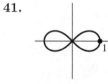

Lemniscate

43.

Lemniscate

45.

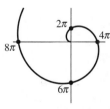

Spiral

47.

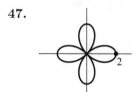

Four-petal rose

49.

Eight-petal rose

51.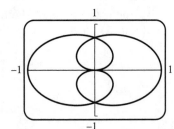

53.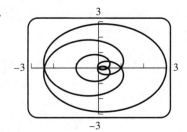

55.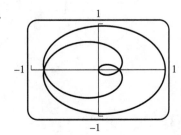

57. (a) $-4\pi < \theta < 4\pi$

59. (a) $r = \dfrac{a}{\cos\theta}, r\cos\theta = a, x = a$ (b) $r\sin\theta = b, y = b$

61. (a) 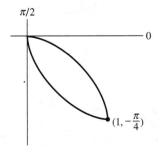 (b)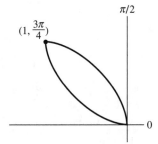

(c) 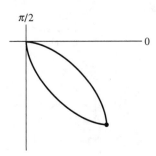 (d)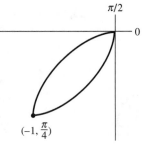

(e)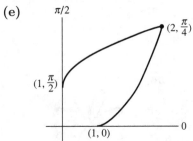

65. (a) $r = 1 + \cos(\theta - \pi/4) = 1 + \dfrac{\sqrt{2}}{2}(\cos\theta + \sin\theta)$

(b) $r = 1 + \cos(\theta - \pi/2) = 1 + \sin\theta$

(c) $r = 1 + \cos(\theta - \pi) = 1 - \cos\theta$

(d) $r = 1 + \cos(\theta - 5\pi/4) = 1 - \dfrac{\sqrt{2}}{2}(\cos\theta + \sin\theta)$

67. Either $r - 1 = 0$ or $\theta - 1 = 0$, so the graph consists of the circle $r = 1$ and the line $\theta = 1$.

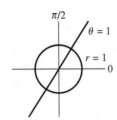

69. $y = r\sin\theta = (1 + \cos\theta)\sin\theta = \sin\theta + \sin\theta\cos\theta$,
$dy/d\theta = \cos\theta - \sin^2\theta + \cos^2\theta = 2\cos^2\theta + \cos\theta - 1 = (2\cos\theta - 1)(\cos\theta + 1)$;
$dy/d\theta = 0$ if $\cos\theta = 1/2$ or if $\cos\theta = -1$;
$\theta = \pi/3$ or π (or $\theta = -\pi/3$, which leads to the minimum point).
If $\theta = \pi/3, \pi$, then $y = 3\sqrt{3}/4, 0$ so the maximum value of y is $3\sqrt{3}/4$ and the polar coordinates of the highest point are $(3/2, \pi/3)$.

71. The width is twice the maximum value of y for $0 \le \theta \le \pi/4$:
$y = r\sin\theta = \sin\theta\cos 2\theta = \sin\theta - 2\sin^3\theta, dy/d\theta = \cos\theta - 6\sin^2\theta\cos\theta = 0$ when $\cos\theta = 0$ or $\sin\theta = 1/\sqrt{6}, y = 1/\sqrt{6} - 2/(6\sqrt{6}) = \sqrt{6}/9$, so the width of the petal is $2\sqrt{6}/9$.

73. (a) Let (x_1, y_1) and (x_2, y_2) be the rectangular coordinates of the points (r_1, θ_1) and (r_2, θ_2) then
$$d = \sqrt{(x_2 - x_1)^2 + (y_2 - y_1)^2} = \sqrt{(r_2\cos\theta_2 - r_1\cos\theta_1)^2 + (r_2\sin\theta_2 - r_1\sin\theta_1)^2}$$
$$= \sqrt{r_1^2 + r_2^2 - 2r_1 r_2(\cos\theta_1\cos\theta_2 + \sin\theta_1\sin\theta_2)} = \sqrt{r_1^2 + r_2^2 - 2r_1 r_2\cos(\theta_1 - \theta_2)}.$$
An alternate proof follows directly from the Law of Cosines.

(b) Let P and Q have polar coordinates $(r_1, \theta_1), (r_2, \theta_2)$, respectively, then the perpendicular from OQ to OP has length $h = r_2\sin(\theta_2 - \theta_1)$ and $A = \frac{1}{2}hr_1 = \frac{1}{2}r_1 r_2\sin(\theta_2 - \theta_1)$.

(c) From Part (a), $d = \sqrt{9 + 4 - 2 \cdot 3 \cdot 2\cos(\pi/6 - \pi/3)} = \sqrt{13 - 6\sqrt{3}} \approx 1.615$

(d) $A = \dfrac{1}{2}2\sin(5\pi/6 - \pi/3) = 1$

75. The tips are located at $r = 1, \theta = \pi/6, 5\pi/6, 3\pi/2$ and, for example,
$$d = \sqrt{1 + 1 - 2\cos(5\pi/6 - \pi/6)} = \sqrt{2(1 - \cos(2\pi/3))} = \sqrt{3}$$

77. $\displaystyle\lim_{\theta \to 0^+} y = \lim_{\theta \to 0^+} r\sin\theta = \lim_{\theta \to 0^+} \dfrac{\sin\theta}{\theta} = 1$, and $\displaystyle\lim_{\theta \to 0^+} x = \lim_{\theta \to 0^+} r\cos\theta = \lim_{\theta \to 0^+} \dfrac{\cos\theta}{\theta} = +\infty$.

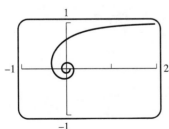

79. Note that $r \to \pm\infty$ as θ approaches odd multiples of $\pi/2$;
$x = r\cos\theta = 4\tan\theta\cos\theta = 4\sin\theta$,
$y = r\sin\theta = 4\tan\theta\sin\theta$
so $x \to \pm 4$ and $y \to \pm\infty$ as θ approaches
odd multiples of $\pi/2$.

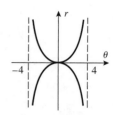

81. Let $r = a\sin n\theta$ (the proof for $r = a\cos n\theta$ is similar). If θ starts at 0, then θ would have to increase by some positive integer multiple of π radians in order to reach the starting point and begin to retrace the curve. Let (r, θ) be the coordinates of a point P on the curve for $0 \le \theta < 2\pi$. Now $a\sin n(\theta + 2\pi) = a\sin(n\theta + 2\pi n) = a\sin n\theta = r$ so P is reached again with coordinates $(r, \theta + 2\pi)$ thus the curve is traced out either exactly once or exactly twice for $0 \le \theta < 2\pi$. If for $0 \le \theta < \pi$, $P(r, \theta)$ is reached again with coordinates $(-r, \theta + \pi)$ then the curve is traced out exactly once for $0 \le \theta < \pi$, otherwise exactly once for $0 \le \theta < 2\pi$. But

$$a\sin n(\theta + \pi) = a\sin(n\theta + n\pi) = \begin{cases} a\sin n\theta, & n \text{ even} \\ -a\sin n\theta, & n \text{ odd} \end{cases}$$

so the curve is traced out exactly once for $0 \le \theta < 2\pi$ if n is even, and exactly once for $0 \le \theta < \pi$ if n is odd.

EXERCISE SET 11.2

1. (a) $dy/dx = \dfrac{2t}{1/2} = 4t$; $dy/dx\big|_{t=-1} = -4$; $dy/dx\big|_{t=1} = 4$

(b) $y = (2x)^2 + 1, dy/dx = 8x, dy/dx\big|_{x=\pm(1/2)} = \pm 4$

3. $\dfrac{d^2y}{dx^2} = \dfrac{d}{dx}\dfrac{dy}{dx} = \dfrac{d}{dt}\left(\dfrac{dy}{dx}\right)\dfrac{dt}{dx} = -\dfrac{1}{4t^2}(1/2t) = -1/(8t^3)$; positive when $t = -1$,
negative when $t = 1$

5. $dy/dx = \dfrac{2}{1/(2\sqrt{t})} = 4\sqrt{t}, d^2y/dx^2 = \dfrac{2/\sqrt{t}}{1/(2\sqrt{t})} = 4, dy/dx\big|_{t=1} = 4, d^2y/dx^2\big|_{t=1} = 4$

7. $dy/dx = \dfrac{\sec^2 t}{\sec t \tan t} = \csc t, d^2y/dx^2 = \dfrac{-\csc t \cot t}{\sec t \tan t} = -\cot^3 t$,
$dy/dx\big|_{t=\pi/3} = 2/\sqrt{3}, d^2y/dx^2\big|_{t=\pi/3} = -1/(3\sqrt{3})$

9. $\dfrac{dy}{dx} = \dfrac{dy/d\theta}{dx/d\theta} = \dfrac{\cos\theta}{1 - \sin\theta}$;
$\dfrac{d^2y}{dx^2} = \dfrac{d}{d\theta}\left(\dfrac{dy}{dx}\right)\Big/\dfrac{dx}{d\theta} = \dfrac{(1 - \sin\theta)(-\sin\theta) + \cos^2\theta}{(1 - \sin\theta)^2}\dfrac{1}{1 - \sin\theta} = \dfrac{1}{(1 - \sin\theta)^2}$;
$\dfrac{dy}{dx}\Big|_{\theta=\pi/6} = \dfrac{\sqrt{3}/2}{1 - 1/2} = \sqrt{3}; \dfrac{d^2y}{dx^2}\Big|_{\theta=\pi/6} = \dfrac{1}{(1 - 1/2)^2} = 4$

11. (a) $dy/dx = \dfrac{-e^{-t}}{e^t} = -e^{-2t}$; for $t = 1, dy/dx = -e^{-2}, (x, y) = (e, e^{-1}); y - e^{-1} = -e^{-2}(x - e)$,
$y = -e^{-2}x + 2e^{-1}$

(b) $y = 1/x$, $dy/dx = -1/x^2$, $m = -1/e^2$, $y - e^{-1} = -\dfrac{1}{e^2}(x - e)$, $y = -\dfrac{1}{e^2}x + \dfrac{2}{e}$

13. $dy/dx = \dfrac{-4 \sin t}{2 \cos t} = -2 \tan t$

    (a) $dy/dx = 0$ if $\tan t = 0$, $t = n\pi$ for $n = 0, \pm 1, \cdots$

    (b) $dx/dy = -\dfrac{1}{2}\cot t = 0$ if $\cot t = 0$, $t = \pi/2 + n\pi$ for $n = 0, \pm 1, \cdots$

15. $x = y = 0$ when $t = 0, \pi$; $\dfrac{dy}{dx} = \dfrac{2 \cos 2t}{\cos t}$; $\left.\dfrac{dy}{dx}\right|_{t=0} = 2$, $\left.\dfrac{dy}{dx}\right|_{t=\pi} = -2$, the equations of the tangent lines are $y = -2x$, $y = 2x$.

17. If $x = 4$ then $t^2 = 4$, $t = \pm 2$, $y = 0$ for $t = \pm 2$ so $(4, 0)$ is reached when $t = \pm 2$. $dy/dx = (3t^2 - 4)/2t$. For $t = 2$, $dy/dx = 2$ and for $t = -2$, $dy/dx = -2$. The tangent lines are $y = \pm 2(x - 4)$.

19. **(a)**

    (b) $\dfrac{dx}{dt} = -3 \cos^2 t \sin t$ and $\dfrac{dy}{dt} = 3 \sin^2 t \cos t$ are both zero when $t = 0, \pi/2, \pi, 3\pi/2, 2\pi$, so singular points occur at these values of t.

21. Substitute $\theta = \pi/6$, $r = 1$, and $dr/d\theta = \sqrt{3}$ in equation (7) gives slope $m = \sqrt{3}$.

23. As in Exercise 21, $\theta = 2$, $dr/d\theta = -1/4$, $r = 1/2$, $m = \dfrac{\tan 2 - 2}{2 \tan 2 + 1}$

25. As in Exercise 21, $\theta = \pi/4$, $dr/d\theta = -3\sqrt{2}/2$, $r = \sqrt{2}/2$, $m = 1/2$

27. $m = \dfrac{dy}{dx} = \dfrac{r \cos \theta + (\sin \theta)(dr/d\theta)}{-r \sin \theta + (\cos \theta)(dr/d\theta)} = \dfrac{\cos \theta + 2 \sin \theta \cos \theta}{-\sin \theta + \cos^2 \theta - \sin^2 \theta}$; if $\theta = 0, \pi/2, \pi$, then $m = 1, 0, -1$.

29. $dx/d\theta = -a \sin \theta (1 + 2 \cos \theta)$, $dy/d\theta = a(2 \cos \theta - 1)(\cos \theta + 1)$

    (a) horizontal if $dy/d\theta = 0$ and $dx/d\theta \neq 0$. $dy/d\theta = 0$ when $\cos \theta = 1/2$ or $\cos \theta = -1$ so $\theta = \pi/3$, $5\pi/3$, or π; $dx/d\theta \neq 0$ for $\theta = \pi/3$ and $5\pi/3$. For the singular point $\theta = \pi$ we find that $\lim\limits_{\theta \to \pi} dy/dx = 0$. There is a horizontal tangent line at $(3a/2, \pi/3), (0, \pi)$, and $(3a/2, 5\pi/3)$.

    (b) vertical if $dy/d\theta \neq 0$ and $dx/d\theta = 0$. $dx/d\theta = 0$ when $\sin \theta = 0$ or $\cos \theta = -1/2$ so $\theta = 0, \pi$, $2\pi/3$, or $4\pi/3$; $dy/d\theta \neq 0$ for $\theta = 0, 2\pi/3$, and $4\pi/3$. The singular point $\theta = \pi$ was discussed in Part (a). There is a vertical tangent line at $(2a, 0), (a/2, 2\pi/3)$, and $(a/2, 4\pi/3)$.

31. $dy/d\theta = (d/d\theta)(\sin^2 \theta \cos^2 \theta) = (\sin 4\theta)/2 = 0$ at $\theta = 0, \pi/4, \pi/2, 3\pi/4, \pi$; at the same points, $dx/d\theta = (d/d\theta)(\sin \theta \cos^3 \theta) = \cos^2 \theta (4 \cos^2 \theta - 3)$. Next, $\dfrac{dx}{d\theta} = 0$ at $\theta = \pi/2$, a singular point; and $\theta = 0, \pi$ both give the same point, so there are just three points with a horizontal tangent.

33.

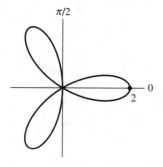

$\theta_0 = \pi/6, \pi/2, 5\pi/6,$
$y = \pm x/\sqrt{3}, x = 0$

35.

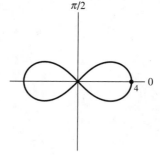

$\theta_0 = \pm\pi/4, y = \pm x$

37.

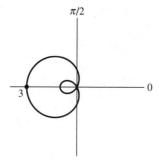

$\theta_0 = 2\pi/3, 4\pi/3, y = \pm\sqrt{3}x$

39. $r^2 + (dr/d\theta)^2 = a^2 + 0^2 = a^2$, $L = \int_0^{2\pi} a\,d\theta = 2\pi a$

41. $r^2 + (dr/d\theta)^2 = [a(1 - \cos\theta)]^2 + [a\sin\theta]^2 = 4a^2\sin^2(\theta/2)$, $L = 2\int_0^{\pi} 2a\sin(\theta/2)d\theta = 8a$

43. $r^2 + (dr/d\theta)^2 = (e^{3\theta})^2 + (3e^{3\theta})^2 = 10e^{6\theta}$, $L = \int_0^2 \sqrt{10}e^{3\theta}d\theta = \sqrt{10}(e^6 - 1)/3$

45. (a) From (3), $\dfrac{dy}{dx} = \dfrac{3\sin t}{1 - 3\cos t}$

(b) At $t = 10$, $\dfrac{dy}{dx} = \dfrac{3\sin 10}{1 - 3\cos 10} \approx -0.46402$, $\theta \approx \tan^{-1}(-0.46402) = -0.4345$

47. (a) $r^2 + (dr/d\theta)^2 = (\cos n\theta)^2 + (-n\sin n\theta)^2 = \cos^2 n\theta + n^2\sin^2 n\theta$
$$= (1 - \sin^2 n\theta) + n^2\sin^2 n\theta = 1 + (n^2 - 1)\sin^2 n\theta,$$
$$L = 2\int_0^{\pi/(2n)} \sqrt{1 + (n^2 - 1)\sin^2 n\theta}\,d\theta$$

(b) $L = 2\int_0^{\pi/4} \sqrt{1 + 3\sin^2 2\theta}\,d\theta \approx 2.42$

(c)

n	2	3	4	5	6	7	8	9	10	11
L	2.42211	2.22748	2.14461	2.10100	2.07501	2.05816	2.04656	2.03821	2.03199	2.02721

n	12	13	14	15	16	17	18	19	20
L	2.02346	2.02046	2.01802	2.01600	2.01431	2.01288	2.01167	2.01062	2.00971

49. $x' = 2t,\ y' = 3,\ (x')^2 + (y')^2 = 4t^2 + 9$

$$S = 2\pi \int_0^2 (3t)\sqrt{4t^2 + 9}\,dt = 6\pi \int_0^4 t\sqrt{4t^2 + 9}\,dt = \frac{\pi}{2}(4t^2 + 9)^{3/2}\Big]_0^2 = \frac{\pi}{2}(125 - 27) = 49\pi$$

51. $x' = -2\sin t\cos t,\ y' = 2\sin t\cos t,\ (x')^2 + (y')^2 = 8\sin^2 t\cos^2 t$

$$S = 2\pi \int_0^{\pi/2} \cos^2 t\sqrt{8\sin^2 t\cos^2 t}\,dt = 4\sqrt{2}\pi \int_0^{\pi/2} \cos^3 t\sin t\,dt = -\sqrt{2}\pi\cos^4 t\Big]_0^{\pi/2} = \sqrt{2}\pi$$

53. $x' = -r\sin t,\ y' = r\cos t,\ (x')^2 + (y')^2 = r^2,\ S = 2\pi \int_0^\pi r\sin t\sqrt{r^2}\,dt = 2\pi r^2 \int_0^\pi \sin t\,dt = 4\pi r^2$

55. (a) $\dfrac{dr}{dt} = 2$ and $\dfrac{d\theta}{dt} = 1$ so $\dfrac{dr}{d\theta} = \dfrac{dr/dt}{d\theta/dt} = \dfrac{2}{1} = 2,\ r = 2\theta + C,\ r = 10$ when $\theta = 0$ so

$10 = C, r = 2\theta + 10.$

(b) $r^2 + (dr/d\theta)^2 = (2\theta + 10)^2 + 4$, during the first 5 seconds the rod rotates through an angle

of $(1)(5) = 5$ radians so $L = \displaystyle\int_0^5 \sqrt{(2\theta + 10)^2 + 4}\,d\theta$, let $u = 2\theta + 10$ to get

$$L = \frac{1}{2}\int_{10}^{20} \sqrt{u^2 + 4}\,du = \frac{1}{2}\left[\frac{u}{2}\sqrt{u^2 + 4} + 2\ln\left|u + \sqrt{u^2 + 4}\right|\right]_{10}^{20}$$

$$= \frac{1}{2}\left[10\sqrt{404} - 5\sqrt{104} + 2\ln\frac{20 + \sqrt{404}}{10 + \sqrt{104}}\right] \approx 75.7\text{ mm}$$

57. (a) The end of the inner arm traces out the circle $x_1 = \cos t, y_1 = \sin t$. Relative to the end of
the inner arm, the outer arm traces out the circle $x_2 = \cos 2t, y_2 = -\sin 2t$. Add to get the
motion of the center of the rider cage relative to the center of the inner arm:
$x = \cos t + \cos 2t, y = \sin t - \sin 2t$.

(b) Same as Part (a), except $x_2 = \cos 2t, y_2 = \sin 2t$, so $x = \cos t + \cos 2t, y = \sin t + \sin 2t$

(c) $L_1 = \displaystyle\int_0^{2\pi}\left[\left(\frac{dx}{dt}\right)^2 + \left(\frac{dy}{dt}\right)^2\right]^{1/2} dt = \int_0^{2\pi}\sqrt{5 - 4\cos 3t}\,dt \approx 13.36489321,$

$L_2 = \displaystyle\int_0^{2\pi}\sqrt{5 + 4\cos t}\,dt \approx 13.36489322$; L_1 and L_2 appear to be equal, and indeed, with the

substitution $u = 3t - \pi$ and the periodicity of $\cos u$,

$L_1 = \dfrac{1}{3}\displaystyle\int_{-\pi}^{5\pi}\sqrt{5 - 4\cos(u + \pi)}\,du = \int_0^{2\pi}\sqrt{5 + 4\cos u}\,du = L_2.$

59. (a) The thread leaves the circle at the point $x_1 = a\cos\theta, y_1 = a\sin\theta$, and the end of the thread
is, relative to the point on the circle, on the tangent line at $x_2 = a\theta\sin\theta, y_2 = -a\theta\cos\theta$;
adding, $x = a(\cos\theta + \theta\sin\theta), y = a(\sin\theta - \theta\cos\theta)$.

(b) $dx/d\theta = a\theta\cos\theta, dy/d\theta = a\theta\sin\theta; dx/d\theta = 0$ has solutions $\theta = 0, \pi/2, 3\pi/2$; and $dy/d\theta = 0$
has solutions $\theta = 0, \pi, 2\pi$. At $\theta = \pi/2, dy/d\theta > 0$, so the direction is North; at $\theta = \pi$,
$dx/d\theta < 0$, so West; at $\theta = 3\pi/2, dy/d\theta < 0$, so South; at $\theta = 2\pi, dx/d\theta > 0$, so East.
Finally, $\displaystyle\lim_{\theta\to 0^+}\frac{dy}{dx} = \lim_{\theta\to 0^+}\tan\theta = 0$, so East.

(c)

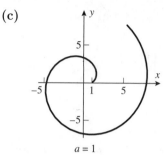

$a = 1$

61. $\tan\psi = \tan(\phi - \theta) = \dfrac{\tan\phi - \tan\theta}{1 + \tan\phi\tan\theta} = \dfrac{\dfrac{dy}{dx} - \dfrac{y}{x}}{1 + \dfrac{y}{x}\dfrac{dy}{dx}}$

$= \dfrac{\dfrac{r\cos\theta + (dr/d\theta)\sin\theta}{-r\sin\theta + (dr/d\theta)\cos\theta} - \dfrac{\sin\theta}{\cos\theta}}{1 + \left(\dfrac{r\cos\theta + (dr/d\theta)\sin\theta}{-r\sin\theta + (dr/d\theta)\cos\theta}\right)\left(\dfrac{\sin\theta}{\cos\theta}\right)} = \dfrac{r}{dr/d\theta}$

63. $\tan\psi = \dfrac{r}{dr/d\theta} = \dfrac{ae^{b\theta}}{abe^{b\theta}} = \dfrac{1}{b}$ is constant, so ψ is constant.

EXERCISE SET 11.3

1. (a) $A = \displaystyle\int_0^\pi \frac{1}{2}4a^2\sin^2\theta\,d\theta = \pi a^2$ **(b)** $A = \displaystyle\int_{-\pi/2}^{\pi/2}\frac{1}{2}4a^2\cos^2\theta\,d\theta = \pi a^2$

3. $A = \displaystyle\int_0^{2\pi}\frac{1}{2}(2 + 2\sin\theta)^2 d\theta = 6\pi$ **5.** $A = 6\displaystyle\int_0^{\pi/6}\frac{1}{2}(16\cos^2 3\theta)d\theta = 4\pi$

7. $A = 2\displaystyle\int_{2\pi/3}^\pi \frac{1}{2}(1 + 2\cos\theta)^2 d\theta = \pi - 3\sqrt{3}/2$

9. area $= A_1 - A_2 = \displaystyle\int_0^{\pi/2}\frac{1}{2}4\cos^2\theta\,d\theta - \int_0^{\pi/4}\frac{1}{2}\cos 2\theta\,d\theta = \pi/2 - \frac{1}{4}$

11. The circles intersect when $\cos\theta = \sqrt{3}\sin\theta, \tan\theta = 1/\sqrt{3}, \theta = \pi/6$, so

$A = A_1 + A_2 = \displaystyle\int_0^{\pi/6}\frac{1}{2}(4\sqrt{3}\sin\theta)^2 d\theta + \int_{\pi/6}^{\pi/2}\frac{1}{2}(4\cos\theta)^2 d\theta = 2\pi - 3\sqrt{3} + 4\pi/3 - \sqrt{3} = 10\pi/3 - 4\sqrt{3}$.

13. $A = 2\displaystyle\int_{\pi/6}^{\pi/2}\frac{1}{2}[9\sin^2\theta - (1 + \sin\theta)^2]d\theta = \pi$

15. $A = 2 \int_0^{\pi/3} \frac{1}{2}[(2 + 2\cos\theta)^2 - 9]d\theta = 9\sqrt{3}/2 - \pi$

17. $A = 2\left[\int_0^{2\pi/3} \frac{1}{2}(1/2 + \cos\theta)^2 d\theta - \int_{2\pi/3}^{\pi} \frac{1}{2}(1/2 + \cos\theta)^2 d\theta\right] = (\pi + 3\sqrt{3})/4$

19. $A = 2 \int_0^{\pi/4} \frac{1}{2}(4 - 2\sec^2\theta)d\theta = \pi - 2$

21.   **(a)** r is not real for $\pi/4 < \theta < 3\pi/4$ and $5\pi/4 < \theta < 7\pi/4$

   (b) $A = 4 \int_0^{\pi/4} \frac{1}{2}a^2 \cos 2\theta \, d\theta = a^2$

   (c) $A = 4 \int_0^{\pi/6} \frac{1}{2}\left[4\cos 2\theta - 2\right] d\theta = 2\sqrt{3} - \frac{2\pi}{3}$

23. $A = \int_{2\pi}^{4\pi} \frac{1}{2}a^2\theta^2 \, d\theta - \int_0^{2\pi} \frac{1}{2}a^2\theta^2 \, d\theta = 8\pi^3 a^2$

25. $r^2 + \left(\dfrac{dr}{d\theta}\right)^2 = \cos^2\theta + \sin^2\theta = 1,$

   so $S = \displaystyle\int_{-\pi/2}^{\pi/2} 2\pi \cos^2\theta \, d\theta = \pi^2.$

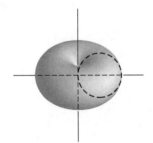

27. $S = \displaystyle\int_0^{\pi} 2\pi(1 - \cos\theta)\sin\theta\sqrt{1 - 2\cos\theta + \cos^2\theta + \sin^2\theta} \, d\theta$

$= 2\sqrt{2}\pi \displaystyle\int_0^{\pi} \sin\theta(1 - \cos\theta)^{3/2} \, d\theta = \frac{2}{5}2\sqrt{2}\pi(1 - \cos\theta)^{5/2}\Big|_0^{\pi} = 32\pi/5$

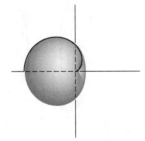

29.   **(a)** $r^3\cos^3\theta - 3r^2\cos\theta\sin\theta + r^3\sin^3\theta = 0, \quad r = \dfrac{3\cos\theta\sin\theta}{\cos^3\theta + \sin^3\theta}$

31.   If the upper right corner of the square is the point (a, a) then the large circle has equation $r = \sqrt{2}a$
   and the small circle has equation $(x - a)^2 + y^2 = a^2$, $r = 2a\cos\theta$, so

   area of crescent $= 2 \displaystyle\int_0^{\pi/4} \frac{1}{2}\left[(2a\cos\theta)^2 - (\sqrt{2}a)^2\right] d\theta = a^2 =$ area of square.

33. $A = \int_0^{\pi/2} \frac{1}{2} 4 \cos^2 \theta \sin^4 \theta \, d\theta = \pi/16$

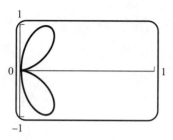

EXERCISE SET 11.4

1. **(a)** $4px = y^2$, point $(1,1), 4p = 1, x = y^2$ **(b)** $-4py = x^2$, point $(3,-3), 12p = 9, -3y = x^2$

(c) $a = 3, b = 2, \dfrac{x^2}{9} + \dfrac{y^2}{4} = 1$ **(d)** $a = 3, b = 2, \dfrac{x^2}{4} + \dfrac{y^2}{9} = 1$

(e) asymptotes: $y = \pm x$, so $a = b$; point $(0,1)$, so $y^2 - x^2 = 1$

(f) asymptotes: $y = \pm x$, so $b = a$; point $(2,0)$, so $\dfrac{x^2}{4} - \dfrac{y^2}{4} = 1$

3. **(a)**
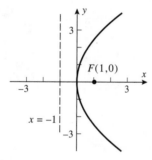

(b)
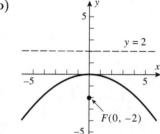

5. **(a)**
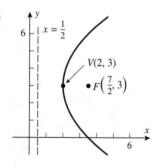

(b)
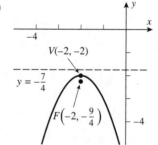

7. **(a)**
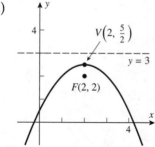

(b)
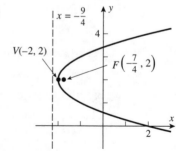

9. (a) $c^2 = 16 - 9 = 7, c = \sqrt{7}$

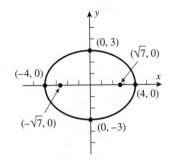

(b) $\dfrac{x^2}{1} + \dfrac{y^2}{9} = 1$

$c^2 = 9 - 1 = 8, c = 2\sqrt{2}$

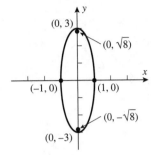

11. (a) $\dfrac{(x-1)^2}{9} + \dfrac{(y-3)^2}{16} = 1$

$c^2 = 16 - 9 = 7, c = \sqrt{7}$

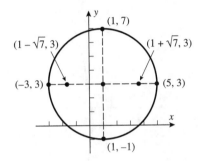

(b) $\dfrac{(x+2)^2}{4} + \dfrac{(y+1)^2}{9} = 1$

$c^2 = 9 - 4 = 5, c = \sqrt{5}$

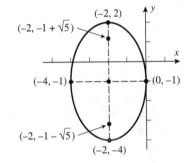

13. (a) $\dfrac{(x+1)^2}{9} + \dfrac{(y-1)^2}{1} = 1$

$c^2 = 9 - 1 = 8, c = 2\sqrt{2}$

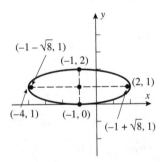

(b) $\dfrac{(x+1)^2}{4} + \dfrac{(y-5)^2}{16} = 1$

$c^2 = 16 - 4 = 12, c = 2\sqrt{3}$

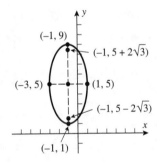

15. (a) $c^2 = a^2 + b^2 = 16 + 9 = 25, c = 5$

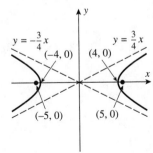

(b) $y^2/4 - x^2/36 = 1$

$c^2 = 4 + 36 = 40, c = 2\sqrt{10}$

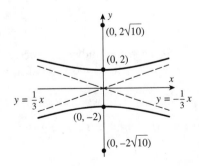

17. **(a)** $c^2 = 9 + 4 = 13, c = \sqrt{13}$

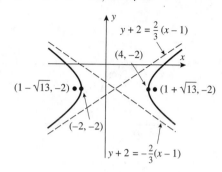

(b) $(y-3)^2/9 - (x-2)^2/4 = 1$
$c^2 = 9 + 4 = 13, c = \sqrt{13}$

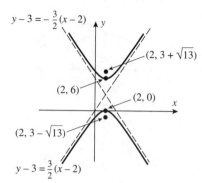

19. **(a)** $(x+1)^2/4 - (y-1)^2/1 = 1$
$c^2 = 4 + 1 = 5, c = \sqrt{5}$

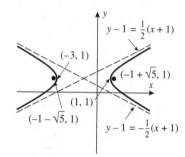

(b) $(x-1)^2/4 - (y+3)^2/64 = 1$
$c^2 = 4 + 64 = 68, c = 2\sqrt{17}$

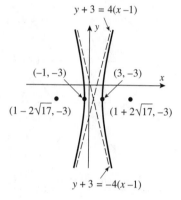

21. **(a)** $y^2 = 4px, p = 3, y^2 = 12x$

(b) $y^2 = -4px, p = 7, y^2 = -28x$

23. **(a)** $x^2 = -4py, p = 3, x^2 = -12y$

(b) The vertex is 3 units above the directrix so $p = 3$, $(x-1)^2 = 12(y-1)$.

25. $y^2 = a(x-h)$, $4 = a(3-h)$ and $2 = a(2-h)$, solve simultaneously to get $h = 1$, $a = 2$ so $y^2 = 2(x-1)$

27. **(a)** $x^2/9 + y^2/4 = 1$

(b) $a = 26/2 = 13, c = 5, b^2 = a^2 - c^2 = 169 - 25 = 144; x^2/169 + y^2/144 = 1$

29. **(a)** $c = 1, a^2 = b^2 + c^2 = 2 + 1 = 3; x^2/3 + y^2/2 = 1$

(b) $b^2 = 16 - 12 = 4; x^2/16 + y^2/4 = 1$ and $x^2/4 + y^2/16 = 1$

31. **(a)** $a = 6, (-3, 2)$ satisfies $x^2/36 + y^2/b^2 = 1$ so $9/36 + 4/b^2 = 1, b^2 = 16/3; x^2/36 + 3y^2/16 = 1$

(b) The center is midway between the foci so it is at $(-1, 2)$, thus
$c = 1, b = 2, a^2 = 1 + 4 = 5, a = \sqrt{5}; (x+1)^2/4 + (y-2)^2/5 = 1$

33. **(a)** $a = 2, c = 3, b^2 = 9 - 4 = 5; x^2/4 - y^2/5 = 1$

(b) $a = 1, b/a = 2, b = 2; x^2 - y^2/4 = 1$

35. **(a)** vertices along x-axis: $b/a = 3/2$ so $a = 8/3$; $x^2/(64/9) - y^2/16 = 1$
vertices along y-axis: $a/b = 3/2$ so $a = 6$; $y^2/36 - x^2/16 = 1$

(b) $c = 5$, $a/b = 2$ and $a^2 + b^2 = 25$, solve to get $a^2 = 20$, $b^2 = 5$; $y^2/20 - x^2/5 = 1$

37. **(a)** The center is at $(3,6)$, $a = 3$, $c = 5$, $b^2 = 25 - 9 = 16$; $(x-3)^2/9 - (y-6)^2/16 = 1$

(b) The asymptotes intersect at $(3,1)$ which is the center, $(x-3)^2/a^2 - (y-1)^2/b^2 = 1$ is the form of the equation because $(0,0)$ is to the left of both asymptotes, $9/a^2 - 1/b^2 = 1$ and $a/b = 1$ which yields $a^2 = 8$, $b^2 = 8$; $(x-3)^2/8 - (y-1)^2/8 = 1$.

39. **(a)** $y = ax^2 + b$, $(20,0)$ and $(10,12)$ are on the curve so $400a + b = 0$ and $100a + b = 12$. Solve for b to get $b = 16$ ft = height of arch.

(b) $\dfrac{x^2}{a^2} + \dfrac{y^2}{b^2} = 1$, $400 = a^2$, $a = 20$; $\dfrac{100}{400} + \dfrac{144}{b^2} = 1$, $b = 8\sqrt{3}$ ft = height of arch.

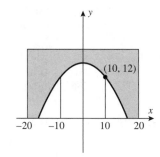

41. We may assume that the vertex is $(0,0)$ and the parabola opens to the right. Let $P(x_0, y_0)$ be a point on the parabola $y^2 = 4px$, then by the definition of a parabola, PF = distance from P to directrix $x = -p$, so $PF = x_0 + p$ where $x_0 \geq 0$ and PF is a minimum when $x_0 = 0$ (the vertex).

43. Use an xy-coordinate system so that $y^2 = 4px$ is an equation of the parabola, then $(1, 1/2)$ is a point on the curve so $(1/2)^2 = 4p(1)$, $p = 1/16$. The light source should be placed at the focus which is $1/16$ ft. from the vertex.

45. **(a)** $P: (b\cos t, b\sin t)$; $Q: (a\cos t, a\sin t)$; $R: (a\cos t, b\sin t)$

(b) For a circle, t measures the angle between the positive x-axis and the line segment joining the origin to the point. For an ellipse, t measures the angle between the x-axis and OPQ, not OR.

47. **(a)** For any point (x,y), the equation $y = b\tan t$ has a unique solution t where $-\pi/2 < t < \pi/2$. On the hyperbola, $\dfrac{x^2}{a^2} = 1 + \dfrac{y^2}{b^2} = 1 + \tan^2 t = \sec^2 t$, so $x = \pm a \sec t$.

(b)

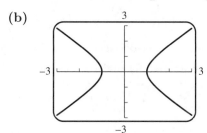

49. $(4,1)$ and $(4,5)$ are the foci so the center is at $(4,3)$ thus $c = 2$, $a = 12/2 = 6$, $b^2 = 36 - 4 = 32$; $(x-4)^2/32 + (y-3)^2/36 = 1$

51. Let the ellipse have equation $\dfrac{4}{81}x^2 + \dfrac{y^2}{4} = 1$, then $A(x) = (2y)^2 = 16\left(1 - \dfrac{4x^2}{81}\right)$,

$V = 2\displaystyle\int_0^{9/2} 16\left(1 - \dfrac{4x^2}{81}\right) dx = 96$

53. Assume $\dfrac{x^2}{a^2} + \dfrac{y^2}{b^2} = 1$, $A = 4\displaystyle\int_0^a b\sqrt{1 - x^2/a^2}\,dx = \pi ab$

55. Assume $\dfrac{x^2}{a^2} + \dfrac{y^2}{b^2} = 1$, $\dfrac{dy}{dx} = -\dfrac{bx}{a\sqrt{a^2 - x^2}}$, $1 + \left(\dfrac{dy}{dx}\right)^2 = \dfrac{a^4 - (a^2 - b^2)x^2}{a^2(a^2 - x^2)}$,

$$S = 2\int_0^a \frac{2\pi b}{a}\sqrt{1 - x^2/a^2}\sqrt{\frac{a^4 - (a^2 - b^2)x^2}{a^2 - x^2}}\,dx = 2\pi ab\left(\frac{b}{a} + \frac{a}{c}\sin^{-1}\frac{c}{a}\right), c = \sqrt{a^2 - b^2}$$

57. Open the compass to the length of half the major axis, place the point of the compass at an end of the minor axis and draw arcs that cross the major axis to both sides of the center of the ellipse. Place the tacks where the arcs intersect the major axis.

59. Let P denote the pencil tip, and let k be the difference between the length of the ruler and that of the string. Then $QP + PF_2 + k = QF_1$, and hence $PF_2 + k = PF_1, PF_1 - PF_2 = k$. But this is the definition of a hyperbola according to Definition 11.4.3.

61. $L = 2a = \sqrt{D^2 + p^2D^2} = D\sqrt{1 + p^2}$ (see figure), so $a = \dfrac{1}{2}D\sqrt{1 + p^2}$, but $b = \dfrac{1}{2}D$,

$$T = c = \sqrt{a^2 - b^2} = \sqrt{\frac{1}{4}D^2(1 + p^2) - \frac{1}{4}D^2} = \frac{1}{2}pD.$$

63. By implicit differentiation, $\dfrac{dy}{dx}\bigg|_{(x_0,y_0)} = -\dfrac{b^2}{a^2}\dfrac{x_0}{y_0}$ if $y_0 \neq 0$, the tangent line is

$y - y_0 = -\dfrac{b^2}{a^2}\dfrac{x_0}{y_0}(x - x_0)$, $a^2 y_0 y - a^2 y_0^2 = -b^2 x_0 x + b^2 x_0^2$, $b^2 x_0 x + a^2 y_0 y = b^2 x_0^2 + a^2 y_0^2$,

but (x_0, y_0) is on the ellipse so $b^2 x_0^2 + a^2 y_0^2 = a^2 b^2$; thus the tangent line is $b^2 x_0 x + a^2 y_0 y = a^2 b^2$, $x_0 x/a^2 + y_0 y/b^2 = 1$. If $y_0 = 0$ then $x_0 = \pm a$ and the tangent lines are $x = \pm a$ which also follows from $x_0 x/a^2 + y_0 y/b^2 = 1$.

65. Use $\dfrac{x^2}{a^2} + \dfrac{y^2}{b^2} = 1$ and $\dfrac{x^2}{A^2} - \dfrac{y^2}{B^2} = 1$ as the equations of the ellipse and hyperbola. If (x_0, y_0) is

a point of intersection then $\dfrac{x_0^2}{a^2} + \dfrac{y_0^2}{b^2} = 1 = \dfrac{x_0^2}{A^2} - \dfrac{y_0^2}{B^2}$, so $x_0^2\left(\dfrac{1}{A^2} - \dfrac{1}{a^2}\right) = y_0^2\left(\dfrac{1}{B^2} + \dfrac{1}{b^2}\right)$ and

$a^2 A^2 y_0^2(b^2 + B^2) = b^2 B^2 x_0^2(a^2 - A^2)$. Since the conics have the same foci, $a^2 - b^2 = c^2 = A^2 + B^2$, so $a^2 - A^2 = b^2 + B^2$. Hence $a^2 A^2 y_0^2 = b^2 B^2 x_0^2$. From Exercises 63 and 64, the slopes of the

tangent lines are $-\dfrac{b^2 x_0}{a^2 y_0}$ and $\dfrac{B^2 x_0}{A^2 y_0}$, whose product is $-\dfrac{b^2 B^2 x_0^2}{a^2 A^2 y_0^2} = -1$. Hence the tangent lines are perpendicular.

67. Let (x_0, y_0) be such a point. The foci are at $(-\sqrt{5}, 0)$ and $(\sqrt{5}, 0)$, the lines are perpendicular if the product of their slopes is -1 so $\dfrac{y_0}{x_0 + \sqrt{5}} \cdot \dfrac{y_0}{x_0 - \sqrt{5}} = -1, y_0^2 = 5 - x_0^2$ and $4x_0^2 - y_0^2 = 4$. Solve to get $x_0 = \pm 3/\sqrt{5}, y_0 = \pm 4/\sqrt{5}$. The coordinates are $(\pm 3/\sqrt{5}, 4/\sqrt{5}), (\pm 3/\sqrt{5}, -4/\sqrt{5})$.

69. Let d_1 and d_2 be the distances of the first and second observers, respectively, from the point of the explosion. Then $t = $ (time for sound to reach the second observer) $-$ (time for sound to reach the first observer) $= d_2/v - d_1/v$ so $d_2 - d_1 = vt$. For constant v and t the difference of distances, d_2

and d_1 is constant so the explosion occurred somewhere on a branch of a hyperbola whose foci are where the observers are. Since $d_2 - d_1 = 2a$, $a = \dfrac{vt}{2}$, $b^2 = c^2 - \dfrac{v^2t^2}{4}$, and $\dfrac{x^2}{v^2t^2/4} - \dfrac{y^2}{c^2 - (v^2t^2/4)} = 1$.

71. (a) Use $\dfrac{x^2}{9} + \dfrac{y^2}{4} = 1$, $x = \dfrac{3}{2}\sqrt{4 - y^2}$,

$$V = \int_{-2}^{-2+h} (2)(3/2)\sqrt{4 - y^2}(18)dy = 54 \int_{-2}^{-2+h} \sqrt{4 - y^2}\,dy$$

$$= 54 \left[\frac{y}{2}\sqrt{4 - y^2} + 2\sin^{-1}\frac{y}{2} \right]_{-2}^{-2+h} = 27 \left[4\sin^{-1}\frac{h - 2}{2} + (h - 2)\sqrt{4h - h^2} + 2\pi \right] \text{ ft}^3$$

(b) When $h = 4$ ft, $V_{\text{full}} = 108\sin^{-1}1 + 54\pi = 108\pi$ ft^3, so solve for h when $V = (k/4)V_{\text{full}}$, $k = 1, 2, 3$, to get $h = 1.19205, 2, 2.80795$ ft or $14.30465, 24, 33.69535$ in.

73. (a) $(x - 1)^2 - 5(y + 1)^2 = 5$, hyperbola

(b) $x^2 - 3(y + 1)^2 = 0$, $x = \pm\sqrt{3}(y + 1)$, two lines

(c) $4(x + 2)^2 + 8(y + 1)^2 = 4$, ellipse

(d) $3(x + 2)^2 + (y + 1)^2 = 0$, the point $(-2, -1)$ (degenerate case)

(e) $(x + 4)^2 + 2y = 2$, parabola

(f) $5(x + 4)^2 + 2y = -14$, parabola

75. distance from the point (x, y) to the focus $(0, -c)$ plus distance to the focus $(0, c) = \text{const} = 2a$,

$$\sqrt{x^2 + (y + c)^2} + \sqrt{x^2 + (y - c)^2} = 2a, x^2 + (y + c)^2 = 4a^2 + x^2 + (y - c)^2 - 4a\sqrt{x^2 + (y - c)^2},$$

$$\sqrt{x^2 + (y - c)^2} = a - \frac{c}{a}y, \text{ and since } a^2 - c^2 = b^2, \frac{x^2}{b^2} + \frac{y^2}{a^2} = 1$$

77. Assume the equation of the parabola is $x^2 = 4py$. The tangent line at $P(x_0, y_0)$ (see figure) is given by $(y - y_0)/(x - x_0) = m = x_0/2p$. To find the y-intercept set $x = 0$ and obtain $y = -y_0$. Thus $Q : (0, -y_0)$. The focus is $(0, p) = (0, x_0^2/4y_0)$, the distance from P to the focus is

$$\sqrt{x_0^2 + (y_0 - p)^2} = \sqrt{4py_0 + (y_0 - p)^2} = \sqrt{(y_0 + p)^2} = y_0 + p,$$

and the distance from the focus to the y-intercept is $p + y_0$, so triangle FPQ is isosceles, and angles FPQ and FQP are equal.

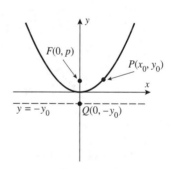

EXERCISE SET 11.5

1. (a) $\sin\theta = \sqrt{3}/2$, $\cos\theta = 1/2$

$x' = (-2)(1/2) + (6)(\sqrt{3}/2) = -1 + 3\sqrt{3}$, $y' = -(-2)(\sqrt{3}/2) + 6(1/2) = 3 + \sqrt{3}$

(b) $x = \dfrac{1}{2}x' - \dfrac{\sqrt{3}}{2}y' = \dfrac{1}{2}(x' - \sqrt{3}y')$, $y = \dfrac{\sqrt{3}}{2}x' + \dfrac{1}{2}y' = \dfrac{1}{2}(\sqrt{3}x' + y')$

$\sqrt{3}\left[\dfrac{1}{2}(x' - \sqrt{3}y')\right]\left[\dfrac{1}{2}(\sqrt{3}x' + y')\right] + \left[\dfrac{1}{2}(\sqrt{3}x' + y')\right]^2 = 6$

$$\frac{\sqrt{3}}{4}(\sqrt{3}x'^2 - 2x'y' - \sqrt{3}y'^2) + \frac{1}{4}(3x'^2 + 2\sqrt{3}x'y' + y'^2) = 6$$

$$\frac{3}{2}x'^2 - \frac{1}{2}y'^2 = 6,\ 3x'^2 - y'^2 = 12$$

(c)

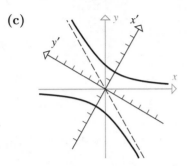

3. $\cot 2\theta = (0-0)/1 = 0,\ 2\theta = 90°,\ \theta = 45°$
$x = (\sqrt{2}/2)(x' - y'),\ y = (\sqrt{2}/2)(x' + y')$
$y'^2/18 - x'^2/18 = 1$, hyperbola

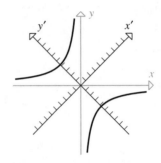

5. $\cot 2\theta = [1 - (-2)]/4 = 3/4$
$\cos 2\theta = 3/5$
$\sin\theta = \sqrt{(1 - 3/5)/2} = 1/\sqrt{5}$
$\cos\theta = \sqrt{(1 + 3/5)/2} = 2/\sqrt{5}$
$x = (1/\sqrt{5})(2x' - y')$
$y = (1/\sqrt{5})(x' + 2y')$
$x'^2/3 - y'^2/2 = 1$, hyperbola

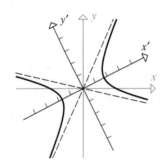

7. $\cot 2\theta = (1 - 3)/(2\sqrt{3}) = -1/\sqrt{3}$,
$2\theta = 120°,\ \theta = 60°$
$x = (1/2)(x' - \sqrt{3}y')$
$y = (1/2)(\sqrt{3}x' + y')$
$y' = x'^2$, parabola

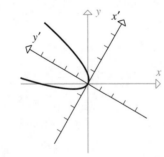

9. $\cot 2\theta = (9-16)/(-24) = 7/24$
$\cos 2\theta = 7/25,$
$\sin \theta = 3/5, \qquad \cos \theta = 4/5$
$x = (1/5)(4x' - 3y'),$
$y = (1/5)(3x' + 4y')$
$y'^2 = 4(x' - 1),$ parabola

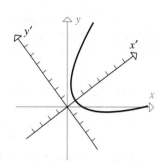

11. $\cot 2\theta = (52-73)/(-72) = 7/24$
$\cos 2\theta = 7/25, \qquad \sin \theta = 3/5,$
$\cos \theta = 4/5$
$x = (1/5)(4x' - 3y'),$
$y = (1/5)(3x' + 4y')$
$(x' + 1)^2/4 + y'^2 = 1,$ ellipse

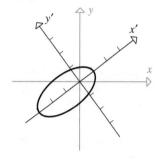

13. $x' = (\sqrt{2}/2)(x+y),\ y' = (\sqrt{2}/2)(-x+y)$ which when substituted into $3x'^2 + y'^2 = 6$ yields $x^2 + xy + y^2 = 3.$

15. Let $x = x'\cos\theta - y'\sin\theta,\ y = x'\sin\theta + y'\cos\theta$ then $x^2 + y^2 = r^2$ becomes
$(\sin^2\theta + \cos^2\theta)x'^2 + (\sin^2\theta + \cos^2\theta)y'^2 = r^2,\ x'^2 + y'^2 = r^2.$ Under a rotation transformation the center of the circle stays at the origin of both coordinate systems.

17. Use the Rotation Equations (5).

19. Set $\cot 2\theta = \dfrac{A-C}{B} = 0, 2\theta = \pi/2, \theta = \pi/4.$ Set $x = (x' - y')\sqrt{2}/2, y = (x' + y')\sqrt{2}/2$ and insert these into the equation to obtain $2x'^2 - 8y' = 0$; parabola, vertex $(0,0)$, focus $(0,1)$, directrix $y = -1$

21. $\cot 2\theta = (9-16)/(-24) = 7/24$ use method of Ex 4 to obtain $\cos 2\theta = \dfrac{7}{25},$ so

$\cos\theta = \sqrt{\dfrac{1+\cos 2\theta}{2}} = \sqrt{\dfrac{1+\frac{7}{25}}{2}} = \dfrac{4}{5}, \sin\theta = \sqrt{\dfrac{1-\cos 2\theta}{2}} = \dfrac{3}{5}.$ Then set

$x = \dfrac{4}{5}x' - \dfrac{3}{5}y',\ y = \dfrac{3}{5}x' + \dfrac{4}{5}y',$ insert these into the original equation to obtain $y'^2 = 4(x' - 1),$ so $p = 1,$ vertex is $(1,0),$ focus $(2,0)$ and directrix $x' = 0.$

23. $\cot 2\theta = (288-337)/(-168) = 49/168 = 7/24$; proceed as in Exercise 21 to obtain $cos\theta = \dfrac{4}{5}, \sin\theta = \dfrac{3}{5},$ so the new equation is $225x'^2 + 400y'^2 - 3600 = 0,\ x'^2/16 + y'^2/9 = 1,$ ellipse, $a = 4,\ b = 3,$ $c = \sqrt{7},$ foci at $(\pm\sqrt{7}, 0),$ vertices at $(\pm 4, 0),$ minor axis has endpoints $(0, \pm 3)$ in the x'-y' plane.

25. $\cot 2\theta = (31-21)/(10\sqrt{3}) = 1/\sqrt{3}, 2\theta = \pi/3, \theta = \pi/6$ and the new equation is $36x'^2 + 16y'^2 + 64y' = 80, 36x'^2 + 16(y' + 2)^2 = 144, x'^2/4 + (y' + 2)^2/9 = 1,$ ellipse, $a = 3, b = 2, c = \sqrt{9-4} = \sqrt{5},$ so vertices at $(0,1), (0,-5),$ foci at $(0, -2\pm\sqrt{5}),$ ends of minor axis at $(\pm 2, -2).$

27. $\cot 2\theta = (1 - 11)/(-10\sqrt{3}) = 1/\sqrt{3}, 2\theta = 2\pi/3, \theta = \pi/3$ and the new equation is
$-4x'^2 + 16y'^2 + 64 = 0, \quad x'^2/16 - y'^2/4 = 1$, hyperbola, vertices $(\pm 4, 0)$, $a = 4, b = 2, c = \sqrt{20} = 2\sqrt{5}$, so foci at $(\pm 2\sqrt{5}, 0)$, asymptotes $y = \pm 2x$.

29. $\cot 2\theta = (32 - (-7))/(-52) = -3/4$; proceed as in Example 4 to obtain

$\cos 2\theta = -3/5, \cos\theta = \sqrt{\dfrac{1 + \cos 2\theta}{2}} = \dfrac{1}{\sqrt{5}}, \sin\theta = \dfrac{2}{\sqrt{5}}$ The new equation is

$-20x'^2 + 45y'^2 - 360y' = -900, 20x'^2 - 45(y' - 4)^2 = 180$, or $\dfrac{x^2}{9} - \dfrac{(y - 4)^2}{4} = 1$, hyperbola,

$a = 3, b = 2, c = \sqrt{13}$, so the vertices are at $(\pm 3, 4)$, the foci at $(\pm\sqrt{13}, 4)$ and the asymptotes are

$y - 4 = \pm\dfrac{2}{3}x$.

31. $(\sqrt{x} + \sqrt{y})^2 = 1 = x + y + 2\sqrt{xy}, (1 - x - y)^2 = x^2 + y^2 + 1 - 2x - 2y + xy = 4xy$, so $x^2 - 3xy + y^2 - 2x - 2y + 1 = 0$. Set $\cot 2\theta = 0$, then $\theta = \pi/4$. Change variables by the Rotation Equations to obtain $2y'^2 - 2\sqrt{2}x' + 1$, which is a parabola.

33. It suffiices to show that the expression $B'^2 - 4A'C'$ is independent of θ. Set
$g = B' = B(\cos^2\theta - \sin^2\theta) + 2(C - A)\sin\theta\cos\theta$
$f = A' = (A\cos^2\theta + B\cos\theta\sin\theta + C\sin^2\theta)$
$h = C' = (A\sin^2\theta - B\sin\theta\cos\theta + C\cos^2\theta)$
It is easy to show that
$g'(\theta) = -2B\sin 2\theta + 2(C - A)\cos 2\theta,$
$f'(\theta) = (C - A)\sin 2\theta + B\cos 2\theta$
$h'(\theta) = (A - C)\sin 2\theta - B\cos 2\theta$ and it is a bit more tedious to show that

$\dfrac{d}{d\theta}(g^2 - 4fh) = 0.$

It follows that $B'^2 - 4A'C'$ is independent of θ and by taking $\theta = 0$, we have $B'^2 - 4A'C' = B^2 - 4AC$.

35. If $A = C$ then $\cot 2\theta = (A - C)B = 0$, so $2\theta = \pi/2$, and $\theta = \pi/4$.

EXERCISE SET 11.6

1. **(a)** $r = \dfrac{3/2}{1 - \cos\theta}, e = 1, d = 3/2$ **(b)** $r = \dfrac{3/2}{1 + \frac{1}{2}\sin\theta}, e = 1/2, d = 3$

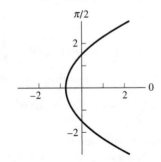

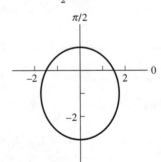

(c) $r = \dfrac{2}{1 + \frac{3}{2}\cos\theta}, e = 3/2, d = 4/3$

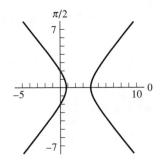

(d) $r = \dfrac{5/3}{1 + \sin\theta}, e = 1, d = 5/3$

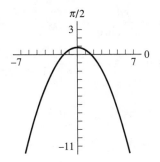

3. (a) $e = 1, d = 8$, parabola, opens up

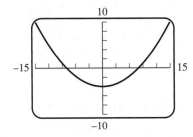

(b) $r = \dfrac{4}{1 + \frac{3}{4}\sin\theta}, e = 3/4, d = 16/3,$

ellipse, directrix 16/3 units
above the pole

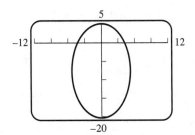

(c) $r = \dfrac{2}{1 - \frac{3}{2}\sin\theta}, e = 3/2, d = 4/3,$

hyperbola, directrix 4/3 units
below the pole

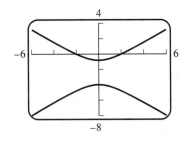

(d) $r = \dfrac{3}{1 + \frac{1}{4}\cos\theta}, e = 1/4, d = 12,$

ellipse, directrix 12 units
to the right of the pole

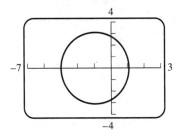

5. (a) $d = 2, r = \dfrac{ed}{1 + e\cos\theta} = \dfrac{3/2}{1 + \frac{3}{4}\cos\theta} = \dfrac{6}{4 + 3\cos\theta}$

(b) $e = 1, d = 1, r = \dfrac{ed}{1 + e\cos\theta} = \dfrac{1}{1 + \cos\theta}$

(c) $e = 4/3, d = 3, r = \dfrac{ed}{1 + e\sin\theta} = \dfrac{4}{1 + \frac{4}{3}\sin\theta} = \dfrac{12}{3 + 4\sin\theta}$

7. (a) $r = \dfrac{ed}{1 \pm e\cos\theta}, \theta = 0 : 6 = \dfrac{ed}{1 \pm e}, \theta = \pi : 4 = \dfrac{ed}{1 \mp e}, 6 \pm 6e = 4 \mp 4e, 2 = \mp 10e,$ use bottom

sign to get $e = 1/5, d = 24, \quad r = \dfrac{24/5}{1 - \cos\theta} = \dfrac{24}{5 - 5\cos\theta}$

(b) $e = 1, r = \dfrac{d}{1 - \sin\theta}, 1 = \dfrac{d}{2}, d = 2, r = \dfrac{2}{1 - \sin\theta}$

(c) $r = \dfrac{ed}{1 \pm e\sin\theta}, \theta = \pi/2 : 3 = \dfrac{ed}{1 \pm e}, \theta = 3\pi/2 : -7 = \dfrac{ed}{1 \mp e}, ed = 3 \pm 3e = -7 \pm 7e, 10 = \pm 4e,$

$e = 5/2, d = 21/5, r = \dfrac{21/2}{1 + (5/2)\sin\theta} = \dfrac{21}{2 + 5\sin\theta}$

9. (a) $r = \dfrac{3}{1 + \frac{1}{2}\sin\theta}, e = 1/2, d = 6,$ directrix 6 units above pole; if $\theta = \pi/2 : r_0 = 2;$

if $\theta = 3\pi/2 : r_1 = 6, a = (r_0 + r_1)/2 = 4, b = \sqrt{r_0 r_1} = 2\sqrt{3},$ center $(0, -2)$ (rectangular

coordinates), $\dfrac{x^2}{12} + \dfrac{(y+2)^2}{16} = 1$

(b) $r = \dfrac{1/2}{1 - \frac{1}{2}\cos\theta}, e = 1/2, d = 1,$ directrix 1 unit left of pole; if $\theta = \pi : r_0 = \dfrac{1/2}{3/2} = 1/3;$

if $\theta = 0 : r_1 = 1, a = 2/3, b = 1/\sqrt{3},$ center $= (1/3, 0)$ (rectangular coordinates),

$\dfrac{9}{4}(x - 1/3)^2 + 3y^2 = 1$

11. (a) $r = \dfrac{3}{1 + 2\sin\theta}, e = 2, d = 3/2,$ hyperbola, directrix 3/2 units above pole, if $\theta = \pi/2 :$

$r_0 = 1; \theta = 3\pi/2 : r = -3,$ so $r_1 = 3,$ center $(0, 2), a = 1, b = \sqrt{3}, -\dfrac{x^2}{3} + (y - 2)^2 = 1$

(b) $r = \dfrac{5/2}{1 - \frac{3}{2}\cos\theta}, e = 3/2, d = 5/3,$ hyperbola, directrix 5/3 units left of pole, if $\theta = \pi :$

$r_0 = 1; \theta = 0 : r = -5, r_1 = 5,$ center $(-3, 0),\;\; a = 2, b = \sqrt{5}, \dfrac{1}{4}(x + 3)^2 - \dfrac{1}{5}y^2 = 1$

13. (a) $r = \dfrac{\frac{1}{2}d}{1 + \frac{1}{2}\cos\theta} = \dfrac{d}{2 + \cos\theta},$ if $\theta = 0 : r_0 = d/3; \theta = \pi, r_1 = d,$

$8 = a = \dfrac{1}{2}(r_1 + r_0) = \dfrac{2}{3}d, d = 12,\;\; r = \dfrac{12}{2 + \cos\theta}$

(b) $r = \dfrac{\frac{3}{5}d}{1 - \frac{3}{5}\sin\theta} = \dfrac{3d}{5 - 3\sin\theta},$ if $\theta = 3\pi/2 : r_0 = \dfrac{3}{8}d; \theta = \pi/2, r_1 = \dfrac{3}{2}d,$

$4 = a = \dfrac{1}{2}(r_1 + r_0) = \dfrac{15}{16}d, d = \dfrac{64}{15}, r = \dfrac{3(64/15)}{5 - 3\sin\theta} = \dfrac{64}{25 - 15\sin\theta}$

(c) $r = \dfrac{\frac{3}{5}d}{1 - \frac{3}{5}\cos\theta} = \dfrac{3d}{5 - 3\cos\theta},$ if $\theta = \pi : r_0 = \dfrac{3}{8}d; \theta = 0, r_1 = \dfrac{3}{2}d, 4 = b = \dfrac{3}{4}d,$

$d = 16/3,\;\; r = \dfrac{16}{5 - 3\cos\theta}$

(d) $r = \dfrac{\frac{1}{5}d}{1 + \frac{1}{5}\sin\theta} = \dfrac{d}{5 + \sin\theta},$ if $\theta = \pi/2 : r_0 = d/6; \theta = 3\pi/2, r_1 = d/4,$

$5 = c = \dfrac{1}{2}d\left(\dfrac{1}{4} - \dfrac{1}{6}\right) = \dfrac{1}{24}d, d = 120,\;\; r = \dfrac{120}{5 + \sin\theta}$

15. A hyperbola is equilateral if and only if $a = b$ if and only if $c = \sqrt{2}a = \sqrt{2}b,$ which is equivalent to $e = \dfrac{c}{a} = \sqrt{2}.$

17. Since the foci are fixed, c is constant; since $e \to 0$, the distance $\dfrac{a}{e} = \dfrac{c}{e^2} \to +\infty$.

19. **(a)** $e = c/a = \dfrac{\frac{1}{2}(r_1 - r_0)}{\frac{1}{2}(r_1 + r_0)} = \dfrac{r_1 - r_0}{r_1 + r_0}$

    (b) $e = \dfrac{r_1/r_0 - 1}{r_1/r_0 + 1}, e(r_1/r_0 + 1) = r_1/r_0 - 1, \dfrac{r_1}{r_0} = \dfrac{1+e}{1-e}$

21. $a = b = 5, e = c/a = \sqrt{50}/5 = \sqrt{2}, r = \dfrac{\sqrt{2}d}{1 + \sqrt{2}\cos\theta}; r = 5$ when $\theta = 0$, so $d = 5 + \dfrac{5}{\sqrt{2}}$,

   $r = \dfrac{5\sqrt{2} + 5}{1 + \sqrt{2}\cos\theta}$.

23. **(a)** $T = a^{3/2} = 39.5^{1.5} \approx 248$ yr

    (b) $r_0 = a(1 - e) = 39.5(1 - 0.249) = 29.6645$ AU $\approx 4{,}449{,}675{,}000$ km

        $r_1 = a(1 + e) = 39.5(1 + 0.249) = 49.3355$ AU $\approx 7{,}400{,}325{,}000$ km

    (c) $r = \dfrac{a(1 - e^2)}{1 + e\cos\theta} \approx \dfrac{39.5(1 - (0.249)^2)}{1 + 0.249\cos\theta} \approx \dfrac{37.05}{1 + 0.249\cos\theta}$ AU

    (d)

25. **(a)** $a = T^{2/3} = 2380^{2/3} \approx 178.26$ AU

    (b) $r_0 = a(1 - e) \approx 0.8735$ AU, $r_1 = a(1 + e) \approx 355.64$ AU

    (c) $r = \dfrac{a(1 - e^2)}{1 + e\cos\theta} \approx \dfrac{1.74}{1 + 0.9951\cos\theta}$ AU

    (d)

27. $r_0 = a(1 - e) \approx 7003$ km, $h_{\min} \approx 7003 - 6440 = 563$ km,

    $r_1 = a(1 + e) \approx 10{,}726$ km, $h_{\max} \approx 10{,}726 - 6440 = 4286$ km

REVIEW EXERCISES, CHAPTER 11

1. (a) $(-4\sqrt{2}, -4\sqrt{2})$ (b) $(7\sqrt{2}/2, -7\sqrt{2}/2)$ (c) $(4\sqrt{2}, 4\sqrt{2})$

    (d) $(5, 0)$ (e) $(0, -2)$ (f) $(0, 0)$

3. (a) $(5, 0.6435)$ (b) $(\sqrt{29}, 5.0929)$ (c) $(1.2716, 0.6658)$

5. (a) $r = 2a/(1 + \cos\theta), r + x = 2a, x^2 + y^2 = (2a - x)^2, y^2 = -4ax + 4a^2$, parabola

    (b) $r^2(\cos^2\theta - \sin^2\theta) = x^2 - y^2 = a^2$, hyperbola

    (c) $r\sin(\theta - \pi/4) = (\sqrt{2}/2)r(\sin\theta - \cos\theta) = 4, y - x = 4\sqrt{2}$, line

    (d) $r^2 = 4r\cos\theta + 8r\sin\theta, x^2 + y^2 = 4x + 8y, (x - 2)^2 + (y - 4)^2 = 20$, circle

7.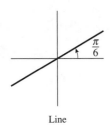

Line

9.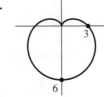

Cardioid

11.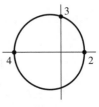

Limaçon

13. (a) $x = r\cos\theta = \cos\theta - \cos^2\theta, dx/d\theta = -\sin\theta + 2\sin\theta\cos\theta = \sin\theta(2\cos\theta - 1) = 0$ if $\sin\theta = 0$ or $\cos\theta = 1/2$, so $\theta = 0, \pi, \pi/3, 5\pi/3$; maximum $x = 1/4$ at $\theta = \pi/3, 5\pi/3$, minimum $x = -2$ at $\theta = \pi$;

    (b) $y = r\sin\theta = \sin\theta - \sin\theta\cos\theta, dy/d\theta = \cos\theta + 1 - 2\cos^2\theta = 0$ at $\cos\theta = 1, -1/2$, so $\theta = 0, 2\pi/3, 4\pi/3$; maximum $y = 3\sqrt{3}/4$ at $\theta = 2\pi/3$, minimum $y = -3\sqrt{3}/4$ at $\theta = 4\pi/3$

15. (a) $dy/dx = \dfrac{1/2}{2t} = 1/(4t); \; dy/dx\big|_{t=-1} = -1/4; dy/dx\big|_{t=1} = 1/4$

    (b) $x = (2y)^2 + 1, dx/dy = 8y, dy/dx\big|_{y=\pm(1/2)} = \pm 1/4$

17. $dy/dx = \dfrac{4\cos t}{-2\sin t} = -2\cot t$

    (a) $dy/dx = 0$ if $\cot t = 0, t = \pi/2 + n\pi$ for $n = 0, \pm 1, \cdots$

    (b) $dx/dy = -\dfrac{1}{2}\tan t = 0$ if $\tan t = 0, t = n\pi$ for $n = 0, \pm 1, \cdots$

19. (a) As t runs from 0 to π, the upper portion of the curve is traced out from right to left; as t runs from π to 2π the bottom portion of the curve is traced out from right to left. The loop occurs for $\pi + \sin^{-1}\dfrac{1}{4} < t < 2\pi - \sin^{-1}\dfrac{1}{4}$.

    (b) $\lim\limits_{t \to 0^+} x = +\infty, \lim\limits_{t \to 0^+} y = 1; \lim\limits_{t \to \pi^-} x = -\infty, \lim\limits_{t \to \pi^-} y = 1; \lim\limits_{t \to \pi^+} x = +\infty, \lim\limits_{t \to \pi^+} y = 1;$

       $\lim\limits_{t \to 2\pi^-} x = -\infty, \lim\limits_{t \to 2\pi^-} y = 1;$ the horizontal asymptote is $y = 1$.

    (c) horizontal tangent line when $dy/dx = 0$, or $dy/dt = 0$, so $\cos t = 0, t = \pi/2, 3\pi/2$; vertical tangent line when $dx/dt = 0$, so $-\csc^2 t - 4\sin t = 0, t = \pi + \sin^{-1}\dfrac{1}{\sqrt[3]{4}}, 2\pi - \sin^{-1}\dfrac{1}{\sqrt[3]{4}},$ $t = 3.823, 5.602$

(d) $r^2 = x^2 + y^2 = (\cot t + 4\cos t)^2 + (1 + 4\sin t)^2 = (4 + \csc t)^2, r = 4 + \csc t$; with $t = \theta$,
$f(\theta) = 4 + \csc\theta$; $m = dy/dx = (f(\theta)\cos\theta + f'(\theta)\sin\theta)/(-f(\theta)\sin\theta + f'(\theta)\cos\theta)$; when
$\theta = \pi + \sin^{-1}(1/4), m = \sqrt{15}/15$, when $\theta = 2\pi - \sin^{-1}(1/4), m = -\sqrt{15}/15$, so the tangent
lines to the conchoid at the pole have polar equations $\theta = \pm\tan^{-1}\dfrac{1}{\sqrt{15}}$.

21. $A = 2\displaystyle\int_0^\pi \frac{1}{2}(2 + 2\cos\theta)^2 d\theta = 6\pi$

23. $= \displaystyle\int_0^{\pi/6} 4\sin^2\theta\, d\theta + \int_{\pi/6}^{\pi/4} 1\, d\theta = \int_0^{\pi/6} 2(1 - \cos 2\theta)\, d\theta + \frac{\pi}{12} = (2\theta - \sin 2\theta)\Big]_0^{\pi/6} + \frac{\pi}{12}$

$= \dfrac{\pi}{3} - \dfrac{\sqrt{3}}{2} + \dfrac{\pi}{12} = \dfrac{5\pi}{12} - \dfrac{\sqrt{3}}{2}$

25.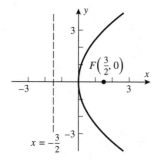

27.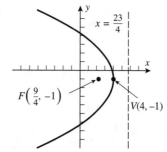

29. $c^2 = 25 - 4 = 21, c = \sqrt{21}$

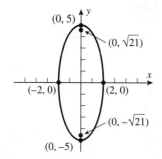

31. $\dfrac{(x-1)^2}{16} + \dfrac{(y-3)^2}{9} = 1$

$c^2 = 16 - 9 = 7, c = \sqrt{7}$

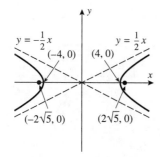

33. $c^2 = a^2 + b^2 = 16 + 4 = 20, c = 2\sqrt{5}$

35. $c^2 = 9 + 4 = 13, c = \sqrt{13}$

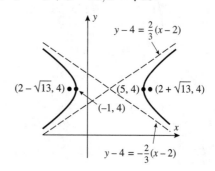

37. **(a)**

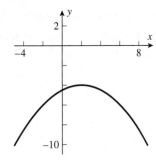

(b)

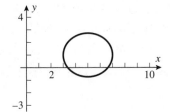

(c)

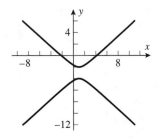

39. $x^2 = -4py$, $p = 4$, $x^2 = -16y$ **41.** $a = 3$, $a/b = 1$, $b = 3$; $y^2/9 - x^2/9 = 1$

43. **(a)** $y = y_0 + (v_0 \sin\alpha)\dfrac{x}{v_0 \cos\alpha} - \dfrac{g}{2}\left(\dfrac{x}{v_0 \cos\alpha}\right)^2 = y_0 + x\tan\alpha - \dfrac{g}{2v_0^2 \cos^2\alpha}x^2$

      (b) $\dfrac{dy}{dx} = \tan\alpha - \dfrac{g}{v_0^2 \cos^2\alpha}x$, $dy/dx = 0$ at $x = \dfrac{v_0^2}{g}\sin\alpha\cos\alpha$,

           $y = y_0 + \dfrac{v_0^2}{g}\sin^2\alpha - \dfrac{g}{2v_0^2\cos^2\alpha}\left(\dfrac{v_0^2\sin\alpha\cos\alpha}{g}\right)^2 = y_0 + \dfrac{v_0^2}{2g}\sin^2\alpha$

45. **(a)** $V = \displaystyle\int_a^{\sqrt{a^2+b^2}} \pi\left(b^2 x^2/a^2 - b^2\right)\,dx$

           $= \dfrac{\pi b^2}{3a^2}(b^2 - 2a^2)\sqrt{a^2+b^2} + \dfrac{2}{3}ab^2\pi$

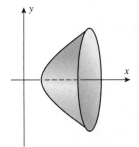

      (b) $V = 2\pi\displaystyle\int_a^{\sqrt{a^2+b^2}} x\sqrt{b^2 x^2/a^2 - b^2}\,dx = (2b^4/3a)\pi$

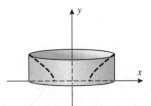

47. $\cot 2\theta = \dfrac{A-C}{B} = 0$, $2\theta = \pi/2$, $\theta = \pi/4$, $\cos\theta = \sin\theta = \sqrt{2}/2$, so

    $x = (\sqrt{2}/2)(x'-y')$, $y = (\sqrt{2}/2)(x'+y')$, $\dfrac{1}{2}x'^2 - \dfrac{5}{2}y'^2 + 3 = 0$, hyperbola

49. $\cot 2\theta = (4\sqrt{5} - \sqrt{5})/(4\sqrt{5}) = 3/4$, so $\cos 2\theta = 3/5$ and thus $\cos\theta = \sqrt{(1 + \cos 2\theta)/2} = 2/\sqrt{5}$ and $\sin\theta = \sqrt{(1 - \cos 2\theta)/2} = 1/\sqrt{5}$. Hence the transformed equation is $5\sqrt{5}x'^2 - 5\sqrt{5}y' = 0$, parabola

51. **(a)** $r = \dfrac{1/3}{1 + \frac{1}{3}\cos\theta}$, ellipse, right of pole, distance $= 1$

(b) hyperbola, left of pole, distance $= 1/3$

(c) $r = \dfrac{1/3}{1 + \sin\theta}$, parabola, above pole, distance $= 1/3$

(d) parabola, below pole, distance $= 3$

53. **(a)** $e = 4/5 = c/a, c = 4a/5$, but $a = 5$ so $c = 4, b = 3$, $\dfrac{(x+3)^2}{25} + \dfrac{(y-2)^2}{9} = 1$

(b) directrix $y = 2, p = 2, (x+2)^2 = -8y$

(c) center $(-1, 5)$, vertices $(-1, 7)$ and $(-1, 3), a = 2, a/b = 8, b = 1/4$, $\dfrac{(y-5)^2}{4} - 16(x+1)^2 = 1$

55. $a = 3, b = 2, c = \sqrt{5}$, $C = 4(3)\displaystyle\int_0^{\pi/2} \sqrt{1 - (5/9)\cos^2 u}\, du \approx 15.86543959$

APPENDIX A
Trigonometry Review

EXERCISE SET A

1. (a) $5\pi/12$ (b) $13\pi/6$ (c) $\pi/9$ (d) $23\pi/30$

3. (a) $12°$ (b) $(270/\pi)°$ (c) $288°$ (d) $540°$

5.

	$\sin\theta$	$\cos\theta$	$\tan\theta$	$\csc\theta$	$\sec\theta$	$\cot\theta$
(a)	$\sqrt{21}/5$	$2/5$	$\sqrt{21}/2$	$5/\sqrt{21}$	$5/2$	$2/\sqrt{21}$
(b)	$3/4$	$\sqrt{7}/4$	$3/\sqrt{7}$	$4/3$	$4/\sqrt{7}$	$\sqrt{7}/3$
(c)	$3/\sqrt{10}$	$1/\sqrt{10}$	3	$\sqrt{10}/3$	$\sqrt{10}$	$1/3$

7. $\sin\theta = 3/\sqrt{10}$, $\cos\theta = 1/\sqrt{10}$ **9.** $\tan\theta = \sqrt{21}/2$, $\csc\theta = 5/\sqrt{21}$

11. Let x be the length of the side adjacent to θ, then $\cos\theta = x/6 = 0.3$, $x = 1.8$.

13.

	θ	$\sin\theta$	$\cos\theta$	$\tan\theta$	$\csc\theta$	$\sec\theta$	$\cot\theta$
(a)	$225°$	$-1/\sqrt{2}$	$-1/\sqrt{2}$	1	$-\sqrt{2}$	$-\sqrt{2}$	1
(b)	$-210°$	$1/2$	$-\sqrt{3}/2$	$-1/\sqrt{3}$	2	$-2/\sqrt{3}$	$-\sqrt{3}$
(c)	$5\pi/3$	$-\sqrt{3}/2$	$1/2$	$-\sqrt{3}$	$-2/\sqrt{3}$	2	$-1/\sqrt{3}$
(d)	$-3\pi/2$	1	0	—	1	—	0

15.

	$\sin\theta$	$\cos\theta$	$\tan\theta$	$\csc\theta$	$\sec\theta$	$\cot\theta$
(a)	$4/5$	$3/5$	$4/3$	$5/4$	$5/3$	$3/4$
(b)	$-4/5$	$3/5$	$-4/3$	$-5/4$	$5/3$	$-3/4$
(c)	$1/2$	$-\sqrt{3}/2$	$-1/\sqrt{3}$	2	$-2\sqrt{3}$	$-\sqrt{3}$
(d)	$-1/2$	$\sqrt{3}/2$	$-1/\sqrt{3}$	-2	$2/\sqrt{3}$	$-\sqrt{3}$
(e)	$1/\sqrt{2}$	$1/\sqrt{2}$	1	$\sqrt{2}$	$\sqrt{2}$	1
(f)	$1/\sqrt{2}$	$-1/\sqrt{2}$	-1	$\sqrt{2}$	$-\sqrt{2}$	-1

17. (a) $x = 3\sin 25° \approx 1.2679$ (b) $x = 3/\tan(2\pi/9) \approx 3.5753$

19.

	$\sin\theta$	$\cos\theta$	$\tan\theta$	$\csc\theta$	$\sec\theta$	$\cot\theta$
(a)	$a/3$	$\sqrt{9-a^2}/3$	$a/\sqrt{9-a^2}$	$3/a$	$3/\sqrt{9-a^2}$	$\sqrt{9-a^2}/a$
(b)	$a/\sqrt{a^2+25}$	$5/\sqrt{a^2+25}$	$a/5$	$\sqrt{a^2+25}/a$	$\sqrt{a^2+25}/5$	$5/a$
(c)	$\sqrt{a^2-1}/a$	$1/a$	$\sqrt{a^2-1}$	$a/\sqrt{a^2-1}$	a	$1/\sqrt{a^2-1}$

21. (a) $\theta = 3\pi/4 \pm n\pi$, $n = 0, 1, 2, \ldots$
(b) $\theta = \pi/3 \pm 2n\pi$ and $\theta = 5\pi/3 \pm 2n\pi$, $n = 0, 1, 2, \ldots$

23. (a) $\theta = \pi/6 \pm n\pi$, $n = 0, 1, 2, \ldots$
   (b) $\theta = 4\pi/3 \pm 2n\pi$ and $\theta = 5\pi/3 \pm 2n\pi$, $n = 0, 1, 2, \ldots$

25. (a) $\theta = 3\pi/4 \pm n\pi$, $n = 0, 1, 2, \ldots$ (b) $\theta = \pi/6 \pm n\pi$, $n = 0, 1, 2, \ldots$

27. (a) $\theta = \pi/3 \pm 2n\pi$ and $\theta = 2\pi/3 \pm 2n\pi$, $n = 0, 1, 2, \ldots$
   (b) $\theta = \pi/6 \pm 2n\pi$ and $\theta = 11\pi/6 \pm 2n\pi$, $n = 0, 1, 2, \ldots$

29. $\sin\theta = 2/5$, $\cos\theta = -\sqrt{21}/5$, $\tan\theta = -2/\sqrt{21}$, $\csc\theta = 5/2$, $\sec\theta = -5/\sqrt{21}$, $\cot\theta = -\sqrt{21}/2$

31. (a) $\theta = \pm n\pi$, $n = 0, 1, 2, \ldots$ (b) $\theta = \pi/2 \pm n\pi$, $n = 0, 1, 2, \ldots$
   (c) $\theta = \pm n\pi$, $n = 0, 1, 2, \ldots$ (d) $\theta = \pm n\pi$, $n = 0, 1, 2, \ldots$
   (e) $\theta = \pi/2 \pm n\pi$, $n = 0, 1, 2, \ldots$ (f) $\theta = \pm n\pi$, $n = 0, 1, 2, \ldots$

33. (a) $s = r\theta = 4(\pi/6) = 2\pi/3$ cm (b) $s = r\theta = 4(5\pi/6) = 10\pi/3$ cm

35. $\theta = s/r = 2/5$

37. (a) $2\pi r = R(2\pi - \theta)$, $r = \dfrac{2\pi - \theta}{2\pi}R$

   (b) $h = \sqrt{R^2 - r^2} = \sqrt{R^2 - (2\pi - \theta)^2 R^2/(4\pi^2)} = \dfrac{\sqrt{4\pi\theta - \theta^2}}{2\pi}R$

39. Let h be the altitude as shown in the figure, then
   $h = 3\sin 60° = 3\sqrt{3}/2$ so $A = \dfrac{1}{2}(3\sqrt{3}/2)(7) = 21\sqrt{3}/4$.

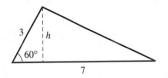

41. Let x be the distance above the ground, then $x = 10\sin 67° \approx 9.2$ ft.

43. From the figure, $h = x - y$ but $x = d\tan\beta$,
   $y = d\tan\alpha$ so $h = d(\tan\beta - \tan\alpha)$.

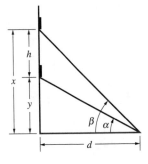

45. (a) $\sin 2\theta = 2\sin\theta\cos\theta = 2(\sqrt{5}/3)(2/3) = 4\sqrt{5}/9$
   (b) $\cos 2\theta = 2\cos^2\theta - 1 = 2(2/3)^2 - 1 = -1/9$

47. $\sin 3\theta = \sin(2\theta + \theta) = \sin 2\theta\cos\theta + \cos 2\theta\sin\theta = (2\sin\theta\cos\theta)\cos\theta + (\cos^2\theta - \sin^2\theta)\sin\theta$
   $= 2\sin\theta\cos^2\theta + \sin\theta\cos^2\theta - \sin^3\theta = 3\sin\theta\cos^2\theta - \sin^3\theta$; similarly, $\cos 3\theta = \cos^3\theta - 3\sin^2\theta\cos\theta$

49. $\dfrac{\cos\theta\tan\theta + \sin\theta}{\tan\theta} = \dfrac{\cos\theta(\sin\theta/\cos\theta) + \sin\theta}{\sin\theta/\cos\theta} = 2\cos\theta$

51. $\tan\theta + \cot\theta = \dfrac{\sin\theta}{\cos\theta} + \dfrac{\cos\theta}{\sin\theta} = \dfrac{\sin^2\theta + \cos^2\theta}{\sin\theta\cos\theta} = \dfrac{1}{\sin\theta\cos\theta} = \dfrac{2}{2\sin\theta\cos\theta} = \dfrac{2}{\sin 2\theta} = 2\csc 2\theta$

53. $\dfrac{\sin\theta + \cos 2\theta - 1}{\cos\theta - \sin 2\theta} = \dfrac{\sin\theta + (1 - 2\sin^2\theta) - 1}{\cos\theta - 2\sin\theta\cos\theta} = \dfrac{\sin\theta(1 - 2\sin\theta)}{\cos\theta(1 - 2\sin\theta)} = \tan\theta$

55. Using (47), $2\cos 2\theta \sin\theta = 2(1/2)[\sin(-\theta) + \sin 3\theta] = \sin 3\theta - \sin\theta$

57. $\tan(\theta/2) = \dfrac{\sin(\theta/2)}{\cos(\theta/2)} = \dfrac{2\sin(\theta/2)\cos(\theta/2)}{2\cos^2(\theta/2)} = \dfrac{\sin\theta}{1 + \cos\theta}$

59. From the figures, area $= \dfrac{1}{2}hc$ but $h = b\sin A$

so area $= \dfrac{1}{2}bc\sin A$. The formulas

area $= \dfrac{1}{2}ac\sin B$ and area $= \dfrac{1}{2}ab\sin C$

follow by drawing altitudes from vertices B and C, respectively.

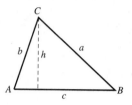

61. (a) $\sin(\pi/2 + \theta) = \sin(\pi/2)\cos\theta + \cos(\pi/2)\sin\theta = (1)\cos\theta + (0)\sin\theta = \cos\theta$
(b) $\cos(\pi/2 + \theta) = \cos(\pi/2)\cos\theta - \sin(\pi/2)\sin\theta = (0)\cos\theta - (1)\sin\theta = -\sin\theta$
(c) $\sin(3\pi/2 - \theta) = \sin(3\pi/2)\cos\theta - \cos(3\pi/2)\sin\theta = (-1)\cos\theta - (0)\sin\theta = -\cos\theta$
(d) $\cos(3\pi/2 + \theta) = \cos(3\pi/2)\cos\theta - \sin(3\pi/2)\sin\theta = (0)\cos\theta - (-1)\sin\theta = \sin\theta$

63. (a) Add (34) and (36) to get $\sin(\alpha - \beta) + \sin(\alpha + \beta) = 2\sin\alpha\cos\beta$ so
$\sin\alpha\cos\beta = (1/2)[\sin(\alpha - \beta) + \sin(\alpha + \beta)]$.

(b) Subtract (35) from (37). **(c)** Add (35) and (37).

65. $\sin\alpha + \sin(-\beta) = 2\sin\dfrac{\alpha - \beta}{2}\cos\dfrac{\alpha + \beta}{2}$, but $\sin(-\beta) = -\sin\beta$ so

$\sin\alpha - \sin\beta = 2\cos\dfrac{\alpha + \beta}{2}\sin\dfrac{\alpha - \beta}{2}$.

67. Consider the triangle having a, b, and d as sides. The angle formed by sides a and b is $\pi - \theta$ so
from the law of cosines, $d^2 = a^2 + b^2 - 2ab\cos(\pi - \theta) = a^2 + b^2 + 2ab\cos\theta$, $d = \sqrt{a^2 + b^2 + 2ab\cos\theta}$.

APPENDIX B
Solving Polynomial Equations

EXERCISE SET B

1. (a) $q(x) = x^2 + 4x + 2, r(x) = -11x + 6$
   (b) $q(x) = 2x^2 + 4, r(x) = 9$
   (c) $q(x) = x^3 - x^2 + 2x - 2, r(x) = 2x + 1$

3. (a) $q(x) = 3x^2 + 6x + 8, r(x) = 15$
   (b) $q(x) = x^3 - 5x^2 + 20x - 100, r(x) = 504$
   (c) $q(x) = x^4 + x^3 + x^2 + x + 1, r(x) = 0$

5.

x	0	1	-3	7
$p(x)$	-4	-3	101	5001

7. (a) $q(x) = x^2 + 6x + 13, r = 20$ (b) $q(x) = x^2 + 3x - 2, r = -4$

9. Assume $r = a/b$ a and b integers with $a > 0$:
   (a) b divides 1, $b = \pm 1$; a divides 24, $a = 1, 2, 3, 4, 6, 8, 12, 24$;
      the possible candidates are $\{\pm 1, \pm 2, \pm 3, \pm 4, \pm 6, \pm 8, \pm 12, \pm 24\}$
   (b) b divides 3 so $b = \pm 1, \pm 3$; a divides -10 so $a = 1, 2, 5, 10$;
      the possible candidates are $\{\pm 1, \pm 2, \pm 5, \pm 10, \pm 1/3, \pm 2/3, \pm 5/3, \pm 10/3\}$
   (c) b divides 1 so $b = \pm 1$; a divides 17 so $a = 1, 17$;
      the possible candidates are $\{\pm 1, \pm 17\}$

11. $(x + 1)(x - 1)(x - 2)$ **13.** $(x + 3)^3(x + 1)$

15. $(x + 3)(x + 2)(x + 1)^2(x - 3)$ **17.** -3 is the only real root.

19. $x = -2, -2/3, -1 \pm \sqrt{3}$ are the real roots.

21. $-2, 2, 3$ are the only real roots.

23. If $x - 1$ is a factor then $p(1) = 0$, so $k^2 - 7k + 10 = 0$, $k^2 - 7k + 10 = (k - 2)(k - 5)$, so $k = 2, 5$.

25. If the side of the cube is x then $x^2(x - 3) = 196$; the only real root of this equation is $x = 7$ cm.

27. Use the Factor Theorem with x as the variable and y as the constant c.
   (a) For any positive integer n the polynomial $x^n - y^n$ has $x = y$ as a root.
   (b) For any positive even integer n the polynomial $x^n - y^n$ has $x = -y$ as a root.
   (c) For any positive odd integer n the polynomial $x^n + y^n$ has $x = -y$ as a root.